Eco-Deconstruction

Eco-Deconstruction

Derrida and Environmental
Philosophy

Matthias Fritsch,
Philippe Lynes,
and David Wood

EDITORS

Fordham University Press | *New York 2018*

Fordham University Press has no responsibility for the persistence or accuracy of URLs for external or third-party Internet websites referred to in this publication and does not guarantee that any content on such websites is, or will remain, accurate or appropriate.

Fordham University Press also publishes its books in a variety of electronic formats. Some content that appears in print may not be available in electronic books. Visit us online at www.fordhampress.com.

Library of Congress Cataloging-in-Publication Data

Names: Fritsch, Matthias, editor.
Title: Eco-deconstruction : Derrida and environmental philosophy / Matthias
    Fritsch, Philippe Lynes, and David Wood, editors.
Description: First edition. | New York, NY : Fordham University Press, 2018. |
    Series: Groundworks: ecological issues in philosophy and theology |
    Includes bibliographical references and index.
Identifiers: LCCN 2017054129 | ISBN 9780823279500 (cloth : alk. paper) |
    ISBN 9780823279517 (pbk. : alk. paper)
Subjects: LCSH: Derrida, Jacques. | Ecology. | Environmental ethics. |
    Deconstruction.
Classification: LCC B2430.D484 E26 2018 | DDC 194—c23
LC record available at https://lccn.loc.gov/2017054129

Printed in the United States of America
20 19 18   5 4 3 2 1
First edition

Contents

List of Abbreviations   *vii*

Introduction
   Matthias Fritsch, Philippe Lynes, and David Wood   1

Part I: Diagnosing the Present

1   The Eleventh Plague: Thinking Ecologically
   after Derrida
    David Wood   29

2   Thinking after the World: Deconstruction
   and Last Things
    Ted Toadvine   50

3   Scale as a Force of Deconstruction
    Timothy Clark   81

Part II: Ecologies

4   The Posthuman Promise of the Earth
    Philippe Lynes   101

5   Un/Limited Ecologies
    Vicki Kirby   121

6   Ecology as Event
    Michael Marder   141

7　Writing Home: Eco-Choro-Spectrography
　　John Llewelyn　165

Part III: Nuclear and Other Biodegradabilities

8　E-Phemera: Of Deconstruction, Biodegradability,
　　and Nuclear War
　　Michael Naas　187

9　Troubling Time/s and Ecologies of Nothingness:
　　Re-turning, Re-membering, and Facing the Incalculable
　　Karen Barad　206

10　Responsibility and the Non(bio)degradable
　　Michael Peterson　249

11　Extinguishing Ability: How We Became
　　Postextinction Persons
　　Claire Colebrook　261

Part IV: Environmental Ethics

12　An Eco-Deconstructive Account of the Emergence
　　of Normativity in "Nature"
　　Matthias Fritsch　279

13　Opening Ethics onto *the Other Shore of Another
　　Heading*
　　Dawne McCance　303

14　Wallace Stevens's Birds, or, Derrida and Ecological
　　Poetics
　　Cary Wolfe　317

15　Earth: Love It or Leave It?
　　Kelly Oliver　339

　　List of Contributors　355

　　Index　361

Abbreviations

Jacques Derrida's Works

A   *Aporias* (Stanford: Stanford University Press, 1993); trans. Thomas Dutoit, from *Apories* (Paris: Galilée, 1996).

AA   *The Animal That Therefore I Am* (New York: Fordham University Press, 2008); trans. David Wills, from *L'animal que donc je suis* (Paris: Galilée, 2006).

AC   *Archaeology of the Frivolous: Reading Condillac* (Lincoln: University of Nebraska Press, 1980); trans. John P. Leavey Jr., from *L'archaeologie du frivole* (Paris: Galilée, 1973).

AE   *Adieu: Emmanuel Levinas* (Stanford: Stanford University Press, 1999); trans. Pascale-Anne Brault and Michael Naas, from *Adieu à Emmanuel Levinas* (Paris: Galilée, 1997).

AI   "Avowing—The Impossible: 'Returns,' Repentance, and Reconciliation," in *Living Together: Jacques Derrida's Communities of Violence and Peace*, ed. Elisabeth Weber (New York: Fordham University Press, 2013), 18–44; trans. Gil Anidjar, from "Avouer—l'impossible: 'Retours,' repentir et réconciliation," in *Le dernier des Juifs* (Paris: Galilée, 2014), 13–66.

AL   *Acts of Literature*, ed. Derek Attridge (London: Routledge, 1992).

ALT   "Alterities," *Parallax* 10, no. 4 (2004); trans. Stefan Herbrechter, from *Altérités* (Paris: Osiris, 1986).

AR   *Acts of Religion*, ed. Gil Anidjar (London: Routledge, 2002).

AV   *Advances* (Minneapolis: University of Minnesota Press, 2018); trans. Philippe Lynes, from "Avances," in Serge Margel, *Le tombeau du dieu artisan* (Paris: Minuit, 1995).

AW   *Arguing with Derrida*, ed. Simon Glendinning (Oxford: Blackwell, 2001).

BS1 *The Beast & the Sovereign*, vol. 1 (Chicago: University of Chicago Press, 2009); trans. Geoffrey Bennington, from *Séminaire: La bête et le souverain*, vol. 1, *2001–2002* (Paris: Galilée, 2008).

BS2 *The Beast & the Sovereign*, vol. 2 (Chicago: University of Chicago Press, 2011); trans. Geoffrey Bennington, from *Séminaire: La bête et le souverain*, vol. 2, *2002–2003* (Paris: Galilée, 2010).

BSD "Biodegradables: Seven Diary Fragments," trans. Peggy Kamuf, *Critical Inquiry* 13, no. 4 (1989): 812–37.

CF   *Of Cosmopolitanism and Forgiveness* (London: Routledge, 2001); trans. Mark Dooley and Michael Hughes, from *Cosmopolites de tous les pays, encore un effort!* (Paris: Galilée, 1997) and *Pardonner: L'impardonnable et l'imprescriptible* (Paris: Galilée, 2012).

CP   *Counterpath: Traveling with Jacques Derrida* (with Catherine Malabou) (Stanford: Stanford University Press, 2004); trans. David Wills, from *La contre-allée: Voyager avec Jacques Derrida* (Paris: La Quinzaine Littéraire–Louis Vuitton, 1999).

D   *Derrida* (with Geoffrey Bennington) (Chicago: University of Chicago Press, 1993); trans. Geoffrey Bennington, from *Derrida* (Paris: Seuil, 2008).

DF   *Demeure: Fiction and Testimony* (Stanford: Stanford University Press, 2000); trans. Elizabeth Rottenberg, from *Demeure: Maurice Blanchot* (Paris: Galilée, 1998).

DN   *Deconstruction in a Nutshell: A Conversation with Jacques Derrida*, ed. John Caputo (New York: Fordham University Press, 1997).

DP1 *The Death Penalty*, vol. 1 (Chicago: University of Chicago Press, 2014); trans. Peggy Kamuf, from *Séminaire: La peine de mort*, vol. 1, *1999–2000* (Paris: Galilée, 2012).

DP2 *The Death Penalty*, vol. 2 (Chicago: University of Chicago Press, 2017); trans. Elizabeth Rottenberg, from *Séminaire: La peine de mort*, vol. 2, *2000–2001* (Paris: Galilée, 2015).

DS   *Dissemination* (London: Athlone Press, 1981); trans. Barbara Johnson, from *La dissémination* (Paris: Seuil, 1972).

E   "Economimesis," *Diacritics* 11 (1981): 3–25; trans. R. Klein, from "Économimesis," in *Mimesis des articulations*, ed. Sylviane Agacinski et. al. (Paris: Aubier-Flammarion, 1972).

EO   *The Ear of the Other: Otobiography, Transference, Translation* (New York: Shocken, 1985); trans. Peggy Kamuf, from *L'oreille de l'autre: Otobiographies, transferts, traductions* (VLB Éditeur: Montréal, 1982).

FF   "La forme et la façon (Plus jamais: Envers et contre tout, ne plus jamais penser ça 'pour la forme')," préface à Alain David, in *Racisme et antisémitisme: Essai de philosophie sur l'envers de concepts* (Paris: Ellipses, 2001).

FK   "Faith and Knowledge: Two Sources of 'Religion' at the Limits of Reason Alone," in *Acts of Religion* (AR), 40–101; trans. Samuel Weber, from *Foi et savoir, suivi de Le siècle et le pardon* (Paris: Seuil, 2000).

FL   "Force of Law," in *Acts of Religion* (AR), 228–98; trans. Mary Quaintance, from *Force de loi* (Paris: Galilée, 1994).

FW   *For What Tomorrow* (Stanford: Stanford University Press, 2004); trans. Jeff Fort, from *De quoi demain . . . dialogue (avec Elisabeth Roudinesco)* (Paris: Fayard et Galilée, 2001).

G   *Glas* (Lincoln: University of Nebraska Press, 1986); trans. John P. Leavey Jr., from *Glas* (Paris: Galilée, 1974).

GD   *The Gift of Death and Literature in Secret*, 2nd ed. (Chicago: University of Chicago Press, 2008); trans. David Wills, from *Donner la mort* (Paris: Galilée, 1999).

GT   *Given Time*, vol. 1, *Counterfeit Money* (Chicago: University of Chicago Press, 1992); trans. Peggy Kamuf, from *Donner le temps*, vol. 1, *La fausse monnais* (Paris: Galilée, 1991).

H   "Hostipitality," *Angelaki* 5, no. 3 (2000): 3–18; trans. Barry Stocker and Forbes Morlock, from "Hostipitalité," *Cogito* 85 (1999): 17–44.

HQ   *Heidegger: The Question of Being & History* (Chicago: University of Chicago Press, 2016); trans. Geoffrey Bennington, from *Heidegger: La question de l'être et l'histoire* (Paris: Galilée, 2013).

LI   *Limited Inc.* (Evanston, Ill.: Northwestern University Press, 1988); trans. Elisabeth Weber, from *Limited Inc.* (Paris: Galilée, 1988).

LL   *Learning to Live Finally: The Last Interview* (New York: Palgrave Mcmillan, 2007); trans. Pascale-Anne Brault and

Michael Naas, from *Apprendre à vivre enfin: Entretien avec Jean Birnbaum* (Paris: Galilée, 2005).

MP   *Margins of Philosophy* (Brighton: Harvester, 1982); trans. Alan Bass from *Marges: De la philosophie* (Paris: Galilée, 1972).

MPD *Memoires for Paul de Man*, rev. ed. (New York: Columbia University Press, 1989); trans. Cecile Lindsay et al. from *Mémoires: Pour Paul de Man* (Paris: Galilée, 1988).

NII *Negotiations: Interventions and Interviews 1971–2002* (Stanford: Stanford University Press, 2002).

OG   *Of Grammatology* (Baltimore: Johns Hopkins University Press, 1974); trans. Gayatri Chakravorty Spivak, from *De la grammatologie* (Paris: Minuit, 1967).

OHG *The Other Heading: Reflections on Today's Europe* (Bloomington: Indiana University Press, 1992); trans. Pascale-Anne Brault and Michael Naas, from *L'autre cap, suivi de La démocratie ajournée* (Paris: Minuit, 1991).

OHY *Of Hospitality: Anne Dufourmantelle Invites Jacques Derrida to Respond* (Stanford: Stanford University Press, 2000); trans. Rachel Bowlby, from *De l'hospitalité: Anne Dufourmantelle invite Jacques Derrida à répondre* (Paris: Calmann-Lévy, 1997).

OS   *Of Spirit: Heidegger and the Question* (Chicago: University of Chicago Press, 1989); trans. Geoffrey Bennington and Rachel Bowlby, from *Heidegger et la question: De l'esprit et autres essais* (Paris: Champs Essais, 2010).

OT   *On Touching—Jean-Luc Nancy* (Stanford: Stanford University Press, 2005); trans. Christine Irizzary, from *Le toucher, Jean-Luc Nancy* (Paris: Galilée, 2000).

P   *Parages* (Stanford: Stanford University Press, 2011); trans. Tom Conley et. al., from *Parages* (Paris: Galilée, 1986).

PC   *The Post Card: From Socrates to Freud and Beyond* (Chicago: University of Chicago Press, 1987); trans. Alan Bass, from *La carte postale: De Socrate à Freud et au-delà* (Paris: Flammarion, 1980).

PF   *The Politics of Friendship* (London: Verso, 2005); trans. George Collins, from *Politiques de l'amitié* (Paris: Galilée, 1994).

PG   *The Problem of Genesis in Husserl's Philosophy* (Chicago: University of Chicago Press, 2003); trans. Marian Hobson, from *Le problème de la genèse dans la philosophie de Husserl* (Paris: Presses universitaires de France, 1990).

PI   *Points . . . Interviews, 1974–1994* (Stanford: Stanford University Press, 1995); trans. Peggy Kamuf et. al., from *Points de suspension: Entretiens* (Paris: Galilée, 1992).

PM   *Paper Machine* (Stanford: Stanford University Press, 2005); trans. Rachel Bowlby, from *Papier machine* (Paris: Galilée, 2001).

PS   *Positions* (Chicago: University of Chicago Press, 1981); trans. Alan Bass, from *Positions: Entretiens* (Paris: Minuit, 1972).

PT   *Philosophy in a Time of Terror: Dialogues with Jürgen Habermas and Jacques Derrida* (Chicago: University of Chicago Press, 2003); ed. Giovanna Borradori, from *Le concept du 11 septembre: Dialogues à New York (octobre–décembre 2001)* (Paris: Galilée, 2004).

P1   *Psyche: Inventions of the Other*, vol. 1 (Stanford: Stanford University Press, 2007); ed. Peggy Kamuf and Elizabeth Rottenberg, from *Psyché: Inventions de l'autre*, vol. 1 (Paris: Galilée, 1997).

P2   *Psyche: Inventions of the Other*, vol. 2 (Stanford: Stanford University Press, 2008); ed. Peggy Kamuf and Elizabeth Rottenberg, from *Psyché: Inventions de l'autre*, vol. 2 (Paris: Galilée, 2003).

QF   "Que faire—de la question 'Que faire?,'" in *Derrida pour les temps à venir*, ed. René Major (Paris: Stock, 2007), 45–62.

R   *Rogues* (Stanford: Stanford University Press, 2005); trans. Pascale-Anne Brault and Michael Naas, from *Voyous* (Paris: Galilée, 2003).

SM   *Specters of Marx* (New York: Routledge Classics, 2006); trans. Peggy Kamuf, from *Spectres de Marx* (Paris: Galilée, 1993).

SQ   *Sovereignties in Question: The Poetics of Paul Celan* (New York: Fordham University Press, 2005).

TP   *The Truth in Painting* (Chicago: University of Chicago Press, 1987); trans. Geoffrey Bennington and Ian McLeod, from *La vérité en peinture* (Paris: Flammarion, 1978).

TS   *A Taste for the Secret* (Cambridge: Polity, 2002).

UG   "Ulysses Gramophone," in *Acts of Literature* (AL); trans. Tina Kendall, from *Ulysse gramophone* (Paris: Galilée, 1987).

VP   *Voice and Phenomenon* (Evanston, Ill.: Northwestern University Press, 2011); trans. Leonard Lawlor, from *La voix et le phénomène* (Paris: Presses universitaires de France, 1967).

WA   *Without Alibi*, ed. Peggy Kamuf (Stanford: Stanford University Press, 2002).

WD   *Writing and Difference* (London: Routledge Classics, 2001);
        trans. Alan Bass, from *L'écriture et la différence* (Paris: Seuil,
        1967).
WM   *The Work of Mourning* (Chicago: University of Chicago
        Press, 2001); ed. and trans. Pascale-Anne Brault and Michael
        Naas, from *Chaque fois unique, la fin du monde* (Paris: Ga-
        lilée, 2003).

Jean-Luc Nancy's Works

AFEC   *After Fukushima: The Equivalence of Catastrophes* (New
        York: Fordham University Press, 2014); trans. Charlotte
        Mandell, from *L'équivalence des catastrophes (Après Fuku-
        shima)* (Paris: Galilée, 2012).
AFT   *A Finite Thinking* (Stanford: Stanford University Press,
        2003); ed. Simon Sparks, from *Une pensée finie* (Paris: Gali-
        lée, 1991).
BSP   *Being Singular Plural* (Stanford: Stanford University Press,
        2000); trans. Robert D. Richardson and Anne E. O'Byrne,
        from *Être singulier pluriel: Nouvelle édition augmentée*
        (Paris: Galilée, 2013).
CW   *The Creation of the World or Globalization* (Albany: SUNY
        Press, 2007); trans. François Raffoul and David Pettigrew,
        from *La création du monde ou la mondialisation* (Paris:
        Galilée, 2002).
SW   *The Sense of the World* (Minneapolis: University of Min-
        nesota Press, 1997); trans. Jeffrey S. Librett, from *Le sens du
        monde* (Paris: Galilée, 2001).

Introduction

Matthias Fritsch,
Philippe Lynes, and David Wood

"We cannot go on like this!" But while the challenges posed by the degradation of the natural environment are pressing—habitat loss, species extinction, and climate change—they also call for serious philosophical and ethical reflection. How can we best think about this degradation? Should it change how we think about being human? And how we might act differently? The field of environmental ethics arose in the English-speaking world in the 1970s in response to these concerns. It principally understood its task to consist in determining the scope of moral considerability: what in nature has value, and why. While the Western tradition centered value on the human being, proposals were put forward to extend the moral realm to higher primates, all sentient animals, all living things, ecosystems, and even to the earth as a whole. Other voices soon emerged, critical of the frameworks and presuppositions of this extensionist or at least moralizing approach, most notably in the schools of deep ecology and ecofeminism.

Continental philosophy, if we continue to use this problematic term,[1] has older and different paths to these themes, so much so that it may be mistaken for a latecomer in relation to Anglo-American environmental philosophy. Hearkening back to nineteenth- and early twentieth-century philosophy of nature, philosophy on the European continent saw environmental philosophy not so much as a question of determining where to draw the line between what is worthy of moral consideration and what is not. Rather, it inquired into the lived experience of humans and other living beings in their relationship with the natural environment, sometimes as the groundwork for ethical deliberation. Alongside industrialization

1

and modernization, the Romantics were already concerned about the ultimately self-destructive reduction of nature to mere instrumental value, and many of their ideas were picked up by Nietzsche, Heidegger, Benjamin, Adorno, Horkheimer, and others in the nineteenth and twentieth centuries.[2]

In Germany, in particular, the critique of the destruction of nature was often wedded to defending an antimodern nationalism—for example, in Ludwig Klages's 1913 *Man and Earth*, which lambasts deforestation and species loss while extolling the virtues of German landscapes and the homeliness of peasant life. Conservationism and conservatism, soil and blood, became a fateful, often anti-Semitic mixture on which fascism could draw.[3] As is well known, in Heidegger this legacy merged into a far-reaching and at times politically troublesome confrontation with modernity and technology,[4] bequeathing Continental environmental philosophy the task of assuming the productive rethinking of the relation to nature while stripping it of its blood-and-soil ideology. The first generation of the Frankfurt School also appropriated elements of Romanticism but stressed the link between the domination of nature and social oppression, especially along class and gender lines.[5] Next to these ontological, philosophical-historical, and social-political preoccupations in the history of Continental environmental thought, the emergence of the normative in nature emerged as an issue, especially in phenomenology.[6]

Today, Continental environmental philosophy is a burgeoning field of study that draws on this long and rich history to address the contemporary crisis.[7] Within it, eco-phenomenology is probably the most well known.[8] Beginning with ecologically oriented readings of German philosopher Edmund Husserl, as well as the many thinkers his work inspired—including Heidegger,[9] Hans Jonas,[10] Emmanuel Levinas,[11] Maurice Merleau-Ponty,[12] and Gilles Deleuze[13]—eco-phenomenology has established itself as an important voice within contemporary philosophical circles in addressing ecological issues. Given that the work of Jacques Derrida formed within, and in critical response to, the phenomenological tradition, it is surprising that his work is only now beginning to garner interest for what it may offer to environmental philosophy.[14] While important steps have been made, in and through Derrida's work, in rethinking the relation to nonhuman animals,[15] no book-length study or collection currently crystallizes the remarkable contribution deconstruction holds in store for addressing the environmental crisis beyond the question of animality on which Derrida seemed to concentrate toward the end of his life. The present volume enthusiastically targets this gap, launching a broader eco-deconstruction, of which rethinking the human-animal-environment relation is only one dimension.

"Eco-deconstruction" is not a familiar expression. It seems to be cousin to constructions such as the "eco-phenomenology" just mentioned or "eco-hermeneutics" (environmental hermeneutics). Legitimate questions can be raised on two different fronts. First, how does the "eco" modify the approach in each case? Is it a narrowing specification of scope or an expansion? Second, is there some sort of historical/dialectical progression from (say) eco-phenomenology through eco-hermeneutics to eco-deconstruction? Would that not be implied by any parallel claim about the movement from phenomenology to deconstruction? In addition to these *ecos* (and there are others—ecofeminism, ecocriticism), there are other philosophical fellow travelers such as deep ecology, social ecology, and ecosophy.

For anyone trying to navigate the waters of environmental philosophy from a broadly Continental perspective, some guidance is surely helpful. But there are conversations, divergences, and disagreements within each of these approaches that would have to be suppressed for a definitive mapping to be made. And it is not obvious that these various orientations all operate at the same level or are (competing) answers to the same question.

There is another reason to be cautious about a taxonomy of eco-positions—one that has to do with how we understand philosophy itself. Philosophy has an important technical dimension, but one way or the other, it is driven by an emancipatory interest. Truth, justice, and happiness are not just philosophical topics, they are its driving passions. And this century's proliferation of *ecos* bears witness to a real and unevenly acknowledged crisis in terrestrial dwelling, the accelerating consequence of our blindness to our relational dependency on the earth's systems. The subtitle of the book *Eco-Phenomenology* (2003), edited by Charles Brown and *Eco-Deconstruction* contributor Ted Toadvine, was *Back to the Earth Itself*, a call to a new shape and manner of thinking and engagement, not just another philosophical program.[16]

Consider first eco-phenomenology. After an inaugural 2001 essay by David Wood,[17] this expression first came to prominence with the aforementioned eponymous collection. It brought together a number of preeminent Continental philosophers to show how broadly phenomenological approaches could enrich environmental thinking. Canonical figures included Husserl, Heidegger, Levinas, and Merleau-Ponty.

Emerging concerns within phenomenology (embodiment, the face-to-face relation, how to think about nature, earth, and world) were made relevant to a broader ecological perspective. Phenomenology comes into its own when acknowledging the centrality of relationality (as in Heidegger's "being-in-the-world" and Nietzsche's "remain faithful to the earth") and intentionality, the way human experience (perhaps animal, too) is central to understanding how the world is *meaningful* to us. And here eco-phenomenology, if successful, productively addresses its main challenge—showing how the naturalism typically associated with ecology can be squared with intentionality.

Derrida once famously remarked that one needed to go through phenomenology (on the way to deconstruction) to avoid philosophical naivety (such as empiricism or positivism).[18] And to make matters more complicated, some have argued for a deconstructive phenomenology, which (after Bataille, Blanchot, and Derrida) would take seriously the experience of the impossibility of experience.[19] Be that as it may, and at the risk of oversimplification, one could treat the movement from phenomenology to deconstruction via hermeneutics as follows.

The hermeneutic take on phenomenology insists on the centrality of language and interpretation over against Husserl's original claim that phenomenology was essentially descriptive and presuppositionless. This promotes a shift away from individual experience to narrative, art, literature, public discussion, and so on. An eco-hermeneutics looks at the interpretive assumptions built into our understanding of nature, the stories we tell ourselves about our relation to the natural world (in the West often starting with Genesis), our constructions of past history and future prospects, and broader questions of sustainability. Eco-hermeneutics is well represented by the 2014 collection *Interpreting Nature: The Emerging Field of Environmental Hermeneutics*.[20] Such an eco- or environmental hermeneutics has its roots in Heidegger, Gadamer, and Ricoeur. The editors embrace an overlap with eco-phenomenology and a shared interest in questions of justice and activism, not least in "help[ing] to create moral communities by breathing life into our moral language."[21] All this opens up and enriches the grounds for and significance of what Ricoeur called "conflicts of interpretation."[22] If there is one watchword for an eco-hermeneutics, it is the idea of context. Hermeneutics was always about context—often, but not exclusively textual and communicative. The prefix "eco-" both stresses the environmental context and directs us more specifically to what is at

stake in our "interpretations of nature," which are always historically situated and culturally permeated, and which in various ways
reflect (or stand back from) our own species interest. And it is with
the idea of context that we can segue into eco-deconstruction.

Eco-hermeneutics, in terms of its practitioners, is only the junior
cousin of the more established field of ecocriticism (and eco-poetics),
firmly established within a broadly interdisciplinary approach to literature and literary theory. Like eco-hermeneutics, which traces at
least its U.S. origins to Thoreau, Muir, and Leopold, ecocriticism
has its roots, albeit problematically, in pastoral literature, Romanticism, and Rachel Carson's *Silent Spring* (1962).[23] These are usually
a springboard for criticism of the political, colonial, gender-based,
speciesist assumptions that such nature writing often betrayed. As
with much contemporary literary study, what is especially significant is its interdisciplinary outreach and, as of 2016, its maturity as
a discipline. *The Ecocriticism Reader: Landmarks in Literary Ecology* dates from 1996,[24] and *The Oxford Handbook of Eco-Criticism*
(2014) marks something of a coming of age.[25] Recent volumes (*Ecocriticism on the Edge: The Anthropocene as a Threshold Concept*
[2015],[26] and *Material Ecocriticism* [2014])[27] demonstrate a lively
engagement with our planetary crisis and the New Materialism, respectively.[28] Malcolm Miles's *Eco-Aesthetics* (2014) more explicitly
broadens ecocriticism to include art and architecture, suggesting
that art can contribute to a more sustainable lifestyle.[29]

Each of these approaches, perhaps following the contextualist
logic of the *eco-*, is driven to expansion of its scope, toward both
different flavors of interdisciplinarity and various kinds of activism
and outreach. It is then not surprising to see strong connections to
animal studies, (eco)feminism, and postcolonial theory, frequently
on the basis of a shared condition of domination and exploitation,
however one constructs the ground of such oppression; patriarchy,
domination of nature, and class oppression are all contenders.

If the prefix *eco-* points to the significance of natural/environ-
mental context, it might be said that the distinctive thrust of eco-
deconstruction is to affirm the significance of context, both "on the
ground," as it were, and methodologically. This is the most plausible reading of Derrida's oft-quoted early claim that *"il n'y a pas
de hors-texte* [there is nothing outside the text (*there is no outside-
text*)]" (OG 158/227). But it needs to be added that no such context
is ever fully saturated—that is to say, specifiable or determinable.
There is no "final analysis," no ultimate frame. It is this idea that

would put eco-deconstruction at odds with at least a certain reading of Deep Ecology, one that went too far in a holistic direction. Eco-deconstruction vigorously maintains a tension between contextual expansion and caution toward any impatient totalizing consolidation of a new frame of reference. This is one way of explaining how eco-deconstruction is not simply another program or approach alongside those previously mentioned but, among other things, a marker of methodological openness (and caution). In this sense, it is entirely appropriate that several of the contributors to this volume found their place in earlier collections devoted to eco-phenomenology, hermeneutics of nature, ecocriticism, and so on.

But there is more to be said about the specific contours of eco-deconstruction. We have suggested that one key task of Continental thought about nature is to welcome the powerful anti-Cartesian rethinking of the human-nature relation that stretches from the Romantics to Heidegger, while being mindful of its often too humanist and too spiritual, too homely and at times nationalistic and anti-Semitic, interpretations. Deconstruction, from Derrida and Paul de Man to Jean-Luc Nancy and others, is well placed to assume the task due to its longstanding critical engagement with this tradition from the vantage point of a nonhumanist philosophy of relational difference. That deconstruction is highly promising in this regard may be evident not only from Derrida's own most specific contributions (*The Animal That Therefore I Am*,[30] *The Beast & the Sovereign*,[31] with some of Derrida's work on life, nature, and ethics still unpublished),[32] but from the fact that deconstruction targets logocentrism. When he first coined the term, Derrida meant by it the privilege granted in most of the Western tradition to speech over writing, presence over absence, the intelligible over the sensible, man over woman, and the human or divine mind over the animal body. As early as *Of Grammatology*,[33] logocentrism names the ultimately self-contradictory attempt to distinguish the human vis-à-vis nature by way of "his properties," especially access to language, history, and death. Deconstruction has sought to show the fragility of these binary hierarchies—and their links to ethnocentrism, sexism, and anthropocentrism—by exposing the ways in which the dichotomies draw on, but disavow, ineradicable differentiation processes operating prior to the oppositions, including the nature-culture divide.

From the beginning, Derrida identified these differentiation processes with mortal life and its evolutionary history. Far from being textualist in a narrow sense, *différance*, he suggested, should be seen as coextensive with mortal life (AW 108; FW 63/106–7)—a life he rewrites as life death, in part to point up its necessary relation to, and contestable distinction from, the nonliving.[34] Deferral and differentiation are at work wherever there are elements in a more or less holistic system—for instance, DNA or organisms in an environment.[35] As the alleged textualism associated with the phrase "there is no outside-text" or "there is nothing outside the text [*il n'y a pas d'hors-texte*]" (OG 158/227) has been a point of much misunderstanding in the history of the reception of deconstruction, to the point of questioning its pertinence in relation to ecology—including in this very volume—we permit ourselves a few further comments.

Following readings of this phrase by Edward Said,[36] Michel Foucault,[37] and Richard Rorty,[38] deconstruction has seemed to some to lock human beings up in their own language and culture. References to the outside—the real, materiality, or life itself—are supposedly rendered impossible by linguistic and textual traces that come to take their place. The "general text," understood as linguistic and cultural webs of meaning, would thus form the ultimate context, the limit beyond which it would be futile to seek to go. The contrast between (cultural-linguistic) "text" and (socioeconomic) "power," dominant in the 1970s and 1980s, has given way, in recent years, to the opposition between "text" and "nature" or its substitutes, the charges of reductionism, skepticism, and relativism remaining more or less constant. Speculative realism or new materialism presents new doubts about this textualism, but there are others too who think it makes deconstruction incapable of taking what has been called "the nonhuman turn"[39] and unhelpful in addressing the environmental crisis.[40] In the words of Claire Colebrook, in this volume, deconstruction's emphasis on the future and the messianic promise is "a future of *thought*" that does not quite manage to reach "the radical nonlinear, nonpresent, inhuman, queer existence beyond thought," whether we call it "nature" or not.[41] Similarly, but more strongly, in his chapter Michael Marder claims deconstruction is "allergic" to ecology.[42] Derrida would, on Marder's view, have confined himself to "the economic domain," defined as "the default organization of our dwellings—whether at the level of the psyche, of the household, or of the planet as a whole."[43] Accused elsewhere by Marder of a hypersymbolism, deconstruction then merely "manages to expose the instability of the code, to which reality has been reduced."[44] The real task today, however, would be to confront the economic domain, the world of the code, with ecology and the "positivity of the event" that, presumably for Marder, falls outside the text, in the sense of emerging only "between economy and ecology."[45] Deconstruction can only ever render the code unstable, or contrast the economic with the aneconomic, without ever reaching the ecological in all its "positivity": "another kind of life," a life claimed to be endowed with the "capacity to dwell . . . [without] securing a piece of territory" and "prior to any appropriative overtures."[46]

Over the years, many have sought to counter this reading of the "no outside-text" thesis in *Of Grammatology*, including Derrida himself.[47] For him, the authority given to language as the ultimate context belonged to the very logocentrism he wished to analyze and problematize. Distancing his own work on the trace and the text from the "linguistic turn" and from de Man's more "'rhetoricist' interpretation of deconstruction,"[48] Derrida argued that the trace or the mark are prior to the human, making possible language but also the relation to an other in general.[49] Human language is only a particular system of marks or traces. The identity of every mark, whether of a sign or other element, demands differential relations to other identities and elements in contexts that shift along with the ongoing processes of relational identification of which contexts are made. This would be true, even if vastly different, for human systems of notation as much as of prelinguistic, nonanthropological marks or modes of existence. Textuality, understood in this sense as a texture of differentially related traces, thus precedes oppositions between economy and ecology, thought and nature, or human and animal or vegetal life. Already *Of Grammatology* sought to borrow and transform the antilogocentric forces associated with the notion of the trace and of writing in the critique of metaphysics in Heidegger, Levinas, Nietzsche, Freud, and "in all scientific fields, notably in biology" (OG 70/103). (This is why Rorty, among others, is wrong to claim that Derrida's "textualism" "adopt[s] an antagonistic position to natural science").[50] Derrida argues that the trace "must be thought before the opposition of nature and culture, animality and humanity, etc.," for, he continues:

> the trace is the opening of the first exteriority in general, the enigmatic relationship of the living [*du vivant*] to its other and of an inside to an outside: spacing. The outside, "spatial" and "objective" exteriority which we believe we know as the most familiar thing in the world, as familiarity itself, would not appear without the grammé, without différance as temporalization, without non-presence of the other inscribed within the

sense of the present, without the relationship with death as the concrete structure of the living present [*comme structure concrète du présent vivant*]. (OG 70–71/103)

As we can see, the deconstruction of phenomenology's "principle of principles," Husserl's "living present" as the site and time of all experience and all reality,[51] ushers in not only a rethinking of presence and time more generally, but also heralds a new thinking of life in its relation to death, a reconceptualization of the living being to its outside, to others and to its environment. The prelinguistic priority of the trace and the text implies first of all that human language is in fact a latecomer in evolutionary history: as Derrida clarified, writing in the narrow sense, as an exteriorized memory, is to be understood against the backdrop of the history of life, beginning with the "'genetic inscription' and the 'short programmatic chains' of the amoeba or the annelid"; both "'instinctive' behavior" and "electronic reading machines" would be forms of the "enlargement of *différance*" and the "exteriorization of the trace" that has always already begun (OG 84/125). The differential structure of the trace is thus the genetic and structural condition of vegetal, animal, human, and cultural life. It is against this background that we can place Vicki Kirby's effort, in this volume, to understand the evolution of the environment as "a force that speciates, a force that individuates *itself*" in habitat, species, and organisms.[52]

But this also means that living beings appropriate from their environments according to their own "codes," literally and figuratively, if we may use this problematic opposition for a moment. As Cary Wolfe argues in this volume (clarifying deconstruction with the help of systems theory and against Object-Oriented Ontology),[53] all organisms recodify and reduce the complexity of their environment. This generates what the previous passage from *Of Grammatology* calls a familiar world—precisely because organisms are not closed systems, but embodied and mortal, interwoven with and dependent on a complex environment not of their own making. Living beings—human as well as nonhuman—live only off their environments, appropriating "their" outside for building an inside, an economic dwelling. But the appropriation is, in a Derridean term recalled by Wolfe and Toadvine in their chapters, an "exappropriation"[54]: necessarily encountering the unpalatable and indigestible (for appropriating it all would be death), appropriation is always incomplete, unstable, to be repeated and recalibrated in an ongoing negotiation begun long before this or that particular organism, form of life, institution, or culture. The instability of the "code" is not just negative or aneconomic, but shows the textual—that is to say, differential, historical, and conflictual—belonging of life to its living and nonliving environment. The aneconomic limit to economic appropriation is at the same time a nonnegative gift to the environment: hence, Derrida's interest in the "non-bio-degradable" as well as in vomit, excrement, and corpses.[55] And that is why, for Derrida, there is no stepping outside the text into the real world of lives without appropriation, without indigestion, without death.

Given the need for differentiation, then, living entities are placed in spatiotemporal contexts or environments to which they are indebted, but from which they also must seek distance to establish a however contorted and relational identity. It is the humanist and logocentric way of conceiving this distance that has become a questionable issue for us today. Given its depth and scope, we cannot just shake it off, but must examine it in its various manifestations. That is why Derrida later expanded the neologism "logocentrism" to yield a "carno-phallogocentrism," by which he means a "metaphysico-anthropocentric axiomatic that dominates, in the West, the thought of just and unjust," and that, he demands in "Force of Law," (1989) we must reconsider today (FL 247/43). This hegemonic axiomatic in Western cultures associates the privilege granted to the voice over writing, to reason over the body, with a virile, carnivorous, animal-domesticating subject who ascribes language, understood as the capacity to respond rather than to merely react, to himself. Of this "schema that dominates the concept of the subject," Derrida claims in "Eating Well" (1988), "The subject does not want just to master and possess nature actively. In our cultures, he accepts sacrifice and eats flesh" (PI 280–81/294–95).

The deconstruction of logocentrism, then, was perhaps always an *eco*-deconstruction, one that tracks and reconsiders not only the subjection of femininity and animality, but our appropriative mastery and alleged sovereignty over nature. The "double science" or "double writing" that deconstruction developed to respond to hierarchical oppositions—both overturning and inscribing them in fields of differences that exceed oppositions and dichotomies[56]—can be fruitfully deployed by environmental philosophy in response to our current ecological impasse. This brief history of deconstruction suggests that there is much to learn from Derrida's corpus in this context, both to render questionable inherited, potentially logocentric assumptions and, at the same time, to propose new ways of doing environmental philosophy and ethics.

To sum up, eco-deconstruction offers an account of a differential relationality explored in a nonfinal, nontotalizable ecological context, both quasiontologically and quasinormatively, with attention to diagnosing our times. The book is divided into four parts: "Diagnosing the Present," "Ecologies," "Nuclear and Other Biodegradabilities," and "Environmental Ethics."

Diagnosing the Present

Deconstruction (as much of Continental philosophy) underscores the often subterranean historical embeddedness of contemporary thought and practice. Situating an issue genealogically in both the light of the present and its opening to the future is a key task. In this case, the issue concerns environmental degradation and its impact on how we should reconceive our time and our self-understanding. The chapters assembled in this part diagnose our time as hollowed out, problematically "scaled," and thus in need of deconstructive dispositions.

David Wood—likely the first to argue that "deconstruction is ripe for going green"[57]—opens the book by arguing that the global climate crisis is to be added to Derrida's list of Ten Plagues from the early 1990s; along with the animal holocaust, it is the signature issue of our time. It demands that key questions of violence, law, and social justice be tied to ecological sustainability. Deconstruction can best be taken forward not as a technical apparatus but as a fourfold disposition: negative capability, patient reading, aporetic schematization, and attention to language, coupled with terminological intervention. Ted Toadvine argues that our times are characterized by a hollowing out of the present that comes to be filled by ecological doomsday scenarios. The evacuation of the present is the result of a fast-paced, ever-accelerating time without meaningful goals independent of productivity increases and profit making. In such a time, all presents are equivalent; no moment, person, or place is allowed its singularity, and the future is there only for a calculative management seeking to maintain the status quo. In response, and with the help of Nancy and Derrida's reflections on "world," we should develop a sense of our times as inhabiting the present in

its singularity, futurity, and geomateriality. "Eco-deconstruction,"
writes Toadvine, "thereby becomes an ethos of the rediscovery of
the world right at the very end or limit of the sense of the world."[58]

Our time is the time, Timothy Clark suggests, in which the time
and space we inhabit should come to be understood as scaled: sustained and upset by long-term, distant effects usually shielded from
view, but now coming to haunt our self-conception. A human being,
for instance, appears much less self-conscious when understood as
part of a global, long-term phenomenon such as climate change, to
which s/he contributes. The scale at which s/he is still taken as a
rational agent and, in particular, the assumption of scale invariance,
are in need of deconstruction, one that can be elaborated following
Derrida's deconstruction of the metaphysics of presence. Only then
can the destructiveness of many seemingly innocuous actions come
to the fore: a scale effect manifests a force of alterity in the present
that Derrida's notions of the trace, the *pharmakon*, or the specter
first helped us to think.

Ecologies

The next part mobilizes the spectral ontology of deconstruction
to think what we might call an "originary environmentality," the
constitutive ecological embeddedness of mortal life. Philippe Lynes
develops Bataille's and Derrida's "general economy" into a "general
ecology" suffused with what he calls the "promise of the earth."
Life cannot be thought without individualization in an ecological
context from which living entities must appropriate in a restricted
economy that is, however, "generalized" by the fact that every appropriation exceeds that individual, calculative restriction. The
excess ties the entity back to its ecological context and invests it
with a promise: the promise to let the other live-on or sur-vive on
the earth. Vicki Kirby, too, reads grammatology as offering an interwoven ecology, but here with a stress on evolution. The history
of life is not a gradual unfolding of complexity, nor is selection a
secondary operation between a given organism and its environment.
Rather, evolution is to be thought as a differential force that individuates itself in organisms, species, and ecosystems—including as
human beings. But this truth regarding the human subject cannot
just be acknowledged by that self without it taking itself to be surviving its self-diagnosis. As quantum physics in particular shows,
in response to this self-survival, radical alterity usurps the force of

agency that we took to be deep inside, so much so that ecology no longer thinks in terms of the model of an environment for a given being and its identity, however much we grant that the human or the living is constituted by and indebted to it. In this sense, taking human responsibility for climate change, however necessary from one perspective, might be a ruse that seeks to reappropriate ecology as an *oikos*, a home, wishing to restore it to lost purity. From an eco-deconstructive perspective, then, the much-berated human exceptionalism should not be seen as something to be overcome, but rethought as ecology individuating itself in a being that takes itself as separate and exceptional.

In contrast to Lynes and Kirby, and as already indicated, Michael Marder argues that Derrida stressed the economical at the expense of the ecological. And yet it becomes increasingly evident that ecology interrupts the economic machine of making a home for the human. The nonappearance of the word "ecology" in Derrida's thought, argues Marder, betrays an "allergy" to the event of ecology. From within the household, whether psychic, economic, or planetary, and especially in our times of rampant economism, ecology is experienced as an event of disorder and the harbinger of crisis, one that is taken to stand in need of swift reintegration into the calculative machine. While Derrida did make room for the aneconomic, most notably in his thinking of general economy and the gift, he did not allow these to be positive events, and thereby maintained economy as the "dominant pole." What needs to be thought for Marder, however, is the ecological as a form of life that builds a home "without appropriation" or securing of territory, the emergence of living singularity "without segregation" and without fencing in, and thus without "pragmatic calculations of the appropriative subject."[59]

John Llewelyn, too, fastens upon the theme of the *"oikos"* in "eco-deconstruction," but in contrast to Marder, he takes deconstruction to take place by taking place away,[60] thereby entailing an unhomeliness in the home. Discussing a range of philosophers in something of a tour de force, Llewelyn marks some of the places of this unhomeliness, such as the *khōra* as differential spacing in Derrida's reading of Plato or spectral justice in the reading of Marx. These discussions suggest that eco-deconstruction consider a "blank ecology" that defers predication and appropriation of things. Blank ecology wishes to permit the appearance of "existents as such in their existence,"[61] so that, with and beyond Levinas's account of being persecuted, we may be struck by the realization that "for any given existent its existence is a good, at least for that existent."[62] Deconstructive ecology then becomes an "eco-choro-spectrography," a writing home displaced by inappropriable *khōra* and haunting specters calling for a justice that cannot but stretch the ethicopolitical imagination.

Nuclear and Other Biodegradabilities

In this part, contributors reflect on the remains, by-products, and disintegrations of human culture: poor literary works without discernable long-term traces in a culture, for example, or lasting masterpieces; waste (above all nuclear); but also environmental destruction, climate change, and species extinctions. In the wake of Derrida's long essay "Biodegradables," the authors cultivate a long-term regard for the intermeshing of both material and cultural dimensions of remaining and decomposition. In this way, biodegradability comes to be a name for eco-deconstructive sustainability. Michael Naas opens this part by offering a close reading of "Biodegradables" in view of the relationship between the biodegradability of things and culture or writing or between organic nature and the products of human artifice. Normally we think only artificial products deserve to be called "biodegradable," nature on this common view being nothing but organic (de)composition. Naas argues that in these concerns with the interrelation of nature and culture, Derrida interrogates the (auto-hetero-)deconstructibility of things and texts in a new way, one that is concerned with life, death, and sustainability, but especially with nuclear waste and nuclear war. For if a cultural artefact is most nonbiodegradable when assimilated by a culture without being fully understood or digested, "unforgettable because irreceivable," then this "secret of survivability" also defines nuclear waste and the threat of nuclear war: they are still with us as the possibility of a radical destruction without remainder.

Karen Barad continues this concern with nuclear war in the context of eco-deconstruction. She turns to quantum field theory to propose a complex entanglement of times and spaces. These entanglements entail that the nuclear bombs of Hiroshima and Nagasaki still affect the present time and in a broader context of colonialism, war, and nuclear-physics research. Barad suggests that the history of Euro-American colonialism is inseparable from the Newtonian assumption of a physical void: *terra nullius* justifies occupation. Early modern history continues in the twentieth century with the facile approval of nuclear testing sites, precursors to atomic bombing, despite their inhabitation by indigenous and nonhuman populations, in the United States and elsewhere. If quantum field theory can show us that the Newtonian assumption is untenable, that instead of the void there is ultimately uncolonizable "spacetimemattering," then we must learn to understand land not as property or territory, but as time-being with its own memory and ecological space to which we humans belong without neat divisions.

Michael Peterson likewise connects the material with the historical-cultural dimension of remains and does so by considering biodegradability with a focus on nuclear waste. Countering the U.S. nuclear waste policy and its account of responsibility, he argues for a strong relation between nuclear half-lives and long-term intergenerational responsibilities. Just as nuclear waste does not stop at generational borders, remaining radioactive for thousands of years, so responsibility cannot be passed on completely from us, its producers, to subsequent generations, no matter how ingenious our management and our warnings. When read alongside deconstructive notions of iterability and biodegradability, it becomes clear that responsibility cannot come to an end—in particular, not when dealing with allegedly exterior environmental material, which ignites responsibility each time anew.

The question of what remains, then, is inseparable from what does not remain, what goes extinct. Claire Colebrook argues that deconstruction's emphasis on survival—the view that the living present is only futural—makes it difficult for it to think literal extinctions. While it stresses disability in every ability and loss in every gain or survival, there is a stronger sense of disability that challenges the deconstructive concept of biodegradability: literal extinctions and the colonization of peoples and species that went along with the creation of a techno-industrial world in which the human could dream that climate has always been stable. But what is to be thought is precisely this paradox of the era of the Anthropocene; the very same technological practices that permitted humans to sustainably protect themselves from the vagaries of nature now render nature less hospitable and sustainable. And this also holds for our ability to diagnose climate change by locating our present in deep geological time, an ability inseparable from technological knowledge systems disabling and even destructive of humanity and the environment. The deconstructive account of autoimmunity can help us not just to understand these countervailing processes, but to take the full measure of the disability and extinction involved, which, Colebrook concludes, requires a counterdeconstructive "perhaps," a "perhaps" that "does not necessarily open, promise, or generate futurity."[63]

Environmental Ethics

As preceding parts have indicated, the concerns with diagnosis, ecology, and biodegradabilities are not free from ethical concerns. The final part turns more explicitly toward environmental ethics. As the earlier reference to Derrida's "Force of Law" intimated, the deconstructive attempt to resituate the human subject in an eco-deconstructive context seeks to uncover a demand for justice, and indeed a justice beyond law, for law is constituted in the modern West as thoroughly humanist, taking as its aim the proper of the human beyond mere mortal life.[64] By contrast, Derrida's work insists that normativity, including human responsibility for suffering beings, emerges precisely as a response to original differentiation and the mortality and unmasterable alterity it installs in living beings.

Matthias Fritsch opens the part by relating this eco-deconstructive normativity to well-known but divergent accounts of the emergence of "value" in nature. Value has been argued to emerge with the individual capacity for suffering, with individual self-valuing, or with holistic ecological entities (species, ecosystems, etc.), these three often seen as at odds with one another. But the differential eco-logics developed in previous chapters help us to see that an entity can become individualized, and thus acquire individual value, only in ongoing confrontations with other beings and the wider environment. Each living being can be seen as valuing its own life, then, only in response to a vulnerable exposure to its environment that it cannot just claim as its own: its self-affirmation necessarily affirms others and its environment. In this way, the three sources are interconnected in an aporetic matrix of ecological normativities.

Dawne McCance similarly relates an eco-deconstructive ethics to utilitarian and deontological approaches, but exposes both as discourses that only seem to have superseded religion; these ethical approaches still rest on an anthropotheological concept of sovereignty that is complicit with speciesism and anthropocentrism. This concept of sovereignty prizes spiritual life above the merely natural life of the biozoological. And spiritual life, assumed to be "worth more than life" (as Derrida put it in "Faith and Knowledge"), is connected, argues McCance, to the allegedly superior capacities of the rational subject that ended up privileging "the white male subject in post-Cartesian culture."[65] Hence, an environmental ethics today, for example, the "water ethics" McCance seeks to sketch here, cannot bypass the deconstruction of the ontotheological heritage as broached with particular relevance in Derrida's texts on religion, sovereignty, and animal life.

Cary Wolfe's contribution continues the concern with critically but affirmatively repositioning the source of ethics. With the help of Wallace Stevens's poetry, the autopoietic, second-order systems theory of biologists Maturana and Varela, and Michel Serres's topology, Wolfe argues that, of the hypotheses discussed in Derrida's seminars *The Beast & the Sovereign*, it is not the claim that man and animal are of the same world that is the most ecological. Rather, that distinction falls to the more controversial, and more difficult, claim that no living being, human or animal, inhabits the same world as another and in fact does not have its own world fully available to it. Wolfe shows that what counts as nature is always a product of the selective practices implemented in the embodied enaction of a particular living being and its engendering of an environment. The source of ethical responsibility, Wolfe suggests, lies precisely in that each world is unique but mortal—that is, uniquely vulnerable. It is the scene of address, among humans and across the human-animal divide, that results from this nonsharing of world but sharing of finitude that Stevens's poems put on stage. We have to act responsibly, then, but without grounding in *the* one world, and without being able to see what is good for all.

Further considering the nonsharing of world, Kelly Oliver concludes the collection with her proposal for what she calls "earth ethics." In the second volume of *Beast & Sovereign*, Derrida shows how islands have functioned as a figure for political, colonial, and individualist sovereignty, including democracy; sovereign agents must seek separation from the world while continuing to inhabit it, like islands in stormy seas that threaten to engulf and swallow them alive. Leaning far toward the side of separation and isolation, as in many utopias set on islands, may give rise to colonial appropriation, as in the case of Robinson Crusoe. Oliver argues that space missions—in particular, the U.S. Apollo mission—have made of the entire earth an island for a humanity thus ambiguously united, with similar effects of attraction and repulsion, love and the desire to leave. Derrida's logic of autoimmunity may help us to understand better both this ambiguity and this conflict. It gives rise to dreams of a common ownership of earth by a united humanity that is in fact divided by bitter strife among nation states competing for sovereign control of the earth. This strife and ambivalence, Oliver concludes, call for a lived acceptance of the ambiguities rather than their erasure in utopian or space-missionary projections: "Earth ethics demands that we not only acknowledge, but also embrace the vital fact that although we may not share a world, we do share a singular bond to the earth."[66]

Eco-deconstruction is here launched as a contribution to environmental thought in diagnostic, eco-ontological, biodegradable, and ethicopolitical ways. We hope this volume will more critically situate the human in its time and space.

Notes

1. See Simon Glendinning, *The Idea of Continental Philosophy* (Edinburgh: Edinburgh University Press, 2006), for a critique of the philosophical meaning and alleged usefulness of the term. For some recent reflections on Continental philosophy, see the special issue of the *Southern Journal of Philosophy* 50, no. 2 (2012), *Continental Philosophy: What and Where Will it Be?*, guest-edited by Ted Toadvine. The special issue contains several articles questioning the unity of the category as well as several arguing for Continental philosophy's unique ability to enact a much-needed rethinking of nature.

2. See, for instance, Kate Rigby, *Topographies of the Sacred: The Poetics of Place in European Romanticism* (Charlottesville: University of Virginia Press, 2004), and Dalia Nassar, "Romantic Empiricism after the 'End of Nature,'" in *The Relevance of Romanticism: Essays on German Romantic Philosophy*, ed. Dalia Nassar (Oxford: Oxford University Press, 2014). With respect to Schelling, in particular, Bruce Matthews makes a strong case for the prescient environmental relevance of German romantic philosophy: "Beyond the obvious, yet unfortunately too often overlooked fact that in destroying nature we harm ourselves, we can see that implicit in Schelling's critique of Descartes and Fichte's treatment of nature is the demand to extend Kant's kingdom of ends to all the kingdoms of nature. This follows clearly from his critique of the alienated subject of modernity, who values the gifts of nature only if they can be transformed into 'beautiful houses and proper furniture' or 'tools and household goods,' since it is only then that 'as a tool of his lust and desire' the world of nature takes on meaning and value . . . exploiting nature by making it subservient to our immediate 'economic-teleological ends' as if it had no inherent value in itself"; Matthews, "The New Mythology," in *The Relevance of Romanticism*, ed. Nassar, 211. On the relevance of Schelling for Adorno's well-known critique of capitalist modernity's destruction

of nature, see Peter Dews, "Dialectics and the Transcendence of Dialectics: Adorno's Relation to Schelling," *British Journal for the History of Philosophy* 22, no. 6 (2014): 1180–1207. Heidegger's lecture courses on Schelling are well known (vols. 42 and 86 of the German *Gesamtausgabe*). See Martin Heidegger, *Schelling: Vom Wesen der menschlichen Freiheit (1809)* (Frankfurt am Main: Vittorio Klostermann, 1988); trans. Joan Stambaugh as *Schelling's Treatise on the Essence of Human Freedom* (Athens: Ohio University Press, 1985); and Martin Heidegger, *Seminare: Hegel— Schelling* (Frankfurt am Main: Vittorio Klostermann, 2011).

3. Ludwig Klages, *Mensch und Erde* (Berlin: Matthes & Seitz, 2013). For a recent study of Klages's philosophy of life and his influence on Heidegger, Adorno, and others, as well as his relation to National Socialism, see Nitzan Lebovic, *The Philosophy of Life and Death: Ludwig Klages and the Rise of Nazi Biopolitics* (Basingstoke: Palgrave Macmillan, 2013). The relation between National Socialism (which passed many laws protecting nature) and environmentalism has been extensively researched; for an overview, see Joachim Radkau and Frank Uekötter, eds., *Naturschutz und Nationalsozialismus* (Frankfurt: Campus, 2003). For English translations of some of the ideas expressed in *Mensch und Erde* (Man and Earth), see Klages, *The Biocentric Worldview: Selected Essays and Poems of Ludwig Klages* (London: Arktos, 2013), and *Cosmogonic Reflections: Selected Aphorisms from Ludwig Klages* (London: Arktos, 2015).

4. Of the many studies, we cite Michael E. Zimmerman, *Heidegger's Confrontation with Modernity: Technology, Politics, and Art* (Bloomington: Indiana University Press, 1990).

5. See, for example, Deborah Cook, *Adorno on Nature* (London: Acumen, 2011). The most comprehensive study of the similarities between Heidegger and Adorno are still Hermann Mörchen, *Macht und Herrschaft im Denken von Heidegger und Adorno* (Stuttgart: Klett-Cotta, 1980), and Mörchen, *Adorno und Heidegger: Untersuchung einer philosophischen Kommunikationsverweigerung* (Stuttgart: Klett-Cotta, 1981); see also Iain Macdonald and Krzysztof Ziarek, eds., *Adorno and Heidegger: Philosophical Questions* (Stanford: Stanford University Press, 2007).

6. See in particular Hans Jonas, *The Phenomenon of Life: Toward a Philosophical Biology* (Evanston, Ill.: Northwestern University Press, 2001); Maurice Merleau-Ponty, *La Nature: Notes de cours du Collège de France*, ed. Dominique Séglard (Paris: Seuil, 1995); trans. Robert Vallier as *Nature: Course Notes from the College de France* (Chicago: Northwestern University Press, 2003). Jonas's book was originally published in 1966, while Merleau Ponty's lectures were given from 1956 to 1960.

7. For a brief overview of Continental environmental thought, see Bruce V. Foltz's entry in John Protevi, ed., *A Dictionary of Continental Philosophy* (New Haven: Yale University Press, 2006); see also Foltz and Robert Frodeman, eds., *Rethinking Nature: Essays in Environmental*

Philosophy (Bloomington: Indiana University Press, 2004), and Zimmerman, *Contesting Earth's Future: Radical Ecology and Postmodernity* (Berkeley: University of California Press, 1997).

8. Cf. Charles S. Brown and Ted Toadvine, eds., *Eco-Phenomenology: Back to the Earth Itself* (Albany: SUNY Press, 2003).

9. Foltz, *Inhabiting the Earth: Heidegger, Environmental Ethics and the Metaphysics of Nature* (Amherst, N.Y.: Humanity, 1995); Ladelle McWhorter and Gail Stenstad, eds., *Heidegger and the Earth: Essays in Environmental Philosophy* (Toronto: University of Toronto Press, 2009); and Casey Rentmeester, *Heidegger and the Environment* (London: Rowman & Littlefield International, 2015).

10. Theresa Morris, *Hans Jonas's Ethic of Responsibility: From Ontology to Ecology* (Albany: SUNY Press, 2014).

11. William Edelglass et al., eds., *Facing Nature: Levinas and Environmental Thought* (Pittsburgh: Duquesne University Press, 2012).

12. Suzanne L. Cataldi and William S. Hamrick, eds., *Merleau-Ponty and Environmental Philosophy: Dwelling on the Landscapes of Thought* (Albany: SUNY Press, 2007); see also Toadvine, *Merleau-Ponty's Philosophy of Nature* (Evanston, Ill.: Northwestern University Press, 2009).

13. Cf. Bernd Herzogenrath, ed., *An [Un]Likely Alliance: Thinking Environment[s] with Deleuze/Guattari* (Newcastle: Cambridge Scholars, 2008), and Herzogenrath, ed., *Deleuze/Guattari and Ecology* (London: Palgrave Macmillan, 2009).

14. Coeditor David Wood wrote one of the first essays on eco-deconstruction: "Specters of Derrida: On the Way to Econstruction," in *Ecospirit: Religions and Philosophies for the Earth*, ed. Laurel Kearns and Catherine Keller (New York: Fordham University Press, 2007).

15. Cary Wolfe, ed., *Zoontologies: The Question of the Animal* (Minneapolis: University of Minnesota Press, 2003); Leonard Lawlor, *This Is Not Sufficient: An Essay on Animality and Human Nature in Derrida* (New York: Columbia University Press, 2007); Matthew Calarco, *Zoographies: The Question of the Animal from Heidegger to Derrida* (New York: Columbia University Press, 2008); Kelly Oliver, *Animal Lessons: How They Teach Us to Be Human* (New York: Columbia University Press, 2009); Anne-Emmanuelle Berger and Marta Segarra, eds., *Demenageries: Thinking (of) Animals after Derrida* (Amsterdam: Rodolpi, 2011); David Farrell Krell, *Derrida and Our Animal Others: Derrida's Final Seminar, "The Beast & the Sovereign"* (Bloomington: Indiana University Press, 2013); Dawne McCance, *Critical Animal Studies: An Introduction* (New York: SUNY Press, 2013); Wolfe, *Before the Law: Humans and Other Animals in a Biopolitical Frame* (Chicago: Chicago University Press, 2013); and Judith Still, *Derrida and Other Animals: The Boundaries of the Human* (Edinburgh: Edinburgh University Press, 2015).

16. Cf. Brown and Toadvine, eds., *Eco-Phenomenology: Back to the Earth Itself*. This subtitle is something of a blend of slogans from Edmund

Husserl ("We must go back to the 'things themselves'") and Friedrich Nietzsche ("*remain faithful to the earth*"). For Husserl, see *Logische Untersuchungen, Zweiter Band: Untersuchungen zur Phänomenologie und Theorie der Erkenntnis, 1. Teil* (Tübingen: Max Niemayer Verlag, 1993), 6; trans. J. N. Findlay as *Logical Investigations* (London: Routledge, 2001), 1:168. For Nietzsche, see *Also Sprach Zarathustra: Ein Buch für Alle und Keine* (Stuttgart: Philipp Reclam, 1994), 10; trans. Adrian Del Caro as *Thus Spoke Zarathustra: A Book for All and None* (Cambridge: Cambridge University Press, 2006), 6.

17. Wood, "What Is Eco-Phenomenology?," in *Research in Phenomenology* 31, no. 1 (2001): 78–95; see also Brown and Toadvine, *Eco-Phenomenology*, 211–33, and Wood, *The Step Back* (Albany: SUNY Press, 2005), 149–68. Wood argues that eco-phenomenology, into which are folded both an ecological phenomenology and a phenomenological ecology, offers us a way of developing a middle ground between phenomenology and naturalism, between intentionality and causality. Our grasp of nature is significantly altered by thinking through four strands of time's plexity— the invisibility of time, the celebration of finitude, the coordination of rhythms, and the interruption and breakdown of temporal horizons. It is also transformed by a meditation on the role of boundaries in constituting the varieties of thinghood. Eco-phenomenology takes up in a tentative and exploratory way the traditional phenomenological claim to be able to legislate for the sciences, or at least to think across the boundaries that seem to divide them. In this way, it opens up and develops an access to nature and the natural, one that is independent both of the conceptuality of the natural sciences and of traditional metaphysics.

18. In *Of Grammatology*, Derrida writes that "the critical movement of the Husserlian and Heideggerian questions must be effectively followed to the very end, and their effectiveness and legibility must be conserved. Even if it were crossed out, without it the concepts of play and writing to which I shall have recourse will remain caught within regional limits and an empiricist, positivist, or metaphysical discourse [*un discours empiriste, positiviste ou métaphysique*]" (OG 50/73). A little later he writes that, while noting "how much transcendental phenomenology belongs to metaphysics," the deconstructive "thought of the trace can no more break [*rompre*] with a transcendental phenomenology than be reduced to it" (OG 62/93).

19. See Wood, "Notes Towards a Deconstructive Phenomenology," in *Step Back*, 131–38.

20. Forrest Clingerman, Brian Treanor, Martin Drenthen, and David Utsler, eds., *Interpreting Nature: The Emerging Field of Environmental Hermeneutics* (New York: Fordham University Press, 2014). A related approach exceeding this more philosophical lens can be found in ecolinguistics. For an introduction, see Alwin Fill and Peter Mühlhäusler, eds., *The Ecolinguistics Reader: Language, Ecology, and Environment* (London: Continuum, 2001).

21. Ibid., 10.

22. Paul Ricoeur, *Le conflit des interprétations: Essais d'herméneutique* (Paris: Seuil, 1969); trans. as *The Conflict of Interpretations: Essays in Hermeneutics* (Evanston, Ill.: Northwestern University Press, 2007).

23. Rachel Carson, *Silent Spring* (Boston: Houghton Mifflin, 1962).

24. Cheryll Glotfelly and Harold Fromm, eds., *The Ecocriticism Reader: Landmarks in Literary Ecology* (Athens: University of Georgia Press, 1996).

25. Greg Garrard, ed., *The Oxford Handbook of Eco-Criticism* (Oxford: Oxford University Press, 2014).

26. Timothy Clark, *Ecocriticism on the Edge: The Anthropocene as a Threshold Concept* (London: Bloomsbury, 2015).

27. Serenella Iovino and Serpil Oppermann, *Material Ecocriticism* (Bloomington: Indiana University Press, 2014).

28. Cf. Diana Coole and Samantha Frost, eds., *New Materialisms: Ontology, Agency, and Politics* (Durham: Duke University Press, 2010). David Wood wonders in his "Specters of Derrida" (282) whether environmentalism might not "provoke a certain materialistic mutation in deconstruction." Contributors Karen Barad and Vicki Kirby's works have both often been associated with the New Materialism; cf. especially Barad's interview in *New Materialism: Interviews and Cartographies*, ed. Rick Dolphijn and Iris Van der Tuin (Ann Arbor: Open Humanities Press, 2012), 48–70.

29. Malcolm Miles, *Eco-Aesthetics: Art, Literature and Architecture in a Period of Climate Change* (London: Bloomsbury, 2013).

30. Jacques Derrida, *L'animal que donc je suis*, ed. Marie-Louise Mallet (Paris: Galilée, 2006); trans. David Wills as *The Animal That Therefore I Am* (New York: Fordham University Press, 2008).

31. Derrida, *Séminaire: La bête et le souverain*, vol. 1, *2001–2002*, ed. Michel Lisse, Marie-Louise Mallet, and Ginette Michaud (Paris: Galilée, 2008); trans. Geoffrey Bennington as *The Beast & the Sovereign*, vol. 1 (Chicago: University of Chicago Press, 2009; Derrida, *Séminaire: La bête et le souverain*, vol. 2, *2002–2003*, ed. Lisse, Mallet, and Michaud (Paris: Galilée, 2010); trans. Geoffrey Bennington as *The Beast & the Sovereign*, vol. 2 (Chicago: University of Chicago Press, 2011).

32. Among others, see Derrida, *La vie la mort* Seminar, Jacques Derrida Papers, University of California at Irvine Critical Theory Archives, MS-C01, Box 12, Folders 10–19. The seminar is currently being edited by Pascale-Anne Brault and Peggy Kamuf, to be published by Galilée in the near future. A translation by Pascale-Anne Brault and Michael Naas is currently in preparation for the University of Chicago Press. Parts of this seminar have already been published, notably its second session, "Logique de la vivante," in "Otobiographies: The Teaching of Nietzsche and the Politics of the Proper Name," in EO 3–38/13–56; its eighth and ninth sessions, "Cause 'Nietzsche'" and "Chaos ('Nietzsche') de l'interprétation," as "Interpreting Signatures (Nietzsche/Heidegger): Two Questions," trans.

Diane Michelfelder and Richard Palmer, in *Dialogue and Deconstruction: The Derrida/Gadamer Encounter* (Albany: SUNY Press, 1989). Its final four sessions were published as "To Speculate: on 'Freud,'" in PC 257–410/275–438.

33. See the discussion of logocentrism throughout *Of Grammatology*, especially the link between ontotheology and logocentrism in the section called "The Hinge/La brisure" (OG 65–74/96–108). Derrida already argued back then that "if the trace, arche-phenomenon of 'memory,' which must be thought before the opposition of nature and culture, animality and humanity, etc., belongs to the very movement of signification, then signification is *a priori* written, whether inscribed or not, in one form or another, in a 'sensible' and 'spatial' element that is called 'exterior'" (OG 70/103). Later, in *The Animal That Therefore I Am*, Derrida could thus write, "Logocentrism is first of all a thesis regarding the animal, the animal deprived of the *logos*, deprived of the can-have-the-*logos*: this is the thesis, position, or presupposition maintained from Aristotle to Heidegger, from Descartes to Kant, Levinas, and Lacan" (AA 27/48). See also the discussion of logocentrism in BS1 343/454–55.

34. Cf. note 32 above.

35. Cf. Francesco Vitale, "The Text and the Living: Derrida between Biology and Deconstruction," *Oxford Literary Review* 36, no. 1 (2014): 95–114.

36. Edward Said, "The Problem of Textuality: Two Exemplary Positions," *Critical Inquiry* 4, no. 4 (1978): 673–714.

37. Michel Foucault, "Mon corps, ce papier, ce feu," in *Dits et écrits*, vol. 2, *1970–1975* (Paris: Gallimard, 1994), 267; trans. Jonathan Murphy and Jean Kalpha as "My Body, This Paper, This Fire," in *History of Madness in the Classical Age* (London: Routledge, 2006), 573.

38. Richard Rorty, "Nineteenth-Century Idealism and Twentieth-Century Textualism," in *Consequences of Pragmatism (Essays 1972–1980)* (Minneapolis: University of Minnesota Press, 1982), 139–59.

39. Richard Grusin, ed., *The Nonhuman Turn* (Minneapolis: University of Minnesota Press, 2015).

40. Cf. Quentin Meillassoux, *Après la finitude: Essai sur la nécessité de la contingence* (Paris: Seuil, 2006); trans. Ray Brassier as *After Finitude: An Essay on the Necessity of Contingency* (London: Bloomsbury, 2009). See also Martin Hägglund's deconstructive response to this text in his "The Arche-Materiality of Time: Deconstruction, Evolution, and Speculative Materialism," in *Theory after "Theory,"* ed. Jane Elliott and Derek Attridge (London: Routledge, 2011), 265–77.

41. Cf. Claire Colebrook, Chapter 11 of this volume.

42. Cf. Michael Marder, Chapter 6 of this volume.

43. Cf. ibid.

44. Michael Marder continues, "Derrida refrains from referring back to the elements or, for that matter, to anything falling in the scope of

the outside-text, hors-texte, that does not exist. . . . By indulging in the excess of codifications, bordering on the indecipherable—in other words by succumbing to a certain temptation of hypersymbolism—Derrida fatally wounds understanding detached from life, but, at the same time, falls short of questioning the primacy of the code or the cipher. . . . But a postdeconstructive break with conventional codification cannot only be theoretical; carrying it out means stepping outside, into a forest, a field, a garden, or a park. There it might be feasible to reconnect with the elements and with the vegetal world, and let the plants express themselves, together with the earth in which they are rooted and which they 'say'"; Marder, "Could Gestures and Words Substitute for the Elements?," in *Through Vegetal Being: Two Philosophical Perspectives*, by Luce Irigaray and Michael Marder (New York: Columbia University Press, 2016), 194–95.

45. Cf. Marder, Chapter 6 of this volume.

46. Cf. ibid.

47. Cf. TS 76; LI 148/273. See also Geoffrey Bennington, "Derridabase," in D 98–105/90–94. For recent discussions of this phrase, see Jonathan Culler, "Text: Its Vicissitudes," in *The Literary in Theory* (Stanford: Stanford University Press, 2006), 99–116, and Sean Gaston, "Punctuations," in *Reading Derrida's "Of Grammatology,"* ed. Sean Gaston and Ian MacLachlan (London: Bloomsbury, 2011), xiii–xxviii.

48. TS 76.

49. For example, in 1996 Derrida clarifies: "The first step for me, in the approach to what I proposed to call deconstruction, was putting into question of the authority of linguistics, of logocentrism. And this, accordingly, was a protest against the 'linguistic turn.' . . . The irony—painful at times—of the story is that often, especially in the United States, because I wrote 'il n'y a pas de hors-texte,' because I deployed a thought of the 'trace,' some people believed they could interpret this as a thought of language (it is exactly the opposite). . . . I do the best I can to mark the limits of the linguistic and the rhetorical—this was the crux of my profound debate with Paul de Man, who had a more 'rhetoricist' interpretation of deconstruction. . . . The notion of trace or of text is introduced to mark the limits of the linguistic turn. This is one more reason why I prefer to speak of 'mark' rather than of language. In the first place the mark is not anthropological; it is prelinguistic; it is the possibility of language, and it is everywhere there is relation to another thing or relation to an other. For such relations, the mark has no need of language"; TS 76. As he adds in a 2001 interview, "Precisely contrarily to this currency of deconstruction that circulates everywhere, according to which deconstruction is a thought of language for which there is no outside of language, while I began by saying precisely the contrary, and continue to do so. For deconstruction, for me in any case, there is an outside of language"; in Marcel Dracht, *L'argent* (Paris: La Découverte, 2004), 208; translation ours.

50. Rorty, *Consequences of Pragmatism*, 139. For Derrida's more detailed analysis of conceptions of life and death in biology, see in particular his 1975 seminar *La vie la mort*, cited in note 32 of this Introduction.

51. Cf., among others, WD 205/244; VP 4/3, 46/59–60, 51/67; OG 139/72; and MP 125/150, 161/192.

52. Cf. Vicki Kirby, Chapter 5 of this volume.

53. Some working in object-oriented ontology, however, have also attempted to bring together deconstruction and ecology. See the work of Timothy Morton, especially "Ecology as Text, Text as Ecology," *Oxford Literary Review* 32, no. 1 (2010): 1–17, and "Ecology without the Present," *Oxford Literary Review* 34, no. 2 (2012): 229–38. Morton writes that "what the new weird realisms such as object-oriented ontology (OOO) do is return to the now somewhat neglected jazz standards of Husserl and Heidegger and rework them within a post-Derridean thinking. To say 'post-Derridean' here means to do philosophy within Derrida's continuation of the Heideggerian project of deconstructing the metaphysics of presence. The implicit truth of *there is no outside-text* is now more true than it was when it could be associated with anti-realism in a facile sense. There is a gigantic coral reef of discrete, unique, irreducible *objects* (OOO's term for any entity whatsoever—a blade of grass, a meteor, a block of staples) that lies beneath the Heideggerian U-boat, at a hitherto unplumbed ontological depth"; ibid., 235. The three years of Derrida's *La Chose* [The Thing] seminar on Francis Ponge, Heidegger, and Blanchot will prove a fertile ground for discussion here; Jacques Derrida Papers, University of California at Irvine Critical Theory Archives, MS-C01, Box 13, Folders 1–2, 11– 17, and Box 14, Folder 8.

54. Cf. GT 81/108, 151/191, and PI 269–71/283–86.

55. Cf. Derrida, "*Ja*, or the *Faux-Bond* II" (PI 30–77/37–82); BS2, *passim*. See also the two seminars constituting parts 2 and 3 of the three-year *Politics of Friendship* seminar *Manger l'autre* [Eating the Other] *(1989–1991)*, Jacques Derrida Papers, University of California at Irvine Critical Theory Archives, MS-C01, Box 20, Folders 1–12, and Derrida, *Rhétoriques du cannibalisme* [Rhetorics of Cannibalism], Jacques Derrida Papers, University of California at Irvine Critical Theory Archives, MS-C01 Box 117, Folders 1–4, and Box 20, Folders 13–15.

56. See MP 329/392; DS 4–6/10–2; and PS *passim*.

57. See in particular Wood, "Specters of Derrida."

58. Cf. Ted Toadvine, Chapter 2 of this volume.

59. Cf. Marder, Chapter 6 of this volume.

60. Cf. John Llewelyn, Chapter 7 of this volume.

61. Cf. ibid.

62. Cf. ibid.

63. Cf. Colebrook, Chapter 11 of this volume.

64. Derrida argues for this humanism of the law not only in "Force of Law" (FL) but also in the recently published *The Death Penalty* (DP1,

DP2). The legal subject is construed as one who values subjection to the law—that is, his autonomy and dignity, his capacity for justice—higher than his life. That is why it belongs to properly human law to demand, at least in principle, the sacrifice of life: in capital punishment, in wars claimed to defend the legal order, and in the form of animal life. As the first volume of *The Beast & the Sovereign* puts it succinctly when speaking of Lacan's view, "With Law and Crime, man begins" (BS1 102/147). On this diagnosis, there is a more than contingent link between the formation of legal orders and the domestication and slaughter of animals.

65. Cf. Dawne McCance, Chapter 13 of this volume.

66. Cf. Kelly Oliver, Chapter 15 of this volume.

PART

I

Diagnosing the Present

The Eleventh Plague: Thinking Ecologically after Derrida

David Wood

Introduction: The Deconstructive Disposition

Derrida has been condemned by some ("Dogs bark at what they do not understand."—Heraclitus) and drawn into empty culture wars by others. Derrida himself hardly ever tried to correct or contain this profligacy. But all of us who have followed Derrida and learned from him at some point or other face the question of inheritance. What is it to inherit the work, the writings, the insights of another? This is not unconnected to "what is it to read?" The problem that arises here is not unfamiliar to admirers of Nietzsche, who cautions against those who would be his followers, calling them *Folgende*, the mathematical expression for zeros. Are we then to follow his proscription against following him? Derrida animates the question of inheritance in *Specters of Marx*, offering a model that would require selection and creative transformation. Moreover, as he insists, a gift sometimes calls for ingratitude. At what level can or should we apply these ideas to reading Derrida himself? Do we have to transform the idea of transformation to avoid just following him? Or would that not be the most faithful, and hence least faithful, response? To be faithful to Derrida, do we have to betray him?

Derrida might not endorse this language, but I propose here a reworking of Heidegger's account of what it is to engage with the work of a great thinker—he speaks of not going counter to the other, but going to their encounter (with Being).[1] And to do this we have to bring our existing passions and commitments to the table. He is saying that we have to address what is most at stake in the other's thinking and writing. And we have to have skin in the game. In glossing Heidegger's claim in this way, I am bypassing the problem that one might seem to have to endorse his thesis about the primacy of the question of Being. A less technical cousin of this claim has legs independent of Heidegger's specific formulation.

Analogously, thinking about how we can *follow* Derrida without falling into aporetic elephant traps, we can draw on a distinction between doing as he does because he says so or does so and doing as he does because it's a smart thing to do. Put less casually, Derrida's ruminations about reading Marx (again) are themselves not completely original, and no worse for that. Context, and the space of concern, change. Marx was not saying just one thing, but drawing together multiple threads from which we, in our time, cannot but select. Licensing ourselves to do this can enhance the transformative creativity of our response. It is in this spirit that I want to advance the idea of a deconstructive disposition. And in response to the ten plagues that Derrida names in *Specters of Marx*, I want to insist on an eleventh plague—our growing global climate crisis.[2] To honor Heidegger's formulation at the same time it would be necessary to formulate this reference to an eleventh plague at something like an ontological level without being caught up in the seductions of ontology. Forging an amalgam from Derrida and Heidegger, we would try to show that the eleventh plague was not just one more plague, but was at the heart of the first ten, or at least intimately implied or caught up in them. In the most summary form, this would be to show that questions of violence, law, and social justice are inseparable from ecological sustainability. A similar move would demonstrate that another candidate for the eleventh plague—the animal holocaust—is closely tied up both with those first ten plagues and with ecological sustainability, perhaps serving as something of a bridge. I will only gesture at such an account here.[3]

What then is meant by a deconstructive disposition? The danger of such an account is that it may seem to dilute what deconstruction has to offer by blurring how it differs from other modes of critical reading. I will address this shortly.

I propose four dimensions to a deconstructive disposition.

Negative Capability

Keats described this as a willingness to tolerate ambiguity and uncertainty.[4] This is not to license intellectual laziness but rather to caution against premature closure. Derrida's reference to going through the undecidable could be understood in this way. And indeed, the broader willingness, even passion, to disturb the sleeping dogs of (often binary) complacency.

Patient Reading

The point of reading (and thinking) is not simply to understand, using the handrail of existing meaning, but to open possibilities. This requires patience, even when we have no time! What does such patience yield? It allows us to restore repressed differences and to expose invisible framings and stagings, even of the very occasions at which issues are being discussed. (See Derrida's prefatory remarks to many of his presentations, raising such questions as, "What is an *international* conference?")

Aporetic Schematization

Thinking often takes an essentially aporetic shape: the past that was never present, the gift that resists gratitude, the supplement to what is already complete, the always already, forgiving the unforgivable. These shapes need to be exposed and worked through, if only to grasp the complex underbelly of intelligibility and coherence.

Attention to Language and Terminological Intervention

Language is not neutral. Words, sometimes limiting or regressive, harbor ways of seeing and being in the world. We can intervene in this invisible process with careful attention to these frames and by actively bending old words and inventing new ones.

It might be said that none of these dispositions is exclusive to deconstruction. So is there not a danger of dilution?

Deconstruction in the late '60s and early '70s was an *event*, an interruption, a challenge, one attuned to the time—structuralism, semiology, a quiescent Marxism, pervasive doubts about the complacencies of humanism, of psychoanalysis, of phenomenology and literary theory. (I am trying to cover here the French and Anglo-American situations, which were different.) But it is no longer an event. Its covert influence has waned, even as the scholarly industry prospers, and some of those strongly influenced by Derrida delight us with their own brilliance and originality. Moreover, it is not entirely a bad thing that deconstruction should have metastasized in many directions, even if its pedigree is less visible. Deconstruction would cease to be deconstruction if it became an idol, an orthodoxy, a citadel to be defended. Last, *this* conference,[5] along with the book we are assembling, is something like an event or renewal, a *repetition* of deconstructive strategies, gestures, and sentiments in the context of a new urgency.

Calculating and naming the inheritance of deconstruction is a thankless and unending task. The most salient threads that specifically address environmental concerns would include language, time, the animal, sovereignty, topological complexity, the new international, inheritance, and death.

I propose to make some remarks here about the first three of these threads.

Unsettling Language

Deconstruction's bad press began with the phrase "there is nothing outside the text," which sounded like linguistic idealism. It was later reworked as the ineliminability of con-text, and the impossibility of ever completely specifying that context. But language itself continued to ground both hesitation and creative response. The normative commitments of words like "parasite," "rogue" (nation), "proper," and "authentic" all rest on structures of asymmetrical binary privilege that can be exposed, and perhaps destabilized, by inserting an *indécidable*.

And as Derrida showed in "Des Tours de Babel," translation, which marks the instability of proper meaning, is a powerful site for deconstructive archaeological excavation.

Consider three classic examples:

1. It is said that the bombing of Hiroshima was ordered after Prime Minister Kantaro Suzuki responded to the demand that the Japanese surrender by using the word *Mokusatsu*, which can mean to "ignore/not pay attention to" or to "refrain from any comment." The former could reasonably be considered as a refusal to surrender. The latter was asking for more time. Could the subsequent loss of some quarter of a million lives in Hiroshima and Nagasaki be put down to a mistranslation of a nuance of meaning?

2. The cult of Mary, the miraculous character of Jesus's birth, rests on the translation of the original Hebrew *almah*, הָעַלְמָ, as virgin when it more accurately meant "maiden" or "young woman."

3. The license given by the translation of *rada* as dominion, in Genesis 1:16/1:26, as God's understanding of the relation between man and the other animals and nature more generally has been argued to be the source of much Western complacency over the destructive and exploitative consequences of man's reign over nature.[6] Some have argued that it should be translated as "hold sway" and others as "rule," with the strong implication of the benign responsibility that might be expected of a thoughtful ruler. Others have pointed out that the literal meaning of *rada* is "a point higher up on the root of a plant." Such a point is where the strength of the plant as a whole is centered, offering a more collaborative sense of "privilege."

This last has direct relevance to eco-deconstruction. The authority of canonical texts is a continuing issue, considering the continuing reverence accorded to the Bible, the Koran, and other religious writings. This would be true even if there were no issues of translation. But the hermeneutic mischief with which they can be treated seems to know no limits. The demonstration that even an authoritative text contains within it competing meanings and possibilities allows other ways of reading it to be opened up. And, of course, this applies not just to the Bible, but to the U.S. Constitution, to the pre-Socratics, to Aristotle, Kant, and so on. Latour's recent treatment of Lovelock's Gaia Hypothesis, in which he attempts to empty it of any residue of political theology, is a good example of how the angle of such readings can make a difference to environmental thinking.[7]

Central too to any eco-deconstruction are the force and meaning of the word "nature," caught up as it is in binary opposition to culture, to man, and to spirit and functioning repeatedly as a transcendental signified, a ground of meaning that would escape the play of language and the very oppositions in which it is inscribed. We are all acquainted with nature, whether it be last year's tornadoes or this year's tomatoes. Nature seems, straightforwardly, to be what's "out there," something we realize we are part of when we feel hungry or get lashed by heavy rain. But nature is not just what is real, what is out there. When placed in opposition to culture, it has played a powerful cognitive role in organizing human life and thought. And one of the hallmarks of early deconstruction was to problematize this simple opposition. It is clear, for example, that we approach nature through all kinds of cultural mediations and constructions, which themselves change through history. And these cultural constructions are not just shaping or distorting lenses; they often lead directly to transformations of nature. (When nature is treated as a resource, a mountain becomes a pile of quarry stone.) Eco-deconstruction reflects our hope that we can get clearer about the complex role that nature plays in our thinking, in our understanding of ourselves, and in our practical existence. This issue is important in academic life, not least because university institutions are constructed on the basis of distinctions between natural sciences, social sciences, and humanities, as if these were separate fields of inquiry, distinctions that depend on how we think about nature. Deconstruction has made it more normal to inspect the boundaries, the frontiers, the contaminations, the difficulties in making these clear-cut distinctions.

While there are those for whom this distancing (from a naïve sense of nature) comes easily,[8] there are others who resist, who recognize the desire to point and say, "That's nature," that striving, pulsing force that precisely escapes description, like Roquentin's black root in Sartre's *Nausea*. The question we are left with is this: is it possible to accept that any concept we have of nature, any meaning we give that word, is culturally constructed, riddled with narrative, and as such burdened, while insisting that there *is* something we are in different ways culturally constructing? Much interesting work has been done critiquing the idea of a return to some original "natural condition," the restoration of a pristine origin, the protection of the purity of wilderness. As Bill McKibben wrote long ago, there is no nature anymore.[9] Nothing with air blowing through it has escaped human influence.

So the analysis is taking place at two different levels. The concept (or sign) of nature cannot escape the cultural conditions of conceptuality. And nature itself, materially, has been contaminated by human activity, destroying the purity by which it could function in opposition to man—all this presupposing that man was not always already part of nature.[10]

Arguably our dominant modes of engagement with the natural world are the reflection of narratives, often what Jean-François Lyotard would call "grand narratives," such as man's God-given sovereignty over the natural world, or man's place in the great chain of being, or the story of enlightenment, in which inferior races, religions, and cultures suffer the same subordinating fate as nonhuman creatures, a fate in which these various disparagements are often roped together. This presents us with an option: either to abandon the whole grand-narrative scene in favor of multiple, local smaller-scale narratives or to continue with narrativity as indispensable while interrupting it. Or replacing an oppressive grand narrative with one with more of a future. Recall that Derrida wrote of the Bin Laden narrative that it "does not open a future."[11]

The upshot of these debates is itself a contested space. Some would use the constructedness of nature as an argument against any critique of technology that would accuse it of sullying our natural condition. Others more reasonably argue that we need criteria other than protecting or restoring purity by which to evaluate our engagement with the earth. These considerations all develop from reflection on the word "nature" and the *desire* attached to it.

It is not always clear whether these issues are linguistic, conceptual, or empirical, but the scope we give to words like "pain," "consciousness," and "person" have direct consequences for ways in which we engage with the natural world. Allowing that animals feel pain qualifies them for our consideration. Crediting them with consciousness bestows further rights. Indigenous peoples often attributed personhood to nonhumans that strongly shaped their engagement with them. And the scope of personhood has major consequences not just for the biopolitics of abortion, but also the rights of corporations to recycle their profits in such a way as to promote climate-change denial through the corruption of political discourse by lies and sophistry. It is as legal persons that Americans protect the freedom of billionaires and multinationals to speak with their wallets.

The issue of whether we are indeed dealing with climate change or global warming is itself contested. Those who deny it tend to call it "climate change," while those who accept it, its anthropogenic cause, and want to do something about it tend to speak of global warming. Moreover, it is surely remarkable that we do not have much of a name for what is likely in store for us, which is climate catastrophe. Here I have in mind Derrida's brilliant remarks about 9/11 in which he argues that our use of a date to name this event reflects an inability to grasp what actually happened, much like a traumatic event.[12] The real event, he suggests, is not what happened on the day, but the hole it blew in our sense of security, making us wonder—what next?

The matter of the name is of deep significance in thinking about global catastrophe. As Lacan showed in his discussion of the symbolic phase, having a name is an ambivalent phenomenon. On the one hand, it functions as a handle by which others can manage us. On the other hand, it is a source of self-integration, recognition, rights, and so on. In light of this, it is of particular interest that the mass species extinction currently underway is taking place with most of the 8 million+ species not even being identified or named. Their very existence is a statistical extrapolation. If there is something tragic about threats to species we know (like the Snail Darter fish in the Tellico Dam on the Little Tennessee River in 1973), there is something beyond tragic in the extinction of species we humans have never even identified as such. Nietzsche laughs at philosophers' supposed concern with truth, telling us that, like every other creature, we are really only concerned with what contributes vitally to our lives.[13] And yet at another level we are deeply committed to knowing what is going on around us, and, reflectively, at least, anonymous extinction is surely a shameful matter. Joni Mitchell's "You don't know what you've got 'til its gone" is the optimistic version. More truthfully, you still don't know—we will never know—even when it's gone. The power of the name is real. The masses of animals industrially slaughtered for food die without names, many even without numbers. Their individuality is pre-eclipsed as stuff. Even chicken no. 2013783456 is a source of chicken stuff. A name is no guarantee of protection. At times it can be a death sentence.[14] But it does draw you in to the symbolic and the possibility of negotiation. Similar issues are raised by concern for future generations of humans, who as yet have no names, and indeed do not exist, and yet arguably have interests that need to be taken into account.

When Derrida talks about animals, three obvious points stand out. First, he claims, the very word "animal" is pretty much a license to eat, or at least to consume in whatever way suits us. It has little if any biological significance, occluding every difference between the creatures it subsumes. And instead of talking of animals, when he is careful, he talks about "those we call animals."[15] He rubs this in by coining the word *animot*, highlighting the way language has here been captured by the anthropological machine. Finally, consider his willingness to speak of animal genocide and link their fate with the holocaust in the face of those who would reserve the latter expression for the singular horror of Nazi concentration camps. This is a choice that shows evidence of having gone through the undecidable. Perhaps it was the sense that acquiescence in the face of those who seek exclusive ownership of this word would privilege one event of silent horror even as another one continues, at the dead end of other country roads, unacknowledged, largely unsung, protected by new alibis, new myopias.

Returning more explicitly to global warming (or whatever we call it), the significance of language in grasping or hiding its significance is hard to exaggerate. So too are the opportunities for deconstructive engagement with language—both everyday language, the common discourses of legitimation, and the language of philosophy itself. While the percent of atmospheric CO_2 rises inexorably, the narratives we construct to justify the things we do that contribute to this rise have continuing legitimacy (George H. W. Bush: "The *American way of life* is not up for negotiation. Period" [1992]; my italics). The discourse in which employment opportunities trump sustainability has a certain independence from the real, especially where that real is fabricated in part from what is still around the corner. Language lags behind the real and often distorts it, even if there are no perfectly proper words. A deconstructive disposition does not see language as a surface phenomenon we can set aside, but as a deep and fundamental part of the problem. It offers many ways of performatively challenging and displacing the language in which key aspects of global warming are often articulated.

Aporetic Temporality

For phenomenology it was important to step back from objective worldly time to the internal time consciousness that makes it possible constitutively. For Derrida this is an attempt to return to a subjective self-presence that is in fact riven by linguistic and temporal difference in a way that essentially undermines the very presence it seeks. Deconstruction on the one hand abandons any attempt at a postmetaphysical *theory* of time, but on the other hand it proliferates a slew of aporetic temporalities we would do well to take seriously.[16] There is an ongoing resistance to any seamless linear time of progress. Derrida replaces this with an im-possible messianic time of hope in which, as with a democracy-to-come, any literal sense of time seems to be converted into a certain (im)possibility and openness, much as happens in Heidegger.[17] If Nietzsche displaces eschatological time with an eternal return that forcibly extinguishes any residues of cosmic teleology, deconstruction is happier with multiple temporalities not subordinated to any sovereign time and with what Albert Hoftstadter once called "strange loops," such as a past that was never present, or a hauntology that can never fully repay its debt to the past and is always haunted by what it imagines it could forget. Derrida draws on such an idea specifically in thinking through just what kind of Marx we could still inherit, a Marx who, though a materialist, still has room for specters. He treats Marx in much the way that Nietzsche urges us to treat history, as a resource for critically received possibilities for the creative furthering of life, even as elsewhere he will contest the very opposition between life and death.

The 9/11 attack perhaps offers the most striking example of a cluster of competing nonlinear temporalities. It was said that it could not have been anticipated, and yet many did anticipate it. While it all happened on that one day on September 11, images of it were relentlessly repeated, all over the world, in the days that followed. And the meaning of this event took a while to sink in. Early Derrida had written that "the future can only be anticipated in the form of absolute danger."[18] It is hard to see how this can be generally true, but it fits rather well the traumatic effect of 9/11. On his reading, 9/11—identified only by a date—is the explosion of any sense of cosmic security. Bad though that event was in itself, its true significance lies in what it portends. If *that* could happen, what next? The events of 9/11 fractured our sense of a benignly unfolding future. A new Pearl Harbor. An event in time that shatters the time frame in which it appears. Time becomes irreversibly complex, and we cannot think it without inhabiting such complexity.

At the same time as we become aware of the future as a potential site of danger and dramatic disruption, it also becomes clear that this is nothing other than the past catching up with us. Our past practices, none intrinsically evil or catastrophic, accumulate like DDT in eagles, until they overflow into dramatic change. What comes at us, seemingly from the outside, like the melting of glaciers, is an indirect product of our own agency.

Again, some ask whether global warming will really happen. Others reply that the future is already here. We are witnessing it without being sure what it is, as if only the eyes of the crocodile had broken the surface. And the crocodile deniers spring up everywhere. For with all this talk of the unpredictable, some of the future has happened, and it is well known that the critical 2 percent rise in average surface temperature, after which real unpredictability is predicted, is already in the pipeline. The "always already" is not just a quasitranscendental mantra. We cannot return to simpler phenomenological times in which retention, current awareness, and protention would be happily interwoven with thematic memory and expectation. Some of the ingredients may be the same, but baking them together need not result in a digestible dish. The real objection to the thought that aporetic time is the new normal is that it was already the old normal, but we just did not realize it. If so, we can no more return to a simpler way of inhabiting time than we can return to simpler times or a lost origin. We can at best acknowledge the shape of such desire (for presence), guard against its seductions, and invent new shapes of dwelling.

It is often said that we should adopt a precautionary principle with respect to future environmental damage. In the absence of complete proof, we should still try to prevent harm by acting on the best evidence. In a sense, this is obvious because we cannot have certainty about a future that has not happened and will only ever happen once. In all sorts of areas, we already act like this, taking out insurance, for example, as a precaution. It is something of a complement to being open to whatever comes. And it captures the spirit of a certain middle voice, or perhaps a double voice, blending or mediating between agency and receptivity. Into this mix we need further to inject the disturbing thought that while we may know what kind of earth *we* would want to inherit, our descendants may not actually conform to our expectations. Our great-grandchildren may not value hiking in the mountains. This raises the question of what it is we should be trying to preserve. But it also raises the question of whether we might cease to care. Both the precautionary principle and a commitment to sustainability presuppose our ability and desire to identify with beings somewhat like us, whether it be narrowly conceived (white middle-class humans) or broadly (complex life forms). What if, like the Atlantic Conveyor, this stream of projective identification ceased to flow? This sounds unlikely, but if one asked again Levinas's question "Are we duped by morality?" and came to see most other humans as irredeemably violent, short-sighted, and self-interested, would we automatically want to encourage this species? What if we went beyond video games and developed simple non-toxic ways of directly stimulating the brain's pleasure centers and everything we now know as culture, indirect means to the same end, were set aside as old school? Would such beings be worth saving? These issues arise within the framework of time in that they affect the way we inhabit our temporal horizons—for example, our capacity for future projection, planning, and hope. They are not issues exclusive to deconstruction. But they expand and extend its sense of the subject not as some transcendental constituter, but as colonized by such desires as are captured by the expression "carnophallogocentrism," open to further colonization.

If the future is a site of profound anxiety, the past is not entirely different. Those who resist the theory of evolution often do so because they cannot stomach the thought that we share a common ancestry with monkeys.[19] And while there are some who would remind us that we are made of stardust, there are others (like me) who think of that as bad (reductive) materialism. In what sense is the history of the earth "our" history? Does this extend through to the emergence of primates from mammals (65 million years ago), or early versions of man in Africa,[20] some 2.8 million years ago, or only as far back as Homo sapiens (250,000/400,000 years)? Or do we see ourselves as part of the stream of life, itself developing from mineral existence, through organic molecules, to the earliest life forms? Is the Big Bang part of our history? It is not difficult to see what gentle pressures have elicited such questions. We used to be, roughly speaking, located within geological history. Now we have hatched the idea of the Anthropocene that would mark the unprecedented impact of a lifeform on the geological forces of the planet (if we exclude the first oxygenating bacteria), a confluence of two quite distinct temporal scales and streams. And we need some such schematization of multiple semi-independent temporalities to think through what we might call the "uneven development of humanity." This is well captured by images of boy soldiers wielding AK-47s. We have productive and destructive powers that outrun both our brain development and the political institutions needed to manage them.[21] Humans are notoriously bad, for example, at risk assessment when thinking about the medium to long-term future, consistently underestimating the risk of high-cost, low-likelihood events. There may be good evolutionary reasons for this. Ancient man had much shorter lifespans. International law (and bodies like the UN) are obvious resources both for preventing conflict and for combating climate change. Yet their power still rests on the support, or at least acquiescence, of individual states. And they are still prey to all the paradoxes of collective action they were designed to overcome. We need better brains and better mechanisms of collective agency to cope with the powers we have developed. Time, as Derrida said, quoting Hamlet, is out of joint.[22]

We Animals

If we (humans) can still use the word, we are and are not animals. Biologically we are animals, and yet "animal" is the name we use for a vast array of nonhuman fellow travelers on the planet. To call an individual human an "animal" is usually to denigrate him (or her) or to set the table for slaughter. Deconstruction takes as its starting point the ways in which our self-understanding as human rests on the construction of the animal as a subordinated other. To the extent that our grasp and treatment of nonhumans is the site of the ongoing and probably interminable operation of the anthropological machine, both our experience of nonhumans and the ways in which we try to think about them require constant vigilance lest we merely use nonhumans as projective screens.[23]

I argued some time ago that there is no such thing as an animal.[24] There are aardvarks, antelopes, armadillos, Australians . . . and there are vast differences between them. Is there not an abyss between man and animal, as Heidegger insists? It would be crazy to deny the gulf, but it too is not one thing, but many and varied.[25] What all this argues against is any hierarchical table of species, any attempt to covertly attribute normative rankings to certain key characteristics. It demands a step back, however difficult that might be, from our understandable tendency to value what we humans think we are good at. To speak of the human subject in terms of carnophallogocentrism is to begin to constitute this being in terms of deep desiring practices, including meat eating.[26] There are obvious dangers in any reductionism, and yet the rewards—including glimpses of new shapes of thought—are quite real. In this vein, one might speculatively propose a new natural history of philosophy—as a rationalization of power, first over other humans, and then specifically as a justification for kinnibalism.[27] Derrida insists that vegetarians do not escape the charge of carnivorousness. We "eat" others in countless other forms of violence, and refraining from meat can be an alibi for a broader blindness to violence.[28] Deconstruction *need* not make any such substantive claim, but it can track (and interrupt) the performative ways in which our use of language and its underlying schematizations sustain such claims.

There may be no one "question of the animal." Derrida's ruminations on his cat in the bathroom, allowing himself to be put in question by its gaze, have implications for humans and animals in general (for example, about who is more naked).[29] But he is insistent that he is speaking of this specific cat (while strangely supplying no name). The ethical may indeed be born from such singular encounters. And yet, as we have seen, and in his earlier arguments for a hyperbolic responsibility for all cats when he feeds his own, Derrida does not hesitate to cast his net more widely, eventually addressing animal genocide and even the ways in which it exceeds the holocaust by breeding animals for them to be killed. Taking all this seriously would give dramatic new life to Heidegger's discussion of being-toward-death. It is important too to take seriously the way deconstruction puts pressure on the distinction between active and passive, the temptation to treat self-conscious agency as the paradigm of responsibility. For the *other* animal genocide, the Sixth Extinction,[30] is not the result of a conspiracy of evil, but of creeping negligence. At first we didn't really know what we were doing, then we didn't want to know. To keep going in the same way—with habitat destruction, with threats to countless species from climate disruption—is culpable negligence. In other words, while deconstruction can contribute to our thinking of animal rights by liberating the gaze of the individual animal, it also lubricates the path to environmental ethics by preventing us from hiding under the bush of individual agential responsibility. We, indeed, are responsible. But it immediately raises the question of who "we" are. And this is not (if it ever could be) just a semantic issue. The "we" question has itself to do with how any such collectivity is constituted, by whom, to what end, and with what powers. And the "we" that might be needed to effect a change in "our" treatment of those "we" call animals, whether direct or indirect, would need to be set aside if "we" came to include nonhumans in a broader we—this time a "we" of interdependence and common fate.

The interdependence of humans and nonhumans is often invisible. Few study the role of beetles or fungi in recycling dead trees and leaves. And yet without them the cycle would cease and life on earth as we know it would grind to a halt. "We must all hang together, or assuredly we shall all hang separately."[31] Might we not come to understand Derrida's democracy-to-come as (impossibly) embracing "animals," perhaps in alliance with Bruno Latour's "parliament of things"?

There is an invasion of the "we" on the horizon; both ethically and ecologically we humans are not as separable from other creatures as we would like to think. Traditionally we have tended to suppose that when it comes to microorganisms (such as bacteria, fungi, viruses) we do need to draw the line just to survive. Disease-causing organisms are simply the enemy. But again, deconstruction is well positioned to articulate the difficulty of this position. It looks increasingly as if an oppositional stance is a miscalculation (as in so many other areas). Consider the impact of antibiotics on human health in recent decades. Over-prescription, failure to complete the course, and the routine use of antibiotics in meat production (faster weight gain), have bred drug-resistant bacteria—a phenomenon allied to that of the autoimmune response in its staging of the collapse of oppositional logic: two different failures of sovereign control. Moreover, the microbiome project makes it clear that what I call "*my* body," the one that must protect itself against the alien invader, is always already a "we," crammed full of benign bacteria and other microorganisms, without which I could not, for example, digest. Broad-spectrum antibiotics bomb wedding parties while being aimed at terrorists. We do not know what we are doing to ourselves because we do not know quite how and how much we are a "we" (or many "we's").

We have followed three major deconstructive motifs to begin at least to show why the environmental crisis deserves to be treated as the eleventh plague. The question we asked earlier was whether this could just be added on to the previous ten, as if it had been overlooked or forgotten. Or was Derrida just counting on his fingers and ran out of numbers? The answer to this question is important.

First, we need to remind ourselves that his ten plagues (SM) are plagues of the New World Order, the one triumphantly celebrated by Fukuyama. In other words, they represent the dark underbelly of the free-enterprise, free-market world that, albeit imperfectly, is said to have brought so much prosperity to so many. Some of the headings are economic (unemployment, economic war, the burden of foreign debt, contradictions of the free market) but not all. There are various other dimensions of the military-industrial complex (nuclear weapons, the arms industry, and interethnic wars), and then there are failures of democracy (phantom states like the Mafia, the exclusion of refugees from the democratic process, and the broken promise of international law and institutions). Let us admit there is much that is left out—for example, persistent poverty, growing inequality, the power of multinationals —this list is neither complete nor homogeneous. But how would global warming fit in? Can it just be added to such a short list of neglected plagues?

One reason to resist such an approach is that there are intimate connections between some of these first ten plagues and the looming climate catastrophe. Joining a disconnected list would be a missed opportunity to pursue those connections. Some examples: employment issues are some of the most urgent with which governments have to deal and are specifically used as reasons not to pass or enforce environmental legislation. Intense global economic competition accentuates the tendency to externalize every possible cost—seeking states and countries with lax waste-dumping laws, precipitating a rush to the bottom. Nature (in the shape of rivers, oceans, and the atmosphere) then picks up the tab. Interethnic war is often fought over scarce natural resources feeding the insatiable monster of development—the spreading demand for a better lifestyle. This then generates the refugees fleeing from war and destruction. Global warming will accelerate these displacements as people abandon desertified land, with all the tragedy of life in camps with poor facilities for those who survive. Foreign debt is a crushing burden on a country that deprives it of the economic surpluses that would enable investment in alternative energy sources. More generally, poor and deprived people have a greater interest in surviving until tomorrow than in embracing sustainable lifestyles. These interconnections and more suggest that merely adding number eleven to the first ten is not the right answer.

But there are two further levels at which we can pursue this question.

1. It is eminently plausible to think that without the kind of transformation—revolution—in the shape of human desires, and hence lifestyles, we are either doomed environmentally or we face a return to feudalism or military subjugation along North Korean lines. These latter would enforce poverty for the masses while the 1 percent live high on the hog, which would cut the average energy footprint. If we accept that the alternative is either doom or dreadful social and political regression, the prospect of real social justice, the realization of so much unfulfilled promise would cease to be some sort of felicity and become a survival necessity. We could begin to envisage a convergence between social (and interspecies) justice and environmental necessity. Much of what deconstruction has done already in terms of welcoming the questioning and re-visioning of borders, boundaries, limits, and identities and exposing the costs of modes of thinking and speaking and dwelling that hide the costs of constitutive exclusion would enable such a convergence.

Derrida's remarks about the animal holocaust and about human suffering and misery are set in the context of our denial, blindness, and refusal to acknowledge these phenomena and the way that human suffering especially represents the contradiction, the hidden waste, produced by an ever more efficiently functioning system.

2. There is, however, another step to be taken, which would make it even clearer that the eleventh plague is not just one more plague, but something of a supplement that completes what seemed already to be complete. The naïve and unguarded way of putting this would be to say that climate catastrophe threatens the material ground of our being and so cannot be compared to the other ten plagues, except perhaps the much more uncertain prospect of nuclear war. But does deconstruction have anything special to say about this, or are we stretching to make a connection? As I see it, climate catastrophe would be the material face of the culmination of an emergent contradiction captured in such key deconstructive concepts as the exclusion of the Other and the autoimmune response. The first tells us that establishing and maintaining binary dominance requires the suppression of the lesser force, an operation that will have unintended consequences. The second tells us that attempts at protecting a strong identity will tend to destroy the very identity they seem to serve. Obviously, these are closely connected. To these two we might add a third, from outside deconstruction—that profits, economic value, on which so much of the world turns—rest not just on the exploitation of wage labor, but on the extraction of finite natural capital and the externalization of the costs of waste disposal onto nature—typically such sinks as sea and air. These three comprise a kind of logic. At this point warning bells go off. The detachment of just such a logic from material historical social conditions was Marx's objection to Hegel: *Geist* is properly understood through "relations of production." Are we succumbing again to a kind of idealism, with deconstruction supplying a quasitranscendental underpinning to historical inevitability? Isn't the lesson of such prognostications that we discount what we cannot anticipate? Capitalism did not wither under its contradictions, nor did it bring about the progressive immiseration of the poor. It bought off its contradictions, at least for a while, and fanned mass consumption into a force for economic growth. And communism did not lead to the withering away of the state.

So, to clarify, this third step, this attempt to show that global catastrophe is not just one plague among others, is not an exercise in counterprovidential history or an attempt to draft deconstruction into the business of prognostication. Rather it marks the site at which, it would seem, some of the fundamental bioexistential parameters for human and nonhuman flourishing might well be breached, for all intents and purposes, irreversibly. And it does so by highlighting the aporetic realities of our thinking and dwelling. Can deconstruction, however, really think about hurricanes, the melting of the ice caps, the release of vast quantities of methane from Siberian permafrost, new disease vectors, mass migration, starvation, and agricultural disruption? Derrida was perhaps more comfortable reminding us that 9/11 terrorists attacked, inter alia, the global communication network that made even their own terrorism possible by broadcasting those images, metastasizing the terror. He is perhaps less comfortable thinking of the earth as a whole, even as a complex system of differences, with all the dangers of totalization that that entails. For all his hesitations about the language of rights and its dependence on traditional notions of the subject, agency, and responsibility, he does in the end strongly defend that discourse. And I cannot think that he would treat with a deconstructive aloofness the prospect that human flourishing might in the course of time be replaced by a life-form with an impoverished trajectory. I am left, then with the question of whether deconstruction is of any direct help in adumbrating a kind of (quasi)transcendental materialism, a materialism that would ground what matters as its condition of possibility, even as it occupies the same one surface on the Moebius strip. It may be that we need an alliance with what has come to be called "New Materialism."

Exploring these three threads only gestures at presenting deconstruction as an eco-friendly disposition. A fuller account would consider hospitality in the face of mass migration, welcoming the Other, the New International, expanding the idea of a democracy-to-come to include nonhumans, the ineliminability of enlightenment values, thinking the im-possible, the paradoxes of both autonomy and collective action, and shared sovereignty. Each of these so-called topics is in fact the name for a dispositional exhortation. Wittgenstein once wrote, "Don't look for the meaning, look for the use." And Heidegger: "This has just been a series of propositions. The point is to follow the movement of showing." Derrida is indeed what Rorty called an "edifying philosopher," recommending—in his case—patience, resisting schematizing formulations, noticing the silent ways in which old binaries frame problems, taking the road less traveled, releasing the power of the repressed Other. Deconstruction is not a method, not an algorithm, not a recipe, not a formula, but a complex disposition—a resource we need when addressing the eleventh plague, anthropogenic climate change.

Notes

1. See Martin Heidegger, *What Is Called Thinking?* [1954] (New York: Harper, 1968), 77.

2. Jacques Derrida, *Specters of Marx: The State of the Debt, the Work of Mourning, and the New International* (New York: Routledge, 1994), 81–84.

3. I address these issues in a somewhat different way in David Wood, "Specters of Derrida: On the Way to Econstruction," in *Ecospirit: Religions and Philosophies for the Earth*, ed. Laurel Kearns and Catherine Keller (New York: Fordham, 2007), 264–90; "Derrida Vert?" *Oxford Literary Review* 36, no. 2 (2014): 319–22; and "Globalization and Freedom," in *The Step Back: Ethics and Politics after Deconstruction* (Albany: SUNY Press, 2005).

4. John Keats, the poet, from a letter to one of his brothers (1817).

5. Acknowledging its status as event, I mark that this paper was originally presented at the *Eco-Deconstruction* conference (Nashville: Vanderbilt University, March 2015).

6. The classic paper is Lynn White, "The Historical Roots of Our Ecological Crisis," *JASA* 21 (June 1969): 42–47.

7. See Bruno Latour's Gifford Lectures (2013), http://www.ed.ac.uk/schools-departments/humanities-soc-sci/news-events/lectures/gifford-lectures/archive/series-2012-2013/bruno-latour.

8. See William Cronon, ed., *Uncommon Ground: Rethinking the Human Place in Nature* (New York: W. W. Norton, 1995).

9. Bill McKibben, *The End of Nature* (New York: Random House, 2006).

10. I retain the word "man" here because of, not despite, its ideologically regressive legacy.

11. In Derrida and Jürgen Habermas, *Le "Concept" du 11 septembre* (Paris: Galilée, 2004); trans. as *Philosophy in a Time of Terror*, ed. Giovanna Borradori (Chicago: University of Chicago Press, 2004).

12. Ibid.

13. Friedrich Nietzsche, "Truth and Falsity in Their Ultra Moral Sense," in *The Complete Works of Friedrich Nietzsche* (London: T. N. Foulis, 1911).

14. The TV program *America's Most Wanted* uses names (and images) to apprehend criminals.

15. See Derrida, *L'animal que donc je suis,* ed. Marie-Louise Mallet (Paris: Galilée, 2006); trans. David Wills as *The Animal That Therefore I Am* (New York: Fordham University Press, 2008).

16. I pursue these ideas more systematically in Wood, *The Deconstruction of Time* (Evanston, Ill.: Northwestern University Press, 1989); in *Time after Time* (Bloomington: Indiana University Press, 2007); and in *Deep Time* (New York: Fordham University Press, forthcoming).

17. I have in mind here the shift from *Being and Time* (1926) to "Time and Being" (1962).

18. Derrida, *Of Grammatology* (Baltimore: Johns Hopkins University Press, 1976), 5.

19. One of the more remarkable comments made by a witness to the Skopes ("Monkey") trial (Dayton, Tennessee [1925]) was that he couldn't see what the fuss was about. He didn't mind being descended from monkeys. But fish? No way!

20. Such as *Homo heidelbergensis, Homo rhodesiensis* or *Homo antecessor, Homo erectus, Homo denisova, Homo floresiensis,* and *Homo neanderthalensis.*

21. It is a tragic but sobering thought that this capacity for (self) destruction might contribute to saving the planet. Genocide to the rescue? Gaia moves in mysterious ways? This surely as obscene a thought as it is sobering.

22. Derrida, *Spectres de Marx* (Paris: Galilée, 2006); trans. Peggy Kamuf as *Specters of Marx* (New York: Routledge, 2006).

23. The anthropological machine is an expression coined by Giorgio Agamben in *The Open: Man and Animal* (Stanford: Stanford University Press, 2002), chap. 9.

24. Wood, "Comment ne pas manger: Deconstruction and Humanism," *Death of the Animal* conference, Warwick, Nov. 1993; see also chap. 9 of Wood, *Thinking after Heidegger* (Cambridge: Polity; Malden, Mass.: Blackwell, 2002).

25. Or as Derrida says, "Betise!" See his *La bête et le souverain,* vol. 1 *(2001–2002)* (Paris: Galilée, 2008); trans. Geoffrey Bennington as *The Beast & the Sovereign,* vol. 1 (Chicago: University of Chicago Press, 2009), *passim.*

26. "Eating Well or the Calculation of the Subject," in *Who Comes after the Subject?* ed. Eduardo Cadava, Peter Connor, and Jean-Luc Nancy (New York: Routledge, 2001); see also Derrida, *Points de suspension: Entretiens* (Paris: Galilée, 1992); trans. Peggy Kamuf as *Points . . . Interviews, 1974–1994* (Stanford: Stanford University Press, 1995).

27. I introduce this term in Wood, "Kinnibalism, Cannibalism: Stepping up to the Plate," https://www.academia.edu/6813639/Kinnibalism _Cannibalism_Stepping_Up_to_the.

28. While this can happen, I have strongly argued against this line of thought. Vegetarianism can just as easily be the leading edge of a broader transformation.

29. See Derrida, *The Animal That Therefore I Am*.

30. See Elizabeth Kolbert, *The Sixth Extinction* (New York: Henry Holt, 2014). The title has a strange ambivalence to it. It announces the geological scale of what we are bringing about. And yet it's hardly unprecedented—there were five previous ones; the earth yawns.

31. Benjamin Franklin at the signing of the *Declaration of Independence* [1776]. Arguably what is now needed is a Declaration of Interdependence.

CHAPTER

2

Thinking after the World: Deconstruction and Last Things

Ted Toadvine

> Do stones have to become human beings before they can become
> subject matter for phenomenology? Or, for *physis* to comprise
> stones, do we at least have to be able to ask sensibly what it would
> be like to be a pile of stones? And must it not be sensible for us to
> ask this if a pile of stones or something very like it—a handful of
> dust, ash, earth—is what all of us are destined by nature to become?
> —*John Llewelyn*, Seeing through God

> We must therefore think this: it is the "end of the world," but we do
> not know in what sense.
> —*Jean-Luc Nancy*, The Sense of the World

Speculations about the end of the world are ubiquitous today. The
popularity of apocalyptic fiction and film reveals the deep pleasure
we take in imagining the world's destruction repeatedly and in every possible variation, a pleasure entangled with genuine anxieties
about what the future holds. Apocalyptic narratives are not new, of
course; they may be as old as civilization itself and probably exist
in some form in every culture. But secular eschatological fiction is
more recent, with Mary Shelley's 1826 novel *The Last Man* usually
considered the first major example. Shelley's novel was dismissed
by critics at the time and soon forgotten, but the genre gained wide
popularity in the 1890s that has continued to the present. At the
turn of the twentieth century, as W. Warren Wagar reports, most
doomsday scenarios imagined that humankind would be wiped out
by natural causes: plagues, earthquakes, floods, giant storms, and
so on. But after WWI, most have imagined us destroying ourselves,

usually in wars with technologically advanced weapons.[1] Since the 1960s, in the wake of Rachel Carson's *Silent Spring*, end-of-the-world fiction has increasingly drawn inspiration from what Wagar calls "fashionable prophesies of ecological disaster."[2] Indeed, since the end of the Cold War, and in certain respects as its legacy or continuation, ecological destruction has become our favored vision of the end, with the currently popular genre of "cli-fi" as its latest flavor.

We have been considering apocalyptic fiction, but actual predictions of ecological disaster also share in the appeal of the apocalyptic narrative: the accumulating biotoxins of *Silent Spring*, the population "bomb," the hole in the ozone layer, biodiversity collapse, genetic engineering gone awry, and so on. In fact, this narrative underwrites environmentalism's efforts to "save the world," and the religious overtones here are not irrelevant. Whether expressed openly as fear of the future or disguised as nostalgia for the past, an eschatological vision of the world is essential to environmentalism. This is readily apparent in the public discourse and cultural imaginary surrounding global climate collapse, so that when journalist and activist Bill McKibben proclaims that stripping the remaining bitumen from Alberta's tar sands will mean "game over for the planet," he is appropriating an intimately familiar narrative.[3] Indeed, the "fashionable prophesies" of environmentalism and the cultural imaginary expressed through speculative fiction share a common eschatological vision with roots in shared cultural sources, and they have fed and borrowed from each other to the point where they can no longer rigorously be distinguished. I will call this the "eco-eschatological" narrative.

Whether presented as empirical prediction or speculative fiction, the eco-eschatological narrative is always a phantasm or fable, a tale that we tell ourselves about the future that reflects our investments and anxieties in the present and that consequently constructs our current identities and institutions. Derrida makes a parallel point in "No Apocalypse, Not Now," presented at a colloquium on Nuclear Criticism five years before the opening of the Berlin Wall, where he characterizes the prospect of total nuclear war as a "phantasm of remainderless destruction," destruction that might extend to humanity in its entirety or even to the earth as a whole (P1 396/372, 393/369–70). By naming nuclear war a phantasm, Derrida was not at all denying the reality of stockpiled weaponry or the plausibility that this weaponry might be deployed with catastrophic consequences. His point was rather that such a war is "fabulously textual"—not only because the weapons themselves rely on codes and texts of all sorts and because the strategies of deterrence were themselves textual games, but most importantly because total nuclear war has never yet taken place, so that it "has existence only by means of what is said of it and only where it is talked about" (P1 393/370). In other words, nuclear catastrophe is a fabulous tale, and nevertheless—or precisely for this reason—it effects a positive construction of present reality, so that the "whole of the human *socius* today," the whole of what Derrida, in 1984, could call the "general institution of the nuclear age" (P1 394/369), would be marked by it directly or indirectly. As a terrifying fable about the future, the prospect of total destruction therefore confronts us—humanists, in particular, insofar as we are "specialists in discourse and in texts" (P1 391/368) and insofar as the imagined destruction threatens uniquely and in particular the juridico-literary archive (P1 400/376-77)—with a contemporary task: "The terrifying 'reality' of nuclear conflict can only be the signified referent, never the real referent (present or past) of a discourse or a text. At least today. And that gives us to think the *today*, the presence of this present in and through this fabulous textuality" (P1 393/369). Ecological disaster, including climate collapse, is fabulously textual in just this sense. Turning our attention, as humanists, to the eco-eschatological narrative as a phantasm implies no skepticism about the very real dangers that we face, but may instead be the only responsible way to think the present insofar as it is constructed through our fables about the future.[4]

One interesting aspect of the eco-eschatological narrative is its construction of the present as suspended between the geologically deep past and an indefinitely distant future. As mentioned previously, Mary Shelley, whose *Frankenstein* is often credited with inaugurating modern science fiction, also authored the first major work of secular eschatology, *The Last Man*, published in 1826.[5] Set in the late twentieth century and putatively based on ancient prophetic writings, the novel tells the tale of the destruction of the human race by a global plague, leaving the last survivor, based autobiographically on Shelley herself, to wander the world alone. It is hardly coincidental that Shelley's novel appeared just as biologists were coming to accept Georges Cuvier's evidence, based on reconstruction of fossilized skeletons including mammoths and mastodons, that the world was once populated with creatures that had subsequently gone extinct. Cuvier intended these findings to "burst the limits of time" just as scientific genius had "burst the limits of space," thereby providing a window into the "former world," a world prior to all human history, that he believed had been catastrophically destroyed.[6] Surely if a natural catastrophe could drive so many other species to extinction and bring their entire world to an abrupt end, then the same could be imagined for our species; our world must be equally precarious and finite, and our days on Earth similarly numbered. This construction of our present as suspended between prehistorical catastrophe and anticipated extinction continues to shape contemporary discussions of climate collapse, as we see for example in astrophysicist Neil deGrasse Tyson's remarks on the National Geographic television series *Cosmos*:

> We're dumping carbon dioxide into the atmosphere at a rate the Earth hasn't seen since the great climate catastrophes of the past, the ones that led to mass extinctions. We just can't seem to break our addiction to the kinds of fuel that will bring back a climate last seen by the dinosaurs, a climate that will drown our coastal cities and wreak havoc on the environment and our ability to feed ourselves. . . . The dinosaurs never saw that asteroid coming. What's our excuse?[7]

For more concrete examples of what I am calling the "temporal suspension" of the present, consider current efforts toward global sustainability that extrapolate from deep-past trends to predict and manage far-future scenarios, thereby tacitly assuming that our responsibility toward future generations is to sustain the world in a state that as much as possible resembles our present. One example would be ongoing efforts to establish permanent repositories for radioactive waste, which must avoid human intrusion and environmental degradation on the scale of tens of thousands or even millions of years. The field of nuclear semiotics emerged from the efforts of the Human Interference Task Force, convened by the U.S. Department of Energy in the 1980s with the charge of devising a warning system to dissuade future generations from tampering with repositories of toxic waste for at least the next 10,000 years, roughly twice the length of written human history.[8] These efforts are ongoing by research teams for the Waste Isolation Pilot Plant in New Mexico, which interestingly have included speculative fiction authors among the experts consulted.[9] Similarly, the Greenland Analogue Project studies the deep history of ice sheets on Greenland's western coast to design the first operational geological repository for high-level radioactive waste, scheduled to open in Olkiluoto, Finland, within the next decade.[10]

A second example would be the economic practice of "discounting the future" to assess how much we should spend today to limit the future effects of climate change. Recent studies hold that greenhouse gas emissions to date have already irrevocably committed us to an altered global climate for at least the next millennium, which will lead to sea-level rise, widespread famines and plagues, mass migrations, and catastrophic weather changes.[11] But if we calculate the future growth of the world's economy on the basis of past trends, then the people of the future will be increasingly richer than we are today. How much should we ask the (relatively) poorer people of today to sacrifice for the (relatively) richer people of tomorrow? The answer will vary depending on the "discount rate" that economists choose to apply, similar to money-market interest rates that track what investors are willing to pay today for a certain level of future benefits.[12] My point is that policy decisions and resource allocations being made today on an international scale rely on what I am calling the "temporal suspension" of the present between the deep past and far future.

To briefly review, first I have suggested that our obsession with the end of the world, in the form of the eco-eschatological narrative that frames speculative fiction as well as environmental prediction, is a phantasm that reflects our desires and anxieties in the present and that leaves its mark, directly or indirectly, on our individual and collective identities, institutions, and sense of the world here and now. Second, I have proposed that this phantasm has a history, that it develops in parallel with the emerging conception of deep time: our awareness of an ancient geological past that precedes us opens our imaginations to an indefinitely distant future after us. And third, I have offered some contemporary examples of efforts to calculate and manage the distant future on the basis of the deep past, which I believe tacitly assume an approach to time that is framed by this eco-eschatological narrative. My conviction is that this raises some profound philosophical questions concerning how we understand the world, time, and responsibility.

To open these questions, I first show how Jean-Luc Nancy's notion of the "catastrophe of equivalence" helps to unpack what is at stake in our doomsday obsession by illuminating the relationship between our eco-eschatological narrative and the suspension of the present. The interdependence that characterizes globalization evacuates the present of its singularity, its nonequivalence, in its efforts to calculatively manage the future. The generalized equivalence of time presupposed by these projections of the future is a consequence of the ecotechnical proliferation of ends and means without final end, a vacuum that can only be filled by phantasms of apocalypse. As an alternative to this general equivalence of time, I suggest an eco-phenomenology of the singular present, but one that, through its encounter with deconstruction, stretches both "eco" and "phenomenology" toward a hyperbolic transformation. This reveals an ambivalence or equivocation, best expressed by Nancy, in how we understand the end of the world: either as the end of the sense of the world as a total horizon of intelligibility or as the rediscovery of this world here *as* sense, without reference to any purpose or meaning beyond itself.

Derrida suggests an alternative to Nancy's account in his description of the death of any singular living thing as the end of the world, which he presents as a radicalization of Husserl's famous thought experiment of the destruction of the world in *Ideas I*. Derrida insists that the death of any living thing is the absolute end of *the* world, rather than the end of *a* world, and confronts us with an impossible responsibility for mourning. This leads Derrida to suggest, in the final session of his last seminar, *The Beast & the Sovereign*, that there is no one world, no common world, even if we must carry on *as if* such a world obtains. The phantasm of the world's destruction, I argue, serves precisely to bolster this pretense of a common and shared world: imagining the world as under threat reinforces our projection that the world exists. This brings us to the relationship between the world, its formation or destruction, and the elements. Just as every stone is outside the relation of life and death, stone remains liminal to the world that it nevertheless makes possible in quasitranscendental fashion. This discloses the world's ongoing liability to elemental materiality and memory, especially the memory of earth and stone. Derrida's investigations of the problem of world in *The Beast & the Sovereign* take us part of the way toward understanding world's liability to the elements. But it is to Nancy that we must turn for an account of our own elemental liability and of the exposure of stone that, beyond life-death, may not *have* a world but nevertheless *is* a world. Through its geological memory, stone therefore offers us a glimpse of how to understand our relationship with the future outside of the regime of general equivalence, outside

of a sustainable management of the future that subjects it to totalizing foreclosure.

This returns us, finally, to our apocalyptic vision of the world,
which approaches everything within the world against a background
of absolute contingency or nothingness. By threatening things with
the specter of their own destruction, we force their presentation into
self-identity. As Nancy notes, "Destruction takes place in the world
and not vice versa,"[13] which is why he enjoins us to "learn to stop
dreaming of the end, to stop justifying it. . . . We need to take our leave
of the romantic-historical mode of thinking that promises an apotheosis or an apocalypse—or both, one in the other" and rediscover the
resistance of existence itself as spacing and permanent revolution.[14]
This requires not a better architecture of the future but a deepen-
ing of our exposure to and within the present. Eco-deconstruction
thereby becomes an ethos of the rediscovery of the world right at the
very end or limit of the sense of the world.

Eco-Phenomenology and Ecotechnics

Unsurprisingly, the world that is under threat in our eco-eschatol-
ogies is typically presented in a naturalistic way—for instance, as
nature or planet Earth. What is genuinely at stake, however, is not
the planet as a collection of physical entities, but rather the world
as we know it, the total horizon of meaning, value, and possibility within which our lives unfold. In other words, what is at stake
is "world" in the phenomenological sense. Phenomenology's the-
matization of the world as a philosophical problem is perhaps its
most significant legacy, teaching us to understand the world not
merely as a given totality of entities or events—our planet, for example, or the universe more broadly—but rather as the nonthematic
referential or horizonal structure that the appearance of anything
whatsoever presupposes. Phenomenology therefore opens a path for
describing the world that is distinct from either the Kantian treatment of it as an a priori form correlated with consciousness or the
speculative metaphysical effort to account for the world in terms
of another being or another world—both of which fall back on an
explanation of the world in worldly terms, by way of what the world
alone makes possible. This is why Eugen Fink, in his famous 1933
Kant-Studien article, describes phenomenology's task as the effort
to uncover "the origin of the world," an origin that could not be anything within the world, outside of the world, or in another world.[15]

Interestingly, Donn Welton explains this phenomenological sense of world—a "nexus of significance" distinct from "something like a natural environment or a socio-historical reality or the totality or whole of all such worlds"—with reference to the collapse of the Twin Towers of the World Trade Center: "It was not a particular fact or a string of facts within the world, but the world itself, the very context and background of our everyday life, that came unraveled on September 11."[16] The end of the world does not, therefore, entail the total destruction of all that factually exists, but rather the disintegration of our horizons of significance and possibility, of the context and background presupposed by any worldly thing or event. What our eco-eschatological imagination projects is such an unraveling without return, without recovery, a final and complete dissolution of meaningful horizons. This poses the phenomenological problem of the world in its most radical form, pushing it to its limit: since every question that can be posed, every future that can be pictured, presupposes the world as its horizon, from what perspective can the total destruction of the world even be imagined?

Deconstruction, in the work of both Derrida and Nancy, carries through this turn toward the end of the world. From his early reading of Husserl, which follows out the paradoxes inherent in phenomenology's account of the origin of the world, to his final seminar, which insistently repeats Heidegger's question—*Was ist Welt?* What is world?—Derrida demonstrates that the concept of the world already implies a relationship to its end, to death and to the nonliving at the heart of life. In fact, if we have learned anything from deconstruction, it is that nothing is less obvious or less certain than that we know what we mean by "world," that we know who or what "has" world, or even that there *is* a world at all. The trajectory of deconstruction is toward the *end* of the world, its goal and its limit, and the question of what, if anything, comes after it. In its claim that every autoaffection is fundamentally and inescapably heteroaffection, that every experience involves a passage through the world and through the other, deconstruction can be understood from its very beginnings as the most radical form of eco-phenomenology, one that stakes the sense of the world on its liability to interruption and dissolution. Coming to terms with our apocalyptic obsession, and in particular with its implications for the sense of the world here and now, requires an eco-phenomenology of the end of the world, but one that, through its encounter with deconstruction, stretches both "eco" and "phenomenology" toward a hyperbolic transformation.

"Eco" here would no longer refer to ecology in either the informal or the strict sense—either as any form of organic community or as the scientific study of organism-environment interactions—but instead evokes what Nancy calls the "ecotechnical," in two senses. First, the ecotechnical refers here to the *technē* of bodies according to which all properness or self-relation is originarily interrupted by transplantation, prosthesis, or foreignness at the very heart of the self.[17] This already entails a transformation of phenomenology, which can no longer understand itself as a return to sense in the interiority of self-contact, even as the contact of one hand touching the other, but must instead suffer the suspension and interruption of sense as the extension and exposure of bodies. A phenomenological heteroaffection, if such an expression is meaningful, must touch on the outside, be touched from the outside, and discover its inside as already outside itself in its vulnerability to being touched. Such a reappropriation of phenomenology and of touch is not without unavoidable and perhaps insurmountable risks, as Derrida compellingly demonstrates.[18] Yet if eco-phenomenology is possible in the wake of deconstruction, if it must assume such risks, then it might parallel Nancy's own path toward a "radical materialism" or, in Derrida's phrase, a "post-deconstructive realism," the touchstone of which would be touch itself.[19]

Second, ecotechnics names the contemporary situation of worldwide technology, the "global structuration of the world as the reticulated space of an essentially capitalist, globalist, and monopolist organization that is monopolizing the world."[20] This situates us beyond any demarcation of the "natural" from the "technological" that could prescribe principles, ends, or norms for environmentalism, and in this respect environmentalist critics are right to see in deconstruction—from Derrida's *Of Grammatology* to Nancy's *After Fukushima*[21]—the subversion of any hope for a foundationalist return to pure distinctions between nature and culture, wilderness and technology, materiality and signification.[22] The ecotechnical in this sense makes salient the global regime of economic, technological, and political interdependence that presumes a general equivalence or interchangeability between all means and ends while pursuing no greater end than its own totalizing expansion.[23] Since existence is essentially technological, we have no recourse here to a more natural or authentic way of life. But the contemporary situation is nevertheless ambivalent or equivocal: on the one hand, ecotechnics raises the specter of absolute closure into the immanence of global sustainability, by which the future is entirely programmed and managed to avoid the real threat of a final destruction, be it nuclear war or climate collapse. On the other hand, precisely in the unhinging of technology from production, in its unworking of the relation between means and ends,[24] there may be an opening toward the sense of the world here and now—no longer as nature or ecology but rather as the exposition of singular bodies in the ever-renewed present.[25] In this latter case, the world no longer *has* a sense—it makes no reference to any creator, to any other world, or to any gathering into a whole—but rather it *is* sense in all of the plural and fragmentary singularity of its material existence. Ecotechnics therefore hovers ambivalently between the end of the world as *mundus* or *cosmos* that would ground a meaning and orientation for our lives and its rediscovery as this world here, absent any purpose or meaning beyond its own existence. As Nancy recognizes, such a rediscovery of the world as sense demands a new philosophy of nature, albeit one that transforms both philosophy and nature.[26] Nature would henceforth be the "exposition of bodies," a "network of confines," or the "there is" of the sensible world that is "neither *for* us nor *because* of us."[27]

The equivocal situation of the ecotechnical is illustrated by Nancy's discussion of the 2011 Fukushima nuclear disaster, where he explains that there can no longer be any purely "natural" catastrophe, since the proliferating and uncontainable repercussions of every catastrophe are inextricably natural, technological, social, economic, and political. Nancy terms this the "equivalence of catastrophes," not to suggest that all disasters are equivalent in terms of their consequences or destructiveness, but because all are marked by the paradigm of nuclear catastrophe, which "remains the one potentially irremediable catastrophe, whose effects spread through generations, through the layers of the earth; these effects have an impact on all living things and on the large-scale organization of energy production, hence on consumption as well" (AFEC 3/11–12). With nuclear catastrophe as their paradigm, all disasters today are equivalent with respect to the pervasive entanglement of their consequences, which is the obverse of the complex and ever-deepening interdependence and interconnection of systems worldwide: ecological, economic, technoscientific, sociopolitical, cultural, logical, and so on. Globalization is precisely the process of this ever-deepening interdependence, which presupposes conversion, translation, substitution, exchange. Starting from Marx's insight that money serves

as a "general equivalent," since every cost and benefit can be translated into economic terms, Nancy finds that contemporary capitalism and technological development have generalized this notion of
equivalence even further, so that "the regime of general equivalence
henceforth virtually absorbs, well beyond the monetary or financial sphere but thanks to it and with regard to it, all of the spheres
of existence of humans, and along with them all things that exist"
(AFEC 5/16). This is characterized by a "limitless interchangeability of forces, products, agents or actors, meanings or values" (AFEC
6/16). If this general equivalence makes greater interdependence of
all our systems possible, then it is the reason catastrophes cannot be
circumscribed in their effects. But more than this, it is the general
equivalence itself that is catastrophic, insofar as it inspires a proliferation of means and ends that are ultimately oriented toward no final
end, no ultimate goal other than their own continued expansion and
proliferation. What takes the place of this final end is precisely our
constant awareness of the possibility of our own self-destruction. As
Günther Anders writes, "Today, since the apocalypse is technically
possible and even likely, it stands alone before us: no one believes
anymore that a 'kingdom of God' will follow it. Not even the most
Christian of Christians."[28]

Nancy recognizes that the regime of general equivalence has implications for how we relate to time and to the future, and the examples that I proposed earlier—the planning of nuclear waste repositories and economic discounting of future climate costs—illustrate his
point perfectly. The absence of any end or goal for our ecotechnical
interdependencies apart from their own self-perpetuation traps us
in a cycle of planning and management of the future in general, and
the extrapolation of the past to calculate the future demonstrates
the sway of general equivalence in our understanding of time, since
each chronological present moment is substitutable for every other
(Nancy mentions "atomic time" in this context [AFEC 31/52]). As
Nancy puts it, "No culture has lived as our modern culture has in
the endless accumulation of archives and expectations. No culture
has made present the past and the future to the point of removing the
present from its own passage" (AFEC 40/67). The alternative here is
to recognize the *nonequivalence* of the singularities, the absolutely
unique and nonsubstitutable events and moments that compose our
quotidian experience, and thereby to deepen our respect for the present. In Nancy's words again, "What would be decisive, then, would
be to think in the present and to think the present. No longer the

end of ends to come, or even a felicitous dispersion of ends, but
the present as the element of the near-at-hand" (AFEC 37/62–63).
Since the aim here is to respect the nonequivalence of the present,
this is a proposal for what we might call "ontological or temporal
justice." Such justice would be absolutely distinct from any respect
for nature in the sense typically encountered within environmental
philosophy, which, for Nancy, would remain within the economy of
valuation and therefore of general equivalence (AFEC 39/66).

This call for temporal justice does not, however, imply any rejection of ecotechnics, but rather its extraction from the general equivalence of capitalism and therefore its return to the *technē* of finitude
or the spacing of existence. As Nancy explains in *A Finite Thinking*, "Technology 'as such' is nothing other than the 'technique' of
compensating for the nonimmanence of existence in the given. Its
operation is the existing of that which *is* not pure immanence. . . .
Insofar as its being *is* not, but is the opening of its finitude, existing
is technological through and through" (AFT 24/44). In other words,
technology in its root sense is a compensation for the original absence of any self-sufficiency of nature, of any given natural order
of means and ends, and is therefore the essence of finite existence.
Nancy adds in *The Sense of the World* that the world of technology
is precisely the world becoming *world*, since "a world is always a
'creation': a *teknnē* with neither principle nor end nor material other
than itself. . . . It is necessary to come to appreciate 'technology'
as the infinite art that supplements a nature that never took place
and will never take place. An ecology properly understood can be
nothing other than a technology" (SW 41/66). Because ecotechnics
characterizes finite existence, as the in-finition or unfinishability of
the finite, it undoes any return not only to a given nature but more
broadly to any pure autoaffection or immanent closure, whether of
life or of sovereignty; against the understanding of life as automain-
taining and autoaffecting, the ecotechnical reveals "the infinitely
problematic character of any 'auto' in general" (CW 94/140), and it
"washes out or dissolves sovereignty" (BSP 137/162). The spacing
or finitude of the world opened as ecotechnics "*is itself the empty
place* of sovereignty. That is, it is the empty place of the end, the
empty place of the common good, and the empty place of the common as good" (BSP 137/162).

Here we reach the heart of what Nancy calls the "terrible ambivalence" of the ecotechnical: "The world, as such, has by definition
the power to reduce itself to nothing just as it has the power to be

infinitely its own sense, indecipherable outside of the praxis of its art" (SW 41/67). Ecotechnics can spiral out of control as the infinite proliferation of means and ends without final end in a growing totalization—with nuclear catastrophe or environmental collapse as its ultimate consequence—or it can take on "the sense of the disruption of all closures of signification, a disruption that opens them up to the coming of (necessarily unprecedented) sense" (SW 102/161; cf. BSP 133–43/158–68). Since technology is the "incessant displacement of ends," it reveals the empty place of all former ends, of the world as having a linear history oriented toward completion; consequently, it has "made possible the modern apocalypse—the modern revelation—of destruction."[29] The hesitation or ambivalence of ecotechnics concerns whether this revelation slides toward infinite mastery or a rediscovery of the finitude of the world and the singularity of sense. In *Corpus*, Nancy expresses both the radicality of this understanding of the world and the revolutionary promise of its alternative:

> Our world is the world of the "technical," a world whose cosmos, nature, gods, entire system is, in its inner joints, exposed as "technical": the world of the *ecotechnical*. The ecotechnical functions with technical apparatuses, to which our every part is connected. But what it *makes* are our bodies, which it brings into the world and links to the system, thereby creating our bodies as more visible, more proliferating, more polymorphic, more compressed, more "amassed" and "zoned" than ever before. Through the creation of bodies the ecotechnical has the *sense* that we vainly seek in the remains of the sky or the spirit.
>
> Unless we ponder without reservation the ecotechnical creation of bodies as the truth of *our* world, and a truth *just as valid* as those that myths, religions, and humanisms were able to represent, we won't have begun to think *this* very world. . . . The ecotechnical deconstructs the system of ends, renders them unsystematizable, nonorganic, even stochastic (*except* through an imposition of the ends of political economy or capital, effectively imposed nowadays on the whole of the ecotechnical, thus relinearizing time and homogenizing all ends, but capital also has to stop presenting a final end—Science or Humanity— and, moreover, the creating of bodies harbors revolutionary force . . .). (*Corpus*, 88–89)

Nancy's account suggests that our fixation on the destruction of the world by technology expresses a genuine intuition into the ecotechnical unworking of any system of ends or ultimate ordering of the world. What ecotechnics reveals is that the world as *cosmos* or *mundus* has already ended, that the world no longer has a sense—a meaning or a direction (cf. SW 4–5/13–15). But our mistake is in interpreting this end of the world as "a cataclysm or as the apocalypse of an annihilation"; the end of the world cannot be given a determinate sense as annihilation or total destruction (SW 4–5/14–15). Indeed, to understand the end in these terms is precisely to remain within the regime of general equivalence that inspires our calculative management of the future at the expense of the ever-unfolding and inestimable present, a present "in which something or someone presents itself: the present of an arrival, an approach," rather than a chronologically ordered and countable present moment (AFEC 38/64). The promise of the ecotechnical is in its ever-renewed and in-finite creation of the finitude of sense, which offers an entirely distinct understanding of the world, no longer as the correlative of a sense, but rather as sense itself: "For as long as the world was essentially in relation to some other (that is, another world or an author of the world), it could *have* a sense. But the end of the world is that there is no longer this essential relation, and that there is no longer essentially (that is, existentially) anything but the world 'itself.' Thus the world *no longer has* a sense, but it *is* sense" (SW 8/19). The world *as* sense is precisely the ever-renewed creation and differentiation of bodies in their exposure to and being-with all others, a world that has unity only as "a differential articulation of singularities that make sense in articulating themselves, along the edges of their articulation" (SW 78/126). The end of all ends of the world is therefore the rediscovery of this world here, with no sense or end beyond itself.

"In a World without World"

On the face of it, this ambivalent interpretation of the end of the world—the end of the world as *cosmos* that is the renewal of the world as sense—would seem as far as possible from Derrida's own treatment of the end of the world. While the theme of the end of the world has its roots in Derrida's earliest writings,[30] it becomes increasingly prominent in his later work, where it is most often associated with death: the death of a friend, for instance, as we see in many of the memorial essays collected in *The Work of Mourning* (the French title of which makes this theme explicit: *Chaque fois unique, la fin du monde*), but also, more generally, the death of any living thing, even including insects, protozoa, and plants.[31] Derrida repeatedly insists on the seemingly paradoxical formulation that each and every death of a unique living thing is *the* end of the world, absolutely and infinitely, and not merely the end of *a* world or of a living thing *within* the world. This insistence is intended to respect the incommensurability and inappropriability of the other as a singular origin of existence, with a unique and untranslatable exposure to experience and time, a respect that Derrida finds already implied by Husserl's recognition that the other can be presented only through analogical appresentation rather than direct perception.[32] As Derrida suggests in "Rams," the claim that death is each time the end of the world must be understood as pushing to its limit Husserl's own thought experiment of the annihilation of the world in paragraph 49 of *Ideas I*. Rather than "weakening this phenomenological radicalization," Derrida instead intends to think through the retreat of the world as "the most necessary, the most logical, but also the most insane experience of a transcendental phenomenology," insofar as what remains after the world is gone is precisely my ethical responsibility to bear the absolute transcendence of the other within myself (SQ 160–61/Béliers 74–76).

It is impossible, however, to assign any fixed sense to what is meant by "world" in Derrida's discussion, and for essential reasons. We should recall here Derrida's cautions about whether one should expect from "big bad words" such as "world" anything like an exact sense (OT 7/17), as well as his commentary on the difficulties Heidegger raises concerning the question of world: "It's a bit like it is for the question of being, we do not know what it is, world, what being it is and therefore in view of what we are questioning. . . . A question about the world is about everything and nothing. About everything, therefore about nothing, it's an empty question that bites the tail of its own presupposition" (BS2 59/97). But the recoil of this question onto its own presupposed horizons becomes even more acute, even more insane, when we follow through the implications of Derrida's discussion of the death of the other, which paradoxically undermines any notion of a common or shared world in the phenomenological sense, and consequently—insofar as any shared meaning presumes a horizon in common—of any meaning at all. This "end of the world" as the withdrawal or liquidation of any sense of a common world, of a *cosmos* or a lifeworld, is, as Derrida presents it, the very situation of ethics. Where Derrida's account pushes us, then, is in the direction of recognizing that not only is the end of the world a fiction in the sense alluded to previously, but so is the world itself a fiction or a phantasm, even if one that is in certain respects necessary and perhaps even what our responsibility toward the other requires.

Derrida first suggests that death is each time the end of the world in a much-discussed passage from "Rams," where it introduces his reading of Celan's poem "Vast, Glowing Vault," especially its final line: *"Die Welt ist fort, ich muß dich tragen"* [The world is gone, I must carry you]. As Derrida writes:

> For each time, and each time singularly, each time irreplace-ably, each time infinitely, death is nothing less than an end of *the* world. Not *only one* end among others, the end of someone or of something *in the world*, the end of a life or of a living be-ing. Death puts an end neither to someone in the world nor to *one* world among others. Death marks each time, each time in defiance of arithmetic, the absolute end of the one and only world, of that which each opens as a one and only world, the end of the unique world, the end of the totality of what is or can be presented as the origin of the world for any unique liv-ing being, be it human or not. (SQ 140/Béliers 23)

As Derrida emphasizes here and elsewhere, the death of any liv-ing thing marks not merely the end of *a* world or of a life *within* the world, but rather the absolute end of *the one and only* world, thereby leaving the survivor "in some fashion beyond or before the world itself . . . responsible without world (*weltlos*), without the ground of any world, thenceforth, in a world without world, as if without earth beyond the end of the world" (SQ 140/Béliers 23). That the death of the other is the end of the one and only world fol-lows from the fact that the other's world is never a mere variation of my world, never analogous with or translatable into my world, but remains an absolute interruption of my world that can never be appropriated or made properly my own. "World" is therefore not a general type of which individual worlds would be specific tokens; in its very singularity, the end of "any" world can only be the end of the world *tout court*. Furthermore, neither is the other's world sim-ply alongside my world as one among others, since my "own" world is not properly mine, since I *am* only in the ethical response, the "I must," of carrying the other, where *to carry* would mean to "*bear oneself toward* the infinite inappropriability of the other": "I only am, I can only be, I *must* only be starting from this strange, dislocated bearing of the infinitely other in me" (SQ 161/Béliers 76). This infinite distance that opens within my world as an irreparable rupture announces my responsibility to carry the other's world within me, to mourn it, after the other's death—but also the melancholic impossibility of my doing so, precisely since I can never contain or encompass this unique and singular opening onto the world. But this means that I do not first "have" a world of "my" own, since what might be called "my world" is from the first only an exposure and bearing toward the infinite distance of the other.

Furthermore, the end of the world does not, strictly speaking, await the actual death of the other, since the melancholic certainty that one friend will survive the other's death interrupts every friendship from its first moment: "From this first encounter, interruption anticipates death, precedes death" (SQ 140/Béliers 22; cf. WM 107/138).[33] More generally, it follows that each and every encounter with each and every living thing already announces the heart of absence interrupting and constituting my world, calling me to respond with a mourning both ineluctable and insufficient. The end of the world therefore haunts every world from within, dissolving its pretense of being "one and only," "unique," the "totality of what is"; we find ourselves "there where the world is no longer between us or beneath our feet, no longer ensuring mediation or reinforcing a foundation for us" (SQ 161/Béliers 76). And so, from the first moment when the ethical injunction "I must carry you" announces itself, the world as *cosmos* or *mundus* has already disappeared. "No world can any longer support us, serve as mediation, as ground, as earth, as foundation or as alibi. . . . I am alone in the world right where there is no longer any world" (SQ 158/Béliers 68). In other words, Derrida's claim that death is each time the absolute end of the one and only world concerns not only an event that comes to pass with each living thing's death, but just as much the structural necessity of the dissolution of any common world—as referential horizon of possibilities—already entailed by our exposure to the other. This is consistent with Derrida's insistence that death is each time the end of *the* world, and not only of *a* world, of one possible world among others. Our ethical responsibility goes beyond the exigency of carrying the other's world within our own, then, but rather situates us already beyond the recourse to any ground, earth, or foundation; finding ourselves from the outset to be survivors, we are always "responsible without world (*weltlos*) . . . as if without earth beyond the end of the world" (SQ 140/Béliers 23).

There are several interesting consequences to be drawn from Derrida's account, the first being that the end of the world is already implicated within and even constitutive of the world itself; the world is not a self-enclosed totality that maintains itself until interrupted from the outside, but rather has its outside on the inside. If this is so, we must think the relation between the world and its coming-to-an-end with more nuance than eschatological thinking allows, as Derrida has so often shown concerning the relationship of life with death. Second, since death is each time the end of the world, there is no perspective from which to compare, evaluate, or hierarchize this end of the world in comparison with, say, total nuclear war or ecological collapse. As Derrida writes in "No Apocalypse, Not Now," "There is no common measure able to persuade me that a personal mourning is less grave than a nuclear war" (P1 403/379). Even if there is no basis for comparison, Derrida nevertheless does distinguish, in the final lecture of *The Beast & the Sovereign*, between the death of an individual as the "absolute end of the world" and the stakes of world war as "the end of the world, the destruction of the world, of any possible world, or of what is supposed to make of the world a *cosmos*, an arrangement, an order, an order of ends, a juridical, moral, political order, an international order resistant to the non-world of death and barbarity" (BS2 260/359).[34] Last, Derrida's account makes clear that the end of the world is an ethical matter and perhaps the very opening of ethics insofar as it first eliminates any recourse to a ground or foundation and confronts us with the monstrousness of the other that is also the monstrousness of the future.[35] Consequently, what our responsibility to the other requires is not holding onto the world, attempting to sustain it at any cost, but precisely letting it go.

Nevertheless, there is an ambiguity or ambivalence in Derrida's approach to the end of the world that becomes clearest in the second year of his final seminar—an ambivalence that may allow us to bring Derrida's position on the end of the world closer to that of Nancy. Responding to Heidegger's famous three theses on world from *Fundamental Concepts of Metaphysics*—the stone is worldless [*weltlos*], the animal is poor in world [*weltarm*], and man is world-forming [*weltbildend*][36]—Derrida frames this year's seminar in the first lecture with three theses of his own, three theses that are apparently incompatible with each other, briefly summarized as follows: (1) animals and humans incontestably inhabit the same "objective" world, even if they do not have the same experience of "objectivity"; (2) animals and humans incontestably do not inhabit the same world, since the human world is not identical with that of nonhuman animals; and (3) no two individuals, whether human or animal, inhabit the same world, and the differences between their worlds are essentially unbridgeable. This third thesis follows from the fact that "the community of the world is always constructed, simulated by a set of stabilizing apparatuses, more or less stable, then, and never natural, language in the broad sense, codes of traces being designed, among all living beings, to construct a unity of the world that is always deconstructible, nowhere and never given in nature" (BS2 8–9/31). Between my world (which, for me, can only be the unique and only world, encompassing all others) and the world of any other, therefore, "there is first the space and the time of an infinite difference, an interruption that is incommensurable with all attempts to make a passage, a bridge, an isthmus, all attempts at communication, translation, trope, and transfer that the desire for a world or the want of a world, the being wanting a world will try to pose, impose, propose, stabilize. There is no world, there are only islands" (BS2 9/31).

Derrida returns to the first and third of these theses in the tenth and final session of the seminar, where he again emphasizes, developing the third claim, that the unity of the world is a merely presumptive construction, a means of reassuring ourselves in the face of the absence of the world. Here, the end of the world—again associated with the line from Celan, *"Die Welt ist fort"*—does not await the death of the other but is instead "the ever unsewn and torn tissue of our most constant and quotidian experience," something that we know "with an undeniable and stubborn, i.e., permanently denied, knowledge" (BS2 266/367). The presumptive unity of the word "world," then, is intended to

> mask our panic . . . , to protect us against the infantile but in-
> finite anxiety of the fact that *there is not the world*, that noth-
> ing is less certain than the world itself, that there is perhaps
> no longer a world and no doubt there never was one as totality
> of anything at all . . . and that radical dissemination, i.e. the
> absence of a common world, the irremediable solitude with-

out salvation of the living being, depends first on the absence
without recourse of any world, i.e. of any common meaning of
the word "world," in sum of any common meaning at all. (BS2
265–66/366)

Here, without mention of death, Derrida draws the full consequences
of his view that the death of the other is the end of the world, which
is that, from the outset, there is no world at all, no world in common, and consequently no common meaning. In other words, we
have moved from treating the *end* of the world as a phantasm to
recognizing that the phantasm is actually *the world itself*, that the
phantasm of the world is intended to mask the *absence* of the world.
Understood in this context, our anxieties about the end of the world,
insofar as they present the world as fragile and vulnerable, precisely
reinforce our belief in its reality. In this situation, according to Derrida, *ich muß dich tragen* can mean one of only two things: either
that, with both of us sharing this knowledge that the world is no
longer, I must carry you into the worldless void; or that what I must
do, "with you and carrying you, is make it that there be precisely a
world, just a world, if not a just world, or to do things so as to make
as if there were just a world, and to make the world come to the
world" (BS2 268/369). On Michael Naas's reading, Derrida places his
hope in the second option, which Naas describes as a poetic making
or remaking of the world ex nihilo in full recognition that there is no
world: "Aware of its own powerlessness, undone by its own ability,
this *poiesis* would be a making *as if* that leaves within the world a
trace of the end or loss of the world."[37]

Here Derrida seems quite close to Nancy's claim, in *The Sense of
the World*, that "there is no longer any world: no longer a *mundus*,
a *cosmos*, a composed and complete order (from) within which one
might find a place, a dwelling, and the elements of an orientation"
(SW 4/13). For Nancy, this does not call for a poetic remaking of the
world in the mode of *as if*, but instead a recognition of the ex nihilo
coming-to-presence of the world as sense. Ex nihilo in Nancy's sense
implies no relationship to a creator, poetic or otherwise, but instead
a growing from nothing, without roots, that would be the "genuine formula of a radical materialism" (CW 51/55; cf. BSP 16/35).
This creation of the world from nothing implies, in the words of
Marie-Eve Morin, "that it has no presupposition or precondition,
no ground or reason, no origin or end."[38] While Derrida and Nancy
seem very close, then, in their account of the end of the world as

shared horizon or common ground, Derrida does not seem to move in Nancy's direction concerning a rediscovery of the world as sense, as the exposure and intersection of the material singularity of bodies, as this world here. Without a poetic making or remaking of the world, a making of the world with and for the other, we remain *weltlos*, worldless, like the stone. This is consistent with Derrida's remarks, in the French preface to *Chaque fois unique, le fin du monde*, distinguishing his view on death as "the end of the world in totality, the end of every possible world, and each time the end of the world as unique totality, thus irreplaceable and thus infinite," from Nancy's account of anastasis in *Noli Me Tangere*, which, according to Derrida, concerns only the end of *a* world, of one possible world among many, and therefore the chance—rejected by Derrida—for replacement, survival, resurrection.[39]

Nevertheless, Derrida's final seminar suggests another reconstructive path for understanding the world—namely, in the first of the three theses introduced in the first session: "Animals and humans inhabit the same world, the same objective world" (BS2 8/31); as living beings, they share in common "the finitude of their life, and therefore, among other features of finitude, their mortality in the place they inhabit, whether one calls that place world or earth (earth including sky and sea) and these places that they inhabit in common" (BS2 10/33). When Derrida returns to this common sense of "world" in the final session, he again stresses that it is the same space of inhabitation or cohabitation, a common habitat, characterized precisely in terms of the elements: "water, earth, air, fire" (BS2 263/363). This reference to the elements, to the inorganic, should give us pause: what is the relationship between the elements and world? Is there a common world of the "organic" and the "inorganic," of lifedeath and the stone?

Eschatology and the Elements

Derrida notes at several points that the departure of the world, *Die Welt ist fort*, exceeds and disrupts Heidegger's three theses on world, that its irreducibility to the categories of *weltlos*, *weltarm*, and *weltbildend* requires us to rethink the very thought of world (SQ 163/Béliers 79; cf. BS2 104/159, 169/243). Nevertheless, even if our situation of carrying the other is irreducible to either of these categories, Derrida repeatedly describes it using Heidegger's category for the worldless stone: "We are *weltlos*" (BS2 9/31–32; cf. SQ 140/

Béliers 23; BS2 177/253; R 155/213). Of course, we are clearly not worldless in the same manner as Heidegger had attributed this to the stone, as Derrida says explicitly (BS2 9/32), but then how are we to think this strange lithic proximity? Recall that, for Heidegger, the stone, as *ein Vorhandenes*, is "absolutely indifferent" insofar as it remains entirely outside or before the difference between being indifferent or not indifferent to its own being (OS 20–21/39–41); it is neither awake nor asleep (AA 148/203); it cannot be deprived of world, since it has absolutely no relationship with other entities, no experience of the sun that shines upon it or the lizard that rests atop it (OS 51–52/79-81; cf. AA 155–56/213). Furthermore, and for Heidegger this is the *Prüfstein*, the touchstone (BS2 173/115), the stone "does not die, because it does not live" (BS2 113/171; cf. AA 154/211); it is finite while lacking finitude (AA 150/206) and therefore entirely outside of the relation between life and death, of mortality or lifedeath. What is the status of this "outside" of the world, which cannot be a simple exclusion? Luckily, that the stone may be without world does not entail that the world is without stone. For Derrida, the death of any *living thing* is each time the end of the world, but could there be a world without stone, without earth, without the elements? Could there be a shared habitat of the living?

Derrida calls attention to the fact that, in Heidegger's theses on world, the stone stands in as the sole example of "material things," of the "lifeless" or the "inanimate": "Why does he take the example of an inanimate thing, why a stone and not a plank or a piece of iron, or water or fire?" (BS2 6/27–28; cf. AA 153/209). By privileging "the" stone as exemplary of the material thing, Heidegger participates in what Jeffrey Cohen calls "a long tradition of mining the philosophical from the lithic," which poses the question of what the stone's ontological exemplariness reveals as well as conceals.[40] For Derrida, the choice of the stone as exemplar serves to cover over the ambiguities of the concept of life, which becomes obvious when one considers where to locate plants, for example—or cadavers—in relation to the general categories of "life" or "material things" (BS2 6/28). But in attending here only to the complications of any pure distinction between what is inside or outside of lifedeath, Derrida never addresses—as he does so well with the general category of "the" animal—the fact that there can be no "the" stone, no "the" material thing, but only a plurality of material singularities—as, for Nancy, matter is "always singular or singularized" (SW 58/97), the very difference and *différance* "through which *something* is possible, as *thing* and as *some*" (SW 57/95).

The stone is, for Heidegger, *vorhanden*, present-at-hand, a mere part of the world in distinction from the human being as one who also *has* world. With this in mind, it is curious that Derrida begins the second year of *The Beast & the Sovereign* with his own example of a stone, a polished pebble found lying on the beach with the inscription "The beasts are not alone" (BS2 5–6/26–28). This stone is a "stumbling block" [*pierre d'achoppement*], an obstacle that "interrupts one's progress and obliges one to lift one's foot," but it is also—like death for Heidegger—a "touchstone" (BS2 6/28). It seems to serve these functions, however, not insofar as it is stone or a stone, but only in terms of its cryptic inscription, an inscription that Derrida soon abandons for being "like this stone, isolated, insularized, forlorn, singularly solitary" (BS2 7/29). Yet the stone is, then, the very figure of an island, of the *Einsamkeit*, the loneliness or solitude, that Heidegger reserves for man as what first brings him into proximity to the world (BS2 30/59–60). Consequently, like the beasts of its inscription, the stone is not, can never be, alone. Can this absolute isolation, beyond any solitude, still constitute a manner of being-*toward*, of *l'être-à*, such that, as Nancy suggests, even if the stone does not "have" a world, it nevertheless *is* a world? "To be sure, the concrete stone does not 'have' a world . . . but it is nonetheless toward or in the world [*au monde*] in a mode of *toward* or *in* that is at least that of *areality*: extension of the area, spacing, distance, 'atomistic' constitution. Let us say not that it is 'toward' or 'in' the world, but that it is world" (SW 62/103, cf. 28/48).

Beyond the distinction between life and death, the stone is consigned to the edge or margin of the world, at or beyond the world's limits, like a pebble washed up on the shores of a deserted island. The pebble on the shore, lying in the sand, knows nothing of its position or its situation, knows nothing of the sand or the beach on which it lies. The pebble may change locations with the waves and tides, but it nevertheless remains unmoved. Of course, the sand, in its turn, consists primarily of smaller stones. And the island also consists of rock, just a stone at a larger scale, bathing in the ocean waters while projecting into the air, immersed in the elements and weathering their changes. Pebble, stone, rock. The English term "pebble," referring to a small stone rounded by the actions of water and sand, has uncertain origins that may relate to the Latin *papula*, a swelling or pistule on the skin, or may echo onomatopoetically the sound of walking on pebbles or of the movement of waters with which they are associated. By comparison, the French equivalent, *galet*, is diminutive of the Old French *gal*, stone, which is cognate with Old Irish *gall*, stone pillar, and *gallán*, large upright stone. "Stone," like the German *Stein*, descends from the Proto-Germanic *stainaz*, related to the Proto-Indo-European *steyh-*, "to stiffen," and cognate with the Greek στία, small stone or pebble, while the French *pierre* follows from the Latin *petra* and Greek πέτρα. "Rock," meanwhile, derived from the Old English *stanrocc*, "stonerock," is cognate with the French *roche*, both from Old French *roke*, cognate with the postclassical Latin *rocca*. While "rock" names the solid mineral matter of the lithosphere that encircles the outer layer of our planet, *a* rock, stone, or pebble is a clast or fragment of this materiality. Pebbles are stones and made of stone; stones are rocks and made of rock. According to the scale for grain size introduced by geologist Charles Wentworth in 1922, pebbles are between 4 and 64 millimeters, larger than *granules* but smaller than *cobbles*. But our geological terms in English have generally not been characterized by precision or consistency. As Cohen notes, "Middle English *ston* could designate any lithic chunk from the smallest pebble to a towering menhir."[41] What, then, of the world of the stone? *The* stone? Is there any "the" stone, any stone in general or as such—or rather pebbles, stones, and rock of unimaginably diverse sorts, sizes, and placements, each one singularly unique, each one a world of its own whose originary spacing is the effective exteriority of all else that exists? Nancy responds to Heidegger's three theses on world by stressing precisely what they exclude of our "effective exteriority": "These statements do not do justice, at least, to this: that the world beyond humanity—animals, plants, and stones, oceans, atmospheres, sidereal spaces and bodies—is quite a bit more than the phenomenal correlative of a human taking-in-hand, taking-into-account, or taking-care-of: it is the effective exteriority without which the very disposition of or to sense would not make . . . any sense" (SW 55–56/92).

The stone is both a part of the world and, as its effective exteriority, constitutive of the *there*, the spacing and material singularity, of the world. As a clast of the lithosphere, of the stony planetal skeleton that undergirds any earthly lifeworld, the stone also recalls or remembers the elemental geomateriality that precedes and exceeds all worlds. Just as creation stories envision the emergence of the world from formless waters and earth, the raging elements are

a recurring motif in our eco-eschatological imagination: rising waters, glaciation, parched sands and storms of dust, hurricanes and
earthquakes. "Some say the world will end in fire, some say in ice."
Whether by fire or ice, our vision of the end of the world is haunted
by its dissolution into elemental materials and forces of sublime
scope and scale. As Levinas notes, "The element comes to us from
nowhere; the side it presents to us does not determine an object,
remains entirely anonymous. It is wind, earth, sea, sky, air."[42] The
stone extracted from the elements to become part of the world remains nevertheless inhabited or haunted by this anonymous elementality from which the world is extracted and to which it must
inevitably return. This is why our imaginations of the world's end
run up against a limit that is, finally, indestructible: the fact that
"there is" something, that existence as such continues, perhaps independently of all subjectivity or even all life, if only in elemental
form: fire and ice, dust and gas, atomic radiation, the stars.

Many stones have proper names, among them the Rosetta Stone,
the Blarney Stone, Plymouth Rock, the Stone of Scone, the Rock of
Gibraltar—and their destruction would, in many cases and in some
sense, be mourned. But stone as such, like the other elements, is not
subject to death or even to destruction; it is a world that survives
the absolute destruction of world, that reveals its peculiar relationship with deep time, both past and future. Stone holds a preeminent
place among the elements precisely because of its peculiar temporality, its geological memory. We owe our conception of the deep
past to this memory of stone, which Buffon in 1778 could call "the
world's archives"; just as we may reconstruct human history from
ancient inscriptions and artifacts, so it is possible to "extract ancient
monuments from the earth's entrails" to "place a certain number of
milestones on the eternal road of time."[43] This archival memory of
stone spans all times and worlds, outstripping and undergirding the
juridico-literary archive. Christopher Tilley demonstrates how Neo-
lithic menhirs embody the traces of prehistoric perceptual worlds,
even as the accumulated geomaterial records of our own lives pass
into the far future in the form of nuclear waste, the stratigraphic
traces of radioactive elements from nuclear blasts, and fossilized
plastiglomerates.[44] This timeless memory of stone situates it both
within the world and beyond it, seesawing at its edge, which makes
it the ideal boundary marker, milestone, or tombstone. As John Sallis writes:

Stone comes from a past that has never been present, a past un-
assimilable to the order of time in which things come and go in
the human world; and that nonbelonging of stone is precisely
what qualifies it to mark and hence memorialize such comings
and goings, births and deaths. As if stone were a sensible image
of timelessness, the ideal material on which to inscribe marks
capable of visibly memorializing into an indefinite future.[45]

The stone is always somehow from another world even as it subsists
in this one, like a meteor, a fossil, or a glacial erratic, haunted by its
immemorial passage across worlds.

Alongside the phantasmic projection of a world in common, a
world of shared meaning that would bridge our separate islands,
then, we must take into account the persistent geomateriality that
grants existence its areal spacing and its temporal span. This is less
a matter of common habitat than of the essential and constitutive
lithic materiality of every living being. As Nancy writes, "A stone
is the exteriority of singularity in what would have to be called its
mineral or mechanical actuality. But I would no longer be a 'human'
if I did not have this exteriority 'in me,' in the form of the quasi-
minerality of bone" (BSP 18/37; cf. SW 60–61/100–102). Our liability
to this minerality is figured in the skeleton as symbol of death, as
the endurance of our own lithic elementality into the rhythm of a
temporality other than or exceeding that of lifedeath, just as the
fossil offers a glimpse of the intersection of the time of life with
the immemorial past of stone. Can we love, can we mourn, a stone,
even the stony skeleton that from within makes possible all our
loving and mourning, that bears us toward the other even as it harbors the obscure memory of its birth among the elements before all
worldly time and anticipates its passage through stone and dust to
other times and other worlds?

To rediscover the world's liability to the elements requires foregoing any eschatology that approaches everything within the world and
the very sense of the world itself against a background of absolute
contingency or nothingness, vulnerable to total destruction. Eco-
eschatology not only deforms our relationship to time, but also prevents us from encountering the materiality of the thing in its absolute singularity. By threatening things with the specter of their own
annihilation and therefore silhouetting them against the screen of
nothingness, we force their presentation into self-identity, positivity,

immanence; they either fully are or fully are not. But, as Merleau-Ponty already points out, taking a pebble as his example, this framing is a denaturing of the thing: "Is not thinking the thing against the background of nothingness a double error, with regard to the thing and with regard to nothingness, and, by silhouetting it against nothingness, do we not completely denature the thing? Are not the identity, the positivity, the plenitude of the thing—reduced to what they signify in the context in which experience reaches them—quite insufficient to define our openness upon 'something'?"[46] Nancy develops this insight by adding that the "regulating fiction" of total destruction always leaves something behind, something indestructible: "the pure being of a world or of 'something' in general," the world or the something that no longer refers to us for its existence or its sense.[47] The true last things of eschatology would then be the *there is* of existence itself:

> The pure *there is* as the indestructible, the gift that cannot be refused (since it has no one to give it), of a space without a subject to arrange it, to distribute it, to give it sense. A *there is* that would be neither *for* us nor *because* of us. Either that or the "sensible" world outside the "sense" given to it by a sentient subject: the very thing that philosophy has never been able to think, still less to touch, even though it has doubtless always been obsessed with or haunted by it. ("Indestructible," 85/245)

There would be little comfort, as Nancy recognizes, in recognizing the indestructibility of being on the verge of our own destruction, but the point is that the end of the world as we know it is not the price for this knowledge, since "we already know this, here and now. 'Being' or the 'there is' or 'existence' is, in us, what happens before us and ahead of us, arising from the very step beyond us."[48] This neither prevents nor justifies our own self-destruction, which remains an ever-present possibility. But it does invite us to "stop dreaming of the end, to stop justifying it" and to attend to what, in the here and now, incommensurably resists our efforts to manage the world toward its own annihilation. "The indestructible measures each one of our destructions, their impotence. Existence resists."[49] To think this resistance, to encounter it as "permanent revolution" in the materiality of things and ourselves, is to move beyond apocalypse or apotheosis and thereby to rediscover the world after the end of the world.

Notes

1. W. Warren Wagar, *Terminal Visions: The Literature of Last Things* (Bloomington: Indiana University Press, 1982), 24.

2. Ibid., 30.

3. McKibben is quoted as attributing this phrase to a conversation with climate scientist James Hansen in Jane Mayer, "Taking It to the Streets," *New Yorker*, November 28, 2011, http://www.newyorker.com/ magazine/2011/11/28/taking-it-to-the-streets. Hansen's own published remarks instead use the phrase "game over for the climate"; see Hansen, "Game Over for the Climate," *New York Times*, May 9, 2012, http://www .nytimes.com/2012/05/10/opinion/game-over-for-the-climate.html.

4. To be clear, in calling attention to the eschatological dimensions of environmental rhetoric, my intention is neither to challenge the credibility of scientific predictions nor to debate their effectiveness for influencing public opinion. There is every reason to believe that climate collapse will result in catastrophic changes for life on our planet, human and nonhuman, and that we have a responsibility to act immediately and collectively to the best of our abilities to curb its effects. But the eschatological framing of this narrative, as obvious or unavoidable as it may seem—and precisely for this reason—goes beyond a simple presentation of the facts or call to action. I am suggesting that closer attention to our eschatological imaginary reveals tacit presuppositions that frame how we think about what it means to act ethically in the present.

5. Mary Shelley, *The Last Man*, ed. Morton D. Paley (Oxford: Oxford Paperbacks, 1998).

6. Georges Cuvier, *Recherches sur les ossements fossiles de quadrupèdes* (Paris: Deterville, 1812), 1:3, 70.

7. Quoted in Chris Mooney, "Finally, Neil deGrasse Tyson and 'Cosmos' Take on Climate Change," *Mother Jones*, May 5, 2014, http://www .motherjones.com/environment/2014/05/neil-tyson-cosmos-global -warming-earth-carbon.

8. See Ted Toadvine, "Apocalyptic Imagination and the Silence of the Elements," in *Ecopsychology, Phenomenology, and the Environment: The Experience of Nature*, ed. Douglas A. Vakoch and Fernando Castrillón (Berlin: Springer, 2014), 211–21, and the essay by Michael Peterson in this volume.

9. On the influence of science fiction on the work of these research teams, see Gregory Benford, *Deep Time: How Humanity Communicates across Millennia* (New York: Harper Perennial, 2000).

10. On the relation between these efforts and deep time, see anthropologist Vincent Ialenti's National Public Radio series from September 2014, http://www.npr.org/tags/347050105/deep-time.

11. See Susan Solomon et al., "Irreversible Climate Change Due to Carbon Dioxide Emissions," *Proceedings of the National Academy of the*

Sciences 106, no. 6 (2009): 1704–9, and Nathan P. Gillette et al., "Ongoing Climate Change Following a Complete Cessation of Carbon Dioxide Emissions," *Nature Geoscience* 4 (2011): 83–87.

12. For an accessible overview of ethical factors involved in setting the discount rate for climate intervention, see John Broome, "The Ethics of Climate Change," *Scientific American*, June 2008, 69–73.

13. See Jean-Luc Nancy, "The Indestructible," trans. James Gilbert-Walsh, in *A Finite Thinking*, ed. Simon Sparks (Stanford: Stanford University Press, 2003), 85; Nancy, "L'indestructible," *Cahiers Intersignes*, nos. 4–5 (1992): 245.

14. Nancy, "Indestructible," 87/247.

15. Eugen Fink, "Die phänomenologische Philosophie Edmund Husserls in der gegenwärtigen Kritik," *Kant-Studien* 38 (1933): 338; Fink, "The Phenomenological Philosophy of Edmund Husserl and Contemporary Criticism," in *The Phenomenology of Husserl*, ed. R. O. Elveton (Chicago: Quadrangle, 1970), 95.

16. Donn Welton, "World as Horizon," in *The New Husserl: A Critical Reader*, ed. Donn Welton (Bloomington: Indiana University Press, 2003), 223, 224, 231.

17. See especially Nancy, *Corpus*, bilingual ed., trans. Richard Rand (New York: Fordham University Press, 2008); Nancy, *L'intrus* (Paris: Éditions Galilée, 2000). In *On Touching—Jean-Luc Nancy*, Derrida repeatedly calls attention to the distinctiveness of Nancy's approach to ecotechnology understood as the *technē* of the body, which he says "singles out Nancy's thinking among all other modern ideas about the body proper, the flesh, touch, or the untouchable, which is to say, the taking into account of technics and technical exappropriation on the very 'phenomenological' threshold of the body proper" (OT 56/70).

18. See, for example, OT 116–30/133–49.

19. Nancy, *The Creation of the World or Globalization*, trans. François Raffoul and David Pettigrew (Albany: SUNY Press, 2007), 51; Nancy, *La création du monde ou la mondialisation* (Paris: Éditions Galilée, 2002), 55 [hereafter, cited textually as CW, with English preceding French pagination]; Derrida, OT 46/60.

20. Nancy, *The Sense of the World*, trans. Jeffrey Librett (Minneapolis: University of Minnesota Press, 1997), 101; Nancy, *Le sens du monde* (Paris: Éditions Galilée, 1993), 159–60 [hereafter, cited textually as SW, with English preceding French pagination].

21. Ian James links Nancy's ecotechnics with the Derridean themes of arche-writing and différance, albeit with essential qualifications, in James, *The Fragmentary Demand: An Introduction to the Philosophy of Jean-Luc Nancy* (Stanford: Stanford University Press, 2006), 147–48.

22. As paradigmatic examples of the environmentalist response to deconstruction, see Edward O. Wilson, *Consilience: The Unity of Knowledge* (New York: Alfred Knopf, 1998), 41, 214–15, 301, and Michael E. Soulé and

Gary Lease, eds., *Reinventing Nature: Responses to Postmodern Decon-struction* (Washington D.C.: Island Press, 1995). Despite the subtitle of the latter work, only a few passing references to Derrida appear here (usually listed alongside such authors as Rorty, Lacan, and Lyotard), and none of his texts is explicitly cited or discussed. Nancy is not mentioned.

23. See Nancy, *A Finite Thinking*, 24–27; *Une pensée finie* (Paris: Édi-tions Galilée, 1990), 43–48 [hereafter cited textually as AFT, with English preceding French pagination]; Nancy, *Being Singular Plural*, trans. Robert Richardson and Anne O'Byrne (Stanford: Stanford University Press, 2000), 132–43; *Être singulier pluriel* (Paris: Éditions Galilée, 1996), 157–68 [here-after, cited textually as BSP, with English preceding French pagination]; CW 94–95/140–43.

24. AFT 26/47–48; Nancy, "Indestructible," 84/244.

25. See Nancy, *After Fukushima: The Equivalence of Catastrophes*, trans. Charlotte Mandel (New York: Fordham University Press, 2015), 37; *L'équivalence du catastrophes (Après Fukushima)* (Paris: Éditions Galilée, 2012), 62–63 [hereafter, cited textually as AFEC, with English preceding French pagination].

26. Nancy, *Corpus*, 36–37; SW 40/64, 62–63/103–4, 157/237–38.

27. Nancy, *Corpus*, 36–37; SW 40/65; "Indestructible," 85/245.

28. Günther Anders, *Le temps de la fin* (Paris: L'Herne, 2007), 115; quoted in AFEC 19–20/38.

29. Nancy, "Indestructible," 84/244.

30. This could be traced back at least to Derrida's discussion of the pos-sibility of the disappearance of truth and of a universal conflagration in his *Edmund Husserl's "Origin of Geometry": An Introduction*, trans. John P. Leavey Jr. (Lincoln: University of Nebraska Press, 1962), 93ff.; *Introduc-tion à "L'Origine de la géométrie" de Husserl* (Paris: Presses Universitaires de France, 1962), 91ff.

31. This is most explicit in the January 10, 2001, session from Derrida's death penalty seminar, where he writes that "the death one makes or lets come in this way is not the end of this or that, this or that individual, the end of a who or a what *in the world*. Each time something dies, it's the end of the world. Not the end of a world, but of the world, of the whole of the world, of the infinite opening of the world. And this is the case for no matter what living being, from the tree to the protozoa, from the mosquito to the human, death is infinite, it is the end of the infinite. The finitude of the infinite"; quoted in Michael Naas, *The End of the World and Other Teachable Moments: Jacques Derrida's Final Seminar* (New York: Ford-ham University Press, 2015), 181n14. See Derrida, *Séminaire La peine de mort*, vol. 2, *2000–2001* (Paris: Galilée, 2015), 118–19.

32. See "Rams," in Derrida, *Sovereignties in Question* (SQ 161); *Béliers: Le dialogue ininterrompu: entre deux infinis, le poème* [hereafter *Béliers*] (Paris: Galilée, 2003), 75–76; OT 190–94/217–20; Naas, *End of the World*, 48.

33. See also WM 107/138 and Naas, *End of the World*, 53.

34. Although this contrast is not further developed in Derrida's final lectures, it might be fruitfully compared to similar remarks in his "No Apocalypse, Not Now" (P1 387–409/363–86) and "Autoimmunity: Real and Symbolic Suicides—A Dialogue with Jacques Derrida" (PT 85–136/133–96).

35. On the relationship between the monstrosity of the other and ethics, see BS1 108/155, as well as BS2 266/367. See also Derrida's famous remark concerning the monstrosity of the future in the Exergue to *Of Grammatology* (OG 5/140).

36. See Martin Heidegger, *The Fundamental Concepts of Metaphysics*, trans. William McNeill and Nicholas Walker (Bloomington: Indiana University Press, 1995), 184ff; Heidegger, *Die Grunbegriffe der Metaphysik: Welt—Endlichkeit—Einsamkeit* (Frankfurt am Main: Vittorio Klostermann, 1992), 272ff.

37. Naas, *End of the World*, 60.

38. Marie-Eve Morin, *Jean-Luc Nancy* (Cambridge: Polity, 2012), 43–44.

39. See Derrida, *Chaque fois unique, la fin du monde* (Paris: Galilée, 2003), 9–11; Nancy, *Noli me Tangere: On the Raising of the Body*, trans. Sarah Clift, Pascale-Anne Brault, and Michael Naas (New York: Fordham University Press, 2008), esp. 18–19, 41, 45; Nancy, *Noli me tangere* (Paris: Bayard, 2003), 33–36, 70, 74–75. Nancy makes reference to this distinction by Derrida in a recent interview, but without mentioning that Derrida had directed this criticism at him; see Nancy and Marcia Sá Cavalcante Schuback, eds., *Being with the Without* (Stockholm: Axl, 2013), 40.

40. Jeffrey Cohen, *Stone: An Ecology of the Inhuman* (Minneapolis: University of Minnesota Press, 2015), 4.

41. Ibid., 14.

42. Emmanuel Levinas, *Totality and Infinity*, trans. Alphonso Lingis (Pittsburgh: Duquesne University Press, 1969), 132; *Totalité et Infini* (The Hague: Martinus Nijhoff, 1971), 139.

43. Georges Buffon, *Histoire naturelle des époques de la nature* (Paris: De l'Imprimerie Royale, 1778), 1.

44. Christopher Tilley, *The Materiality of Stone: Explorations in Landscape Phenomenology* (Oxford: Berg, 2004). On plastiglomerites, see Patricia L. Corcoran, Charles J. Moore, and Kelly Jazvac, "An Anthropogenic Marker Horizon in the Future Rock Record," *GSA Today* 24, no. 6 (June 2014): 4–8.

45. John Sallis, *Stone* (Bloomington: Indiana University Press, 1994), 26.

46. Maurice Merleau-Ponty, *The Visible and the Invisible*, trans. Alphonso Lingis (Evanston, Ill.: Northwestern University Press, 1968), 162; *Le visible et l'invisible* (Paris: Gallimard, 1964), 213.

47. Nancy, "Indestructible," 85/245.

48. Ibid., 85/246.

49. Ibid., 86/246.

Scale as a Force of Deconstruction

Timothy Clark

There are many ways in which "they have eyes but do not see."
—*David Wood, "What Is Eco-Phenomenology?," in*
Eco-Phenomenology: Back to the Earth Itself, *215*

All this, this open and non-self-identical totality of the world is
deconstruction.
—*Derrida,* Negotiations: Interventions and Interviews, *193*

"Scale critique": this term has recently been proposed by Derek
Woods to name an emerging tendency within the environmental
humanities.[1] Dipesh Chakrabarty, Mark McGurl, Ursula Heise, and
others have found that attempting to engage environmental issues of
a dauntingly global nature entails a reconsideration of the concept of
scale and that this leads to a surprising reconceptualization of once
seemingly familiar issues.[2] Thinkers have started to map out an intellectual practice in which we "need to think human agency over
multiple and incommensurable scales at once."[3] "Scale critique"
would also include David Wood's work to expand eco-phenomenology beyond the scales of normal human perception and intuition.[4]

"Scale critique" is a response to one of the most challenging aspects in thinking the so-called Anthropocene: the way it highlights
anew the question of the ontology of the human. The term should
also embrace Baird Callicott's work when he argues that the global
environmental crisis has a root cause in a scalar disjunction between
human activity and the rest of the natural world.

What renders strip mines, clear-cuts, and beach developments un-
natural is not that they are anthropogenic—for, biologically speaking,

81

Homo sapiens is as natural a species as any other—but that they occur at temporal and spatial scales that were unprecedented in nature
until nature itself evolved another mode (the Lamarckian mode) of
evolution: cultural evolution.[5]

So it is not just that the technical and the prosthetic are, paradoxically, inherent to what the human is, but that the speed of change in
these things ("cultural evolution") means that the human inhabits a
very different time scale from the vast majority of the natural world.
In this sense, a latently destructive scalar discrepancy underlies human inventiveness. The human names the particular site of a capi-
talization of information and energy far speedier than in the rest of
nature, making it a capricious force of interference in earth-system
processes, notably in the exploitation of fire and combustion.

Callicott's argument raises further questions. Is the Anthropocene,
with its attendant collapse of biosystems across the world, still an
accident—that is, something that might be mitigated or addressed
by cultural measures, by the putative construction or retrieval of a
supposedly more "ecological" human culture, as some eco-thinkers
like to argue? Or is it in some sense inherent or inevitable in the
kind of creature we are? If the latter, then any viable new environmental ethic can only arise out of a recognition of that tragic reality, rather than the advocacy of more nostalgic forms of humanism.
Or is it something else again, a new event not susceptible to being
conceived or traced as the revelation, extrapolation, development of
anything that preexisted it?

Let us return to the specific issue of scale. Writing in 1997 to illustrate a social constructionist understanding of geographic scale,
David Delaney and Helga Leitner wrote that "'scale' is not simply
an external fact awaiting discovery but a way of framing conceptions of reality."[6] This statement led to an investigation, characteristic of the 1990s, into the way various geographical scales, as they
correlate to objects of study or of government, are not given but are
produced by various social and political factors, which they deeply
influence in turn. Such an emphasis now seems apt but incomplete.
The concept of scale now at issue is at least as much ontological as
epistemological in emphasis, and its provocation is often to hover
undecidably between the two.

Scale does not constitute some sort of background to experience:
it inheres in and effects its basic structure, categories, and openness
to phenomena. The factors that determine the nature of a phenomenon considered at one scale may give way to quite different ones,

in a nonlinear way, at a larger scale, just as the properties of a crowd differ from those of an individual, the potential of a complex processor from that of one microchip, or the structural properties of a plane from those of its small-scale copy as a model. The issue of scale is something that particularly marks ecology, where longer-term processes of climatic or geographical change emerge as decisive over a longer time scale but that yet may have seemed incidental or contingent at a lesser one, where immediate issues of survival of offspring and rates of predation seem to hold the stage.

Scale must thus be a crucial issue in any proposed eco-phenomenology that attempts, in David Wood's words, to think those "non- or pre-intentional characteristics of nature"[7] that yet structure or condition the intentionality of immediate human experience. Human intentionality is neither reducible to nor independent of such physiological facts as those that structure the sensorium at certain scales: no insect could be scaled up to the size of an average human without suffocating, nor could the functions of the human brain exist in a body with the dimensions of an insect. The expanded sense of the natural that emerges in such eco-phenomenology entails "a world in which the relation between present experience and the complexity of what is being experienced has always been deeply complex and stratified,"[8] something "hidden from view by our ordinary experience,"[9] such as the way the nature of trees, encompassing processes of growth over decades or even centuries, cannot appear in the immediate snapshot of the present.

A given scale is not part of a world in the sense of the latter as some recognized nexus of significances or field of meaning; rather, it inheres in any world as a dimensionality that is all-structuring. It is a kind of grammar whose presence is overlooked in our habitual attention to individual things, to the semantics, so to speak. It both eludes and conditions normative concepts of consciousness as self-presence, as this "deflects us from proper acknowledgment of structures within the heart of our situated openness to the world that cannot be reduced to what is 'at present' alive or 'immediately' available to those who are at home in it."[10]

"Scale variance means that the observation and operation of systems are subject to different constraints at different scales due to real discontinuities."[11] Logically, it is assumptions of scale *invariance* that would be the first object of scale critique. What Derrida called "Western metaphysics," or the metaphysics of "presence" or "proximity," may now appear anew in terms of false norms and assumptions of scale invariance.[12] Scale, ontologically considered, is a constitutive feature of all temporalization, which now clearly cannot be reduced to the homogenous continuity of some hypostatized present. Heidegger's argument on the vulgar concept of time as "the homogenous medium in which the movement of daily existence is reckoned and organized" (MP 35/38) and on the ontotheological determination of being as presence—as that which is near, which persists, exposed to vision, at hand [*vorhanden*], etc., entailed what Derrida's "Ousia and Gramme" described as "this strange epoche of Being hiding itself in the very movement of its presentation" (MP 34/36). Scale is likewise an element hiding itself in the "givenness" of any phenomenon, a facet of its necessary finitude, an inherent element of contingency, unreliability, and incalculable metamorphosis.

Scale Effects

Many forms of thinking previously taken for granted, or certain norms in politics, ethics, and in relation to physical or other systems, must increasingly be recognized as operating only under certain, usually unexamined assumptions about their scale. Jim Dator writes, "Environmental, economic, technological and health factors are global, but our governance systems are still based on the nation state, while our economic system ('free-market' capitalism) and many national political systems (interest group 'democracy') remain profoundly individualistic in input, albeit tragically collective in output."[13]

The way issues of scale have been imposing themselves in numerous fields would yet be another instance,[14] fifty years after Derrida first saw the symptom in the proliferation of notions of writing of our living in an epoch that "seems to be approaching what might be called its own *exhaustion*" (OG 8/18). To be aware of scale variance and the ontological discontinuities entailed by varying determinations of scale is to find unsurprising Derrida's project of "determining the possibility of meaning on the basis of a 'formal' organization which in itself has no meaning" (MP 104/161). This last phrase could be as succinct a definition of the role of scale in this context as one is likely to find and describes simultaneously its challenges to anthropocentrism and narrow forms of humanism.

Many forms of environmentalism are loose manifestations of scale critique—they highlight now disjunctive or anachronistic as-

sumptions of scale in, say, treating environmental problems solely
as a result of underinformed consumer choice, and try to engage
the counterintuitive nature of thinking of climate change in relation to any individual taken as such. They often consider current
forms of human behavior, such as deforestation or pollution, and
extrapolate from present trends into a broader time scale. The result
is that, along with patently destructive practices such as deforestation or overfishing, some activities that seemed normal or relatively
innocuous or even praiseworthy in the present (increased prosperity, more car ownership, even increased longevity) acquire another
destructive face at broader scales. This discrepancy in scale is not
just the object of an intellectual exercise, but is now a source of day-
to-day evasions, contradictions, and tensions, as when, for instance,
an environmental scientist weighs the benefits of her attending a
conference on another continent against the environmental impact
of air travel, or, more controversially, when the reproductive rights
of a young couple are juxtaposed against the increasingly destructive
effects of population pressure.

What of scale effects specifically? A change in scale may constitute the element of metamorphosis in emergent properties. Again,
animal ecology offers the most graphic instance: reproduction rates,
survival of offspring, and so on may meet a scalar threshold at which
the mere factor of increasing numbers becomes newly significant,
either positively self-reinforcing in the form of swamping out possible competitors or autoimmune in the form of exhausting available
sources of self-sustenance—or, at different time scales, both.

If issues of scale are becoming more prominent, it is primarily
because *scale effects* are an elusive and underconceptualized form
of agency. Their elusiveness is at work in the way increasingly dangerous and latently destructive levels of carbon dioxide in the atmosphere are traceable not only immediately to visible problems such
as deforestation but also to the emergent effects of innumerable
actions that are in themselves insignificant and largely innocent.
The so-called Anthropocene could itself be in part defined by the
threshold or scale effect whereby, at the counterintuitive scale of
the whole earth, even once environmentally insignificant behaviors
(emitting an amount of a pollutant, chopping down a tree, and even
having a baby) now feed into nonlinear material processes that have
become both problematically decisive and incalculable.

Scale effects manifest themselves primarily in effects of interference, discontinuity, the unexpected, the multivarious. They add a

supplementary and potentially sinister dimension to what Wood describes as that "essential temporal articulatedness of things" that is "not itself obviously presented in their immediate temporal appearance."[15] In effect, it is through or as scale effects that newly emergent, destructively powerful agencies, conditioned by the finitude of the earth, become relatively autonomous in relation to the individual human acts or natural events that might have first seemed to comprise them.[16] My own singular carbon footprint only matters because there are so many other footprints of various sizes, now and in the past and the future, with incalculable scope and effect—like the destructive and seemingly self-generating vibrations that can emerge as if from nowhere when, for example, a certain number of people happen to walk over a metal bridge at the same time, threatening its structural integrity.

Scale effects manifest materially the priority of alterity within iteration. They would thus add themselves to the series of related but not identical terms in Derrida's work, "*différance*," "the trace," the *pharmakon*, etc., each operating in a slightly different context. A scale effect is a thing, or force, or more precisely a difference of things and forces, with decisive effects while remaining "that which in the presence of the present does not present itself" (PI 83/89).

This brings us to a terminological point most readers will have anticipated. Would "scale critique" be better named "scalar deconstruction"? Because we are largely dealing with effects of discontinuities in nature and discordances and interferences in human thinking and policy, "critique" may not be the best term here, with its stricter philosophical sense looking back to Kant and the aim of a systematic and skeptical/rational inquiry into the limits, presuppositions, and conditions of a concept. Rather, scale effects entail events that do not "await the deliberation, consciousness, or organization of a subject, or even of modernity. *It deconstructs itself. It can be deconstructed [Ça se deconstruit]*" (P2 4/12). "Deconstruction is not a critical operation; it takes critique as its object; deconstruction, at one moment or another, always aims at the trust confided in the critical, critico-theoretical agency, that is, the deciding agency, the ultimate possibility of the decidable; deconstruction is a deconstruction of critical dogmatics" (PI 54/60).[17]

In issues of scale, the kinds of discontinuities, refusal of homogenous sense, and mockery of conceptual synthesis that intellectuals have long engaged in the texts of Derrida, Maurice Blanchot, Gilles Deleuze, and others become increasingly fraught and contentious practical issues in daily life. Thinking through the slippery issue of scale effects in environmental politics also helps to overcome the inadequacies of a term like "environment" or "nature," projecting as they do some sort of external and even homogenously encompassing setting or reliable context for a central, usually human, reference point. The relative contingency, seeming caprice, incalculability, and counterintuitive nature of scale effects make them less easy to accommodate to talk of reestablishing some kinship between the human and nature. The latter relation becomes less a groundedness than an element of deeper uncertainty, a sliding of the once seemingly solid or reliable.[18]

One reason environmental issues are so difficult and fraught is that scale effects entail spectral agencies that present no easily identified target or simple object in empirical reality for politics or law—hence the tendency for environmental arguments to move in two polarized directions, either toward an evasive identification of green politics with individual lifestyle choices, a kind of moralized consumerism, becoming a new and often mildly irritating code of correct behaviors, or, on the other hand, toward a sweepingly general diagnostic of the crisis in terms of such monoliths as cultural anthropocentrism, patriarchy, or modern capitalism. It is not to dismiss the value of any of these approaches to suggest that the stridency and uncertainty of much environmental politics are precisely because of the importance of scale effects, the resistance of such spectral agencies to being resolved into empirically identifiable people ("spectral" here meaning something quite different from "illusory," as in Derrida). In environmental politics and ethics, scale effects may inhabit an area of particular discomfort or incalculability, "places where discourse can no longer dominate, judge, decide: between the positive and the negative, the good and the bad, the true and the false" (PI 86/92), for changes of scale may entail the emergence of different or even clashing determinations of an entity's nature or value.

As Nigel Clark writes, "Researchers in the social sciences and humanities still tend to treat natural and social agency as sliding points on a linear scale."[19] Yet scale effects are something beyond any one individual horizon, perception, or even calculation, also resisting the model of eco-phenomenology given by Ted Toadvine and Charles S. Brown in 2003 that, "for phenomenologists, experience must be the starting point and ultimate court of appeal for all philosophical evidence."[20] Scale effects also correspond more to the thought of a materiality that exceeds and resists conceptualization than to classic Marxist materialist arguments about "ultimate determination by the mode of production."[21]

In sum, to delineate a scalar deconstruction is to add force, specificity, and clarity to critiques over the past fifty years of the deep presupposition that reality itself is continuous or that being can admit of synthesis in some overarching conceptual unity: "Why should not man, supposing that the discontinuous is proper to him and is his work, reveal that the *ground of things*—to which he must surely in some way belong—has much to do with the demand of discontinuity as it does with that of unity?"[22]

Methodological Individualism
and the Anthropocene as a Scale Effect

Callicott's argument on cultural evolution underlines how deeply notions of scale are crucial to definitions of the human. This is an issue that the so-called Anthropocene or "epoch of the human" makes newly salient and unavoidable.

Francesco Vitale, in an essay on Derrida's mid-1970s seminar on the biology of François Jacob, describes the genetic realm in the animal kingdom as one cybernetic-biological program operating at varying levels of complexity, with the far greater complexity of the human brain having entailed a jump in capability. In effect, Vitale is reading Derrida's work on Jacob as describing the human mind as a scale effect of such programs:

> [A] difference in degree [of complexity] consists in the greater or lesser flexibility of the program, in its greater or lesser opening to possible variations in response, that is, in the ability to integrate the possible choices dictated by the influence of the environment and the group on the individual. But these possibilities are ultimately inscribed in the structure of the program. In the annelid and the amoeba, mentioned by both Leroi-Gourhan and Derrida, the program and its execution are very restricted because of the extreme simplicity of the nervous system; in man, the program is very open because of the great complexity of the system and the brain, which is able to operate a much greater number of connections than the brain of the other animals.[23]

Vitale contrasts two stages of complexity in the passage just quoted, between the annelid and the amoeba and the human (with doubtless innumerable other creatures and scale effects in cognitive possibility in between). Beyond the two levels of complexity that Vitale highlights in Jacob (between the annelid, the amoeba, and the human), does not a further level of complexity insist itself beyond the visible boundaries of the organism, at scales that acknowledge the fact that all creatures are constituted in the fact of living in groups, as these in turn interact with other groups? It has become a common gesture in posthumanist arguments and related kinds of ecocriticism to refer to the very small—to the fact that nonhuman bacteria and microorganisms are what make up a large percentage of any person's body mass, or to the way a multispecies genetic landscape underlies the minute biosemiotics of biological functioning. Less considered, but now unavoidable, is the issue of how human nature and its originary environmentality must be reconceived if one changes scale *in the other direction.* If "the living is a text which produces texts in order to survive in relation to its environment,"[24] then why privilege so strongly the level of the individual and its genetic makeup when defining the nature of a living entity?

Why, when trying to define the human, is the focus almost always on that relative abstraction, an individual person and his/her capabilities? What would it look like if the primary scale for thinking human nature were instead ecological, the larger scale of populations and ecological dynamics as distinct from individual or group behaviors, scales that may be global and encompass decades or millennia? Of supplementarity, Derrida writes:

This property [*propre*] of man is not a property of man: it is the very dislocation of the proper in general: it is the dislocation of the characteristic, the proper in general, the impossibility—and therefore the desire—of self-proximity; the impossibility, and therefore the desire of pure presence. That supplementarity is not a characteristic or property of man does not mean only, and in an equally radical manner, that it is not a characteristic or property; but also that its play precedes what one calls man and extends outside of him. Man *calls himself* man only by drawing limits excluding his other from the play of supplementarity: the purity of nature, of animality, primitivism, childhood, madness, divinity. (OG 244/347)

In effect, whether one defines "the human" with primary reference to the capacities and propensities of an individual human being (as Derrida is still doing provisionally in his work on the human/animal distinction) or ecologically, understanding the individual as in part an abstraction from the functions of large groups over various scales of space and time—any such definition entails an element of decision as a matter of pragmatic convenience, cultural prejudice, or political norms (forms of individualism, for instance). Thinking of the Anthropocene must affect this, as we now live in a time in which we know that we cannot not consider things on the larger temporal and spatial scale, even if we have as yet no adequate inherited ethics or even language for doing so. A scalar deconstruction, whether as a practice or as modes of thought that simply impose themselves in these newly realized contexts, highlights the contingency and hidden epistemic and ethical assumptions and decisions at work in thinking or acting absolutely on any one scale.

One obvious effect of the change of scale involved is to render newly questionable the presumed human/animal distinction. At the larger scale—the spectacle of human populations expanding to fill available ecological niches, to maximize energy consumption, or acquire new territory—is very plausibly still that of a manifestly "animal" dynamic, and each member of the human community, however individually rational, responding rather than mechanically reacting, is still living as part of that. It is a scenario in which innumerable people, to quote Eileen Crist, now "appear blind to the meaning and significance of their activities and interactions, and the production of their behaviors . . . [and how these are] determined by forces beyond their control and comprehension."[25] In fact, this quotation is from Crist's overview of accounts of the human/animal distinction and is describing *the latter, not the former*. This highlights the contingent nature of the distinction and suggests that the seeming privilege of the human in comprehension, response, etc., is in part a matter of prejudging the scale at which these issues are framed.

To question methodological individualism is also to query Vitale's statement that the possibilities of behaviors and interaction "are ultimately inscribed in the structure of the program." For individual behavior is not coherently conceived as the externalization of an inner script that is supposedly activated by external inputs. Vitale's summary here contrasts markedly with work in modern anthropology, as well as with Bernard Stiegler's Derridean argument that the essence of the human has never been separable from technicity and that the history of the human is one of the exteriorization of memory or, more specifically, of memory becoming such only through exteriorization, in sign-systems, archives, tools, infrastructure.[26] Qualities and capabilities become human only in this same originary supplementation and prosthesis. This is an exteriorization that is not preceded by an interiority but that determines it, or, in Derrida's words, "The dis-junction *is the relation*, the essential juncture" between so-called living interiority and technics.

Tim Ingold offers a related argument: that individual human behavior, recognized now as a fairly abstract or abstracted object of analysis, exists within a constrained space that is an emergent property of biosocial relations, inherited "traditions," or sets of practices inscribed in variously stable material environments:

> We can no longer think of the organism, human or otherwise, as a discrete, bounded entity, set over against an environment. It is rather a locus of growth within a field of relations traced out in a flow of materials. As such, it has no "inside" or "outside." It is perhaps better imagined topologically, as a knot or tangle of interwoven lines, each of which reaches onward to where it will tangle with other knots.[27]

Ingold describes a "logic of inversion" in the widespread and still dominant misconstruing of human agency as a central, sovereign determinant of events. This model inverts what are really outer multiple, contextual factors into the supposed act or nature of a unitary human agent whose actions are then understood as the outer expression of an interiority, intention, or character.[28]

Callicott, Ingold, and Stiegler's different but mutually supporting arguments offer a suggestive context in which to highlight Derek Woods's incisive point against now widespread accounts of the Anthropocene as a change in the earth's large-scale systems brought about by human activity. Woods does not contest the fact of this change, but argues that ascribing it solely to the agency of the human or to humans as a terraforming species is deeply misleading: "Scale critique shows the sleight of hand whereby accounts of human agency from familiar (individual, novelistic, small groups) scales get scaled up across disjunctures to become (in the present case) terraforming subjects."[29]

"The 'human' appears as a grammatical subject, agent, or population of agents that wields technology as an instrument,"[30] in a context in which it has actually become a complex and sometimes even marginal part of a broader assemblage, one lever of geomorphic change along with cattle, microbes, melting tundra, ocean currents, and so on. So it makes little sense "to narrate the subject of the Anthropocene as a species identity that works across scale domains."[31] Many of the factors matter in turn as scale effects that come to possess a perplexing autonomy in relation to any simple or initial human—or other—cause.[32] Yet it is difficult and perhaps too disconcerting to get away from discussing human action as the sole factor.

Read in terms of the biocybernetic conception of the human developed by Callicott (on cultural evolution) or Stiegler (on technics), the Anthropocene emerges as a culminating manifestation of the human as *anachronistic* in the strict sense—as living in disjunctive time scales. But it is also the moment of a profound dehumanization, for the Anthropocene is also the emergence of a new, dispersed, or nonunitary form of agency arising through the capricious force of scale effects of which human agency is only one part. The global scale is not simply that at which the human species, entwined in its various technical and infrastructural prostheses, becomes an unintended force of disruption in the earth's systems. The bizarre fact is that it is the accelerating development of technics, of cultural evolution, that is breaking down the distinction between human and nonhuman, on which it may seem to have been based.

So what is happening at the world scale clearly is only partially captured in Immanuel Wallerstein's well-known thesis of a relative unification of human societies under the hegemony of capitalist systems of production.[33] This account still effectively assumes a homogenous entity at the human scale growing into an expanded area. What in fact also happens at the world scale is that many of the constitutive elements of concepts of a world itself dissolve—not just because human cultures remain plural, with capitalism being absorbed or resisted in so many different ways, but because the once-foregrounded thinking of human agency gives space increasingly to more impersonal forms of dynamic. The Anthropocene should not then be seen as the external manifestation of something already "in" humanity, some essence or fundamental characteristic, but as a contamination, interference, and working together of various forms of happenstance (the finitude of the earth, short-term simplistic systems of economics, the absorptive properties of the atmosphere and oceans, hierarchy and exploitation in human societies, the unknown thresholds of various ecosystem functions, population growth, the unmeant effects of innumerable small human actions).

To return then to the question at the start of this essay: is the destructiveness associated with humanity an accident in some sense, is it inherent to what human beings are, or is it a new event not susceptible to being conceived or traced as the revelation, extrapolation, development of anything that preexisted it? One would now argue that it is a new event that cannot be adequately conceived or traced as the revelation, extrapolation, and development of anything that preexisted it. This presents a newly tragic and imponderable context for environmental ethics, for, effectively, human nature, if considered on a scale that eschews methodological individualism, has changed. It cannot be adequately the object of inherited systems of ethics developed for an entity at lesser scales.

Conclusion

Modern humanity lives increasingly in a counterintuitive unworld of scale effects. This context of excessive resource use, pollution, the incalculable or catalytic effects of the sheer and growing numbers of other people and material impacts involved means that we now live in a changed environment in which old, once innocent norms—on expectations of personal space, resource use, and even the desirability of children—acquire an autoimmune quality.

Any "green" thinker making appeals for a more truly human relation to the natural world, for living on a human scale, or being more fully human is confronted with a sense of how empty such uses of the word "human" are now becoming. A person living in 2017 may well be living a life very much like that of his or her parent a generation before—working, consuming energy and resources, producing waste, occupying so much space—but nevertheless, in this changed context of so many other people now doing the same thing and increasing sections of the world devoted to rampant capitalist accumulation, what this person is actually doing now is significantly different from anyone in the previous generation because of various and often opaque threshold and scale effects.

The specific force of a putative scalar deconstruction would be first to highlight the assumptions of scale inherent in given ways of thinking that grant them a spurious coherence—whether this is how certain conceptions of space, of what presents itself to us, are conditioned by a certain time scale, or, conversely, how conceptions of unfolding events or processes (the working of a "national economy," for instance) achieve coherence and noncontradiction only through kinds of unacknowledged scale framing. Second and more pointedly, scalar deconstruction highlights the contradictory determinations of being and judgments of value that emerge as scales are shifted.

The Anthropocene is so morally and politically dangerous because it is a situation in which the mere continued existence of some entities (e.g., seven billion human beings) has become already incompatible with that of the existence of others, so that many of the players, human as well as nonhuman, can only be pushed over the edge.[34] Increasing recognition of this situation must induce a heightened sense of responsibility, even as its stress already challenges any ethics of care based on notions of interdependence or of recognition of the other.

Notes

1. Derek Woods, "Scale Critique for the Anthropocene," *Minnesota Review* 83 (2014): 133–42.

2. Dipesh Chakravarty, "The Climate of History: Four Theses," *Critical Inquiry* 35 (2009): 197–222; Mark McGurl, "The Posthuman Comedy," *Critical Inquiry* 38 (2012): 533–53; Ursula Heise, *Sense of Place and Sense of Planet: The Environmental Imagination of the Global* (Oxford: Oxford University Press, 2008).

3. Chakravarty, "Climate of History," 1.

4. David Wood, "What Is Eco-Phenomenology?," in *Eco-Phenomenology: Back to the Earth Itself*, ed. Charles S. Brown and Ted Toadvine (Albany: SUNY Press, 2003), 215.

5. Baird Callicott, "Lamarck Redux: Temporal Scale as the Key to the Boundary between the Human and the Natural Worlds," in *Nature's Edge: Boundary Explorations in Ecological Theory and Practice*, ed. Charles S. Brown and Ted Toadvine (Albany: SUNY Press, 2007), 36.

6. David Delaney and Helga Leitner, "The Political Construction of Scale," *Political Geography* 16 (1997): 94–95.

7. Wood, "What Is Eco-Phenomenology?," 212.

8. Ibid., 213.

9. Ibid., 217.

10. Simon Glendinning, *In the Name of Phenomenology* (London: Routledge, 2007), 183.

11. Woods, "Scale Critique for the Anthropocene," 133.

12. Derrida, in arguing that "this is my starting point. No meaning can be determined out of context, but no context permits saturation" (P 108/125) is effectively denying the authority of scale invariance. As a

mode of reading, deconstruction is often a refusal to take things in "normal" proportion—the context at issue is simultaneously miniaturized, with the focus on a minor aside in text, a footnote or a mere snippet such as Joyce's "he war" from *Finnegans Wake* (see Derrida, *Ulysse gramophone: Deux mots pour Joyce* [Paris: Galilee, 1987]), and vastly expanded, to embrace that large-scale entity known as the "Western metaphysics of presence."

13. Jim Dator, "Assuming 'Responsibility for our Rose,'" in *Environmental Values in a Globalizing World: Nature, Justice and Governance*, ed. Ian Lowe and Jouni Paavola (London: Routledge, 2005), 215–16.

14. "For example, distinctions are made in biology between micro- and macro-evolution, in economics between micro- and macro-economics, and in history between micro- and macro-historical processes. The physical sciences use the term 'scale' to denote a regime or domain. Astronomers refer to 'the large-scale structure of the Universe,' while particle physicists talk about the sub-atomic scale. Scale is implicit in all measurements, e.g. the temperature scale, weight scale, Richter scale, etc. In this sense it is related to the concept of units, and the invention and adoption of various systems of units of measurement have many manifestations in the physical sciences as well as multiple implications for social, political and historical analysis. In each disciplinary area, there are deep questions about the relationships between different scales and whether principles and laws describing behaviour at the micro level can account for macro-level phenomena. Knowledge and understanding are often more certain at smaller scales than at larger scales. Is the apparent similarity in problems of scale real or superficial?"; "Introducing Scale," theme for 2016–17, Institute for Advanced Study, Durham University https://www.dur.ac.uk/ias/themes/scale/; accessed August 1, 2015.

15. Wood, "What Is Eco-Phenomenology?," 215.

16. The inhuman semiautonomy of scale effects, arising in part out of the total force of human actions, correlates with what David Wood says of the emergence of rhythms in the happening of natural events, such as the emergence of some species of cicada over a period of seventeen years. These are processes in which "time acquires sufficient autonomous efficacy to generate its own relational differentiation" (ibid., 216), something that is not "the result of the synthetic or constitutive activity of any kind of subject, nor any simple causal mechanism" (ibid., 217).

17. See also PI 212/226.

18. Wood, "What Is Eco-Phenomenology?," 215.

19. Nigel Clark, "Geo-Politics and the Disaster of the Anthropocene," *Sociological Review* 62 (2014): 26.

20. Brown and Toadvine, "Eco-Phenomenology: An Introduction," in Brown and Toadvine, *Eco-Phenomenology*, xi.

21. Frederic Jameson, *The Political Unconscious: Narrative as a Socially Symbolic Act* (Ithaca: Cornell University Press, 1981), 45.

22. Maurice Blanchot, *The Infinite Conversation*, trans. Susan Hanson (Minneapolis: University of Minnesota Press, 1993), 9. Compare Gilles Deleuze and Félix Guattari: "When a multiplicity [of the kind of a rhizome] changes direction, it necessarily changes in nature as well, undergoes a metamorphosis"; A *Thousand Plateaus: Capitalism & Schizophrenia* (Minneapolis: University of Minnesota Press, 1987), 21.

23. Francesco Vitale, "The Text and the Living: Jacques Derrida between Biology and Deconstruction," *Oxford Literary Review* 36, no. 1 (2014): 102. It is suggestive to ask what the earlier reception of Derrida's work would have been like, with the accusations of textual idealism, etc., if Derrida had chosen not "writing" in the 1960s as the significant term for the workings and unworkings of *différance*, but one that did not have a potentially misleading implication of an exclusively human/cultural reference, such as "reproduction," "replication," or "duplication." Such terms would legibly undermine at a stroke fantasies of human autonomy by stressing the human's constitution through an incalculable physical and informational embeddedness in tangled networks of chemical and energy exchange, microbiological, chemical, meteorological, and geological—in effect an originary environmentality. In some ways "originary environmentality" might seem a stronger term of deconstruction than the familiar "originary trace": its immediate inference is multidimensional, a circle or environing sphere of relations, whereas "trace" still suggests a residually linear figure, a trail, marked line, or path. This misleading implication, at odds with Derrida's actual argument, is perhaps behind the reductiveness in Vitale's reading of Derrida on "'an exteriorization of the trace, that has always already begun, and yet, becomes larger and larger'" (ibid., 101)—a way of putting this that risks still sounding like a homogenous and continuous process, not one involving scale discontinuities and unpredictable retroactive effects.

24. Ibid., 111.

25. Eileen Crist, *Images of Animals: Anthropomorphism and Animal Mind* (Philadelphia: Temple University Press, 1999), 5. The admittedly sometimes simplistic talk about the "population bomb" in the 1960s and 1970s, when world population was half current numbers, arose in part from applying to human beings results derived from the study of other animals, producing Paul Ehrlich's and others' "population ecology"; see Sabine Höhler, "The Law of Growth: how Ecology Accounted for World Population in the 20th Century," *Distinktion: Scandinavian Journal of Social Theory* 14 (2007): 45–64.

26. Bernard Stiegler, *Technics and Time*, vol. 1, *The Fault of Epimetheus*, trans. Richard Beardsworth and George Collins (Stanford: Stanford University Press, 1998).

27. Tim Ingold, "Prospect," in *Biosocial Becomings: Integrating Social and Biological Anthropology*, ed. Tim Ingold and Gisli Palsson (Cambridge: Cambridge University Press, 2013), 10.

28. Diana Coole endorses Nietzsche's argument about the way concepts of the subject as unitary and intentional agent are a kind of pragmatic illusion made credible when life forces are turned back upon themselves, "internalized," in a fold that effects a notion and seeming experience of interiority and reflexivity: "Modern ideas of agency have elided a series of phenomenal processes—such as consciousness, meaning-generation, reflexivity, will, reasoning—that are then unified in the figure of the onto-logical individual or transcendental subject"; Coole, "Rethinking Agency: A Phenomenological Approach to Embodiment and Agentic Capacities," *Political Studies* 53 (2005): 128.

29. Woods, "Scale Critique for the Anthropocene," 137.

30. Ibid., 138.

31. Ibid., 139.

32. Woods suggests that "as subject of the Anthropocene the species concept can . . . do little work here"; ibid., 139.

33. For Wallerstein's work, see *Immanuel Wallerstein and the Problem of the World: System, Scale, Culture,* ed. David Palumbo-Liu, Bruce Robbins, and Nirvana Tanoukhi (Durham, N.C.: Duke University Press, 2011).

34. See, for example, the report of a study by the UN Environment Programme of 2007, finding that "the human ecological footprint is on average 21.9 hectares per person. Given the global population, however, the Earth's biological capacity is just 15.7 hectares per person"; https://www.newscientist.com/article/dn12834-unsustainable-development-puts-humanity-at-risk/, accessed August 1, 2015. It should be noted that the maximum biological capacity here for human life on earth is a concept that acknowledges no right for nonhuman life also to exist except in the service of humanity.

PART

II

Ecologies

The Posthuman Promise of the Earth

Philippe Lynes

Derrida announces what I'll risk calling the "posthuman promise of the earth" in a little-read 1995 text called *Advances*, only published in English translation in 2017,[1] where he writes of

> a finite promise *of* the world, as world: it is up to "us" to make the world survive; and we cannot say this question is not urgently important today; it always is and always will have been, any time it can be a matter—or not—of giving oneself death, that is to say the end of the world; it is thus up to "us" to make what "we" inadequately call the human earth survive, an earth that "we" know is finite, that it can and must exhaust itself in an end. But "we" will have to change all these names, beginning with "ours," "we" know and sense this more than ever. (AV 47–48/39)

In the last interview before his death, Derrida perhaps surprised some of his readers when he claimed the following: "All the concepts that have helped me in my work, and notably that of the trace or of the spectral, were related to this 'surviving' as a structural and rigorously originary dimension" (LL 26/26), later adding that "deconstruction is always on the side of the *yes*, on the side of the affirmation of life. Everything I say . . . about survival as a complication of the opposition life/death proceeds in me from an unconditional affirmation of life" (LL 51–52/54). These would in fact be the last words he ever wrote, and were read by his family at his funeral. "Always prefer life and endlessly affirm survival. . . . I love you and am smiling at you from wherever I am."[2] Our current ecological crisis now enjoins us, with each day more urgently than ever, to understand what Derrida may have meant with these words.

To affirm survival, then, but of whose, of "ours"? Of the "human" earth? Or does the present state of "our" earth not signal the necessity of hearing this affirmation—which is perhaps already a promise—from elsewhere, from beyond the beyond, indeed through the experience of a certain transcendence, without (lived) experience, an experience of the impossible, a certain ecological *end of the human* opening onto what Cary Wolfe has called the posthumanist challenge of "sharing the planet with nonhuman[s] . . . and treating them justly."[3] Borrowing a formulation from Bataille and Blanchot, Derrida has often stated that deconstruction is precisely an experience of the impossible, and "what is impossible is to inscribe this transcendence (the 'relation that goes beyond the world') within immanence, to wit the presence, proximity, immediacy one necessarily associates with the word 'experience'" (FF 26t)—to inscribe this transcendence within the proper, within the *oikos*, then, of what I'll call a "restricted ecology."

Justice, as Derrida puts it, "is above life, beyond life or the life drive in a sur-viving of which the *sur*, the transcendence of the 'sur'—if it is a transcendence—remains to be interpreted" (DP1 271/366). What I'll attempt to develop here as general ecology will thus attempt to think this aneconomic excess and expropriation of survival over transcendental, ontological, psychoanalytic, and dialectical notions of life. General ecology must further problematize the presumed irreducibility of the ontological "is" to the normative or ethical "ought" proclaimed in the naturalistic fallacy. It *must* do so, as Derrida might put it, both as an ontological necessity and an ethical duty, in a change that "will have to affect the very sense and value of these concepts (the ontological and the ethical)" (FW 64/108). The syntax in Derrida's claim that "il faut bien vivre ensemble" hints at this ontological ethics; it is an ontological fact that we must *of course* live together, and it is an ethical duty that we must live together *well*.[4]

"Economical indeed ecological," Derrida writes in "Faith and Knowledge"—"and one would have had to associate the motif of ecology, this large and new dimension of 'living together,' with the motif of economy" (AI 39/61). And so aneconomic as an-ecological? But doesn't sharing the earth more justly with its other living beings precisely require some proper housekeeping, maintaining the law of the *oikos*, the (h)earth? Is not the radical unprecedentedness of our current ecological crisis precisely when and where the transcendence of a relation that goes beyond the world ought to be jettisoned? Perhaps, but this assumes the circulation of a certain propriety that was never that of ecology, even if it will always remain necessary to count and to calculate with. To live together ethically on earth requires the interruption of what Blanchot and Levinas call a "separation," what I'll later refer to as a "relationality without relationality." Derrida illustrates the difficult consequences of thinking this interrupted relationality with respect to ecology as follows:

> "Living together" supposes, therefore, an interrupting excess . . . with regard to *symbiosis*, to a symbolic, gregarious, or fusional living together . . . all "living together" that would limit itself to the symbiotic or that would be regulated according to a figure of the symbiotic or the organic is a first lapse of the sense [*un manquement au sens*] and of the "one must" [*il faut*] of "living together." (AI 26–7/33–34t)

In this sense, the "one must" of living together, along with the question of what must be done with the earth, "is not simply an *ecological* question, even if it remains at the horizon of what ecology *could* have as its most ambitious and most radical today" (QF 49t).

Derrida's notion of survivance must be thought in terms of his well-known notions of the trace or *différance*. Such notions, as he puts it, "can introduce the economy of a new configuration into the immanence of the living being. The interruption itself belongs to the field of what is genetically or biologically possible" (FW 40/73). I'll examine later how this relation of interruption figures into Derrida's notion of "the living in general" and how this allows the generality of "general ecology" to be heard in its particularity.[5] As Derrida elaborates,

> Différance means at once *the same* (the living being, but deferred, relayed, replaced with a substitutive supplement, by a prosthesis, by a (technological) supplementation) [this is what I'm calling "restricted ecology"] and *the other* (absolutely heterogeneous, radically different, irreducible and untranslatable, the aneconomic, the wholly-other or death). The differantial interruption is both reinscribed into the economy of the same and opened to an excess of the wholly other [*tout autre*]. (FW 40/74t)

What I will develop here as general ecology will therefore attempt to think the relation between two indissociable movements. On the one hand, it must think, count, and calculate with a restricted ecology, a dialectics of life and death, openness and closure, exposure and reappropriation, activity and passivity within the autoaffection of transcendental subjectivity and intersubjectivity—and all the anthropocentric consequences of this economy. On the other hand, general ecology must think a radical expropriation without return, an unconditional affirmation of survivance without sovereignty, the promise of sharing the earth with all its others where *tout autre est tout autre*, every other (one) is every (bit) other,[6] whether animal, fungal, vegetal, bacterial, viral, technomachinic, organic or inorganic, God or human, whole or part, in their radical singularity. This will require "rethinking the limits between the human and the animal, the human and the natural, the human and the technical. For the question of animality, that of the earth, of what we may mean by 'life' in general also makes up the promise of this bond" (NII 241). The promise of this bond is itself the promise of the earth, an ethical-ontological promise to *let life live-on* and *let the earth be the earth*.

This chapter will attempt to articulate the promise of the earth through the expropriation of three temporalities of life: Husserl's living present, Heidegger's autoaffective temporalization of *Dasein*, and Freud's death drive. As I will show, all three rest upon a certain decision concerning the restricted ecological reappropriations of life and death and their relations to the subject, the human, and the individual: Husserl's living present as the pure form of absolute subjectivity, Heidegger's autoaffective temporalization as the proper of the human and its death, and Freud's death drive as the detour within the ecology of immanence in the individual living organism. I will conclude by returning to the an-archic temporality of the promise of the earth (which is itself, perhaps, a matter of thinking the eternal return otherwise) to show how deconstruction opens onto another thought of sharing the earth in the life death we share with its other living beings.

I

Derrida's early investigations of Husserl, which he recalls are at the heart of his thinking of nonhuman life and death in *The Animal That Therefore I Am*,[7] can be summarized as describing the structures by which the purity, propriety, immanence, and self-presence of transcendental life and subjectivity find themselves contaminated, indeed expropriated, by the transcendences of temporalization, alterity, the material world, and death. Phenomenology is a philosophy of life in a restricted sense, for Derrida, not only because a transcendental life remains after the bracketing of empirical life and the pure Form of the *living* present as *absolute subjectivity* is itself shown to constitute transcendental subjectivity, but because death is nothing other than an empirical and worldly accident befalling life for phenomenology.[8] The ultratranscendental concept of life put forth by Derrida thus intends to think together transcendental life and its others, beyond any common or biological definitions of life, and "perhaps calls for *another name*" (VP 13/14). As early as his 1953–54 dissertation *The Problem of Genesis in Husserl's Philosophy*, Derrida was engaged in complicating Husserl's concept of the living present through the latter's own descriptions of passive genesis: the structures by which transcendental subjectivity is passively constituted through time and the other and lead back to a "pre-egological and pre-subjectivist zone" (PI 263/277).[9] As he would later explain, the indissociable alterities of time and the other not only structure the dialectic of autoaffection at the heart of every individual living being,[10] but are also the very condition for any relation between living beings—I would add, of any ecology. Derrida's work would in fact eventually seek to articulate this restricted economy, or dialectical, autoaffective complication of the absolute subjectivity of the living present within a more general structure, upon a certain heteroaffection. As he explains in *Voice and Phenomenon*, the flux of the living present is not unnamable as it is for Husserl, who resigns himself to saying, "*Für all das fehlen uns die Namen*" [for all this, we lack names]. Rather, the living present is entirely determined according to a subjectivist metaphysics of presence: "*Its proof is that we do not put into question its being as absolute subjectivity*" (VP 72n1/94n1t). For Derrida, the movements of temporalization and intersubjectivity must be thought on the basis of *différance* rather than a dialectical economy of presence and subjectivity: "There is no constituting subjectivity. And one must deconstruct all the way down to the concept of constitution" (VP 72n1/94n1t). I suggest that this occurs in the proximity of a certain encounter with Levinas—albeit a proximity that installs itself as separation—and their similar thoughts of the trace, radical passivity, and a past that has never been present. These notions, I believe, allow one to enter

into relation with *"on the one hand,* my other living present (retained or anticipated by an indispensible movement of retention and
protention) [this is what I'm calling 'restricted ecology'], and, *on the
other hand,* wholly other, the present of the other, the temporality
of which cannot be reduced, included, assimilated, introjected, appropriated into mine, cannot even resemble it or be similar to it"
(BS1 271–72/364). General ecology, to jump ahead a little, would be
the logic that thinks together these two intertwinements of time
and the other, the economic and the aneconomic.

In *Adieu to Emmanuel Levinas,* Derrida suggests that ethics constitutes phenomenology's own autointerruption, the interruption of
itself from within by itself,[11] a claim he would echo in *Specters of
Marx* in claiming that no justice is possible without a disjunction in
the living present. Deconstruction, I would say, would come down
to a *beyond-within* of the living present: *"beyond-within:* transcendence in immanence" (AE 76/138). Levinas's philosophical project
also begins with a study of Husserl, albeit opposing what the former reads as the ethical spirit of phenomenology in its descriptions
of temporalization, intentionality, and passive genesis, to its letter in
the Husserlian corpus. In fact, Derrida precisely sees a contestation
of the living present at the source of the Levinasian problematic in
Paper Machine.[12] For Levinas, as is well- known, phenomenology is
ultimately a philosophy of violence, since the structure of analogical
appresentation can only account for the other as a modification of
the ego and the same and cannot thereby respect the transcendence
of the other. For Derrida, however, it is precisely this structure of
analogical appresentation that preserves the separation and dissocia-
tion necessary for respecting the infinite alterity of the other: what
Levinas and Blanchot call a "relation without relation" as the condition of the social bond.[13] However, it is difficult, perhaps impossible
to argue that the Other for Levinas constitutes anything other than a
human being or subjectivity. The situation is obviously much more
complicated for Derrida, who recognizes the necessity of an ethics
that opens onto the beyond of the human, where *tout autre est tout
autre,* problematizing the distinctions between "the 'nonliving,' the
'vegetal,' the 'animal,' 'man,' or 'God'" (PI 292/269). As he says elsewhere, "The unrecognizable . . . is the beginning of ethics, and not
of the human. So long as it remains human, among men, ethics remains dogmatic, narcissistic, and not yet thinking" (BS1 155/108).

In a sense, Derrida's reading of the living present can be read as
a deconstruction of Form itself, of a certain formalism, *eidos,* or

idealism, indeed an immanentism and vitalism in transcendental phenomenology that one could read at the heart of not only Euro-centric views of history and science, but also in anthropocentrism more generally. Derrida seems to pose an originary finite materiality to transcendental teleology, reinscribing what calls itself the "hu-man subject" within an-other thought of history before and beyond the humanism of transcendental teleology, in a relation to an im-memorial past beyond even its evolutionary history. In this *beyond-within*, this self-excess and autotranscendence of the living present, "a process of deconstruction is in progress, which is no longer a te-leological process or even a simple event in the course of history" (AE 80/146). What we might anticipate in the promise of the earth, an ecological democracy-to-come, "the Parliament of the Living," as David Wood has called it,[14] can precisely never be a utopian fu-ture democracy, since "their inaccessibility would still retain the temporal form of a *future present*, of a future modality of the *living present*" (SM 81/110). This teleological structure, moreover, would thoroughly annul and neutralize the unknowability, incalculability, and transcendence of the alterity of the other in my ecological re-lationality. How far must this democracy-to-come extend, Derrida asks? To humans, to all nonhuman living beings? To the nonliving beyond the living present, past or to-come? "To the dead, to ani-mals, to trees and to rocks?" (R 54/82).

II

Derrida's debt to Heidegger in attempting to think beyond the hu-manism and anthropologism of transcendental subjectivity cannot be overstated. Indeed, Derrida also reads in the transcendence of *Da-sein* the notion of a presubjective transcendental field. Heidegger himself, he notes, challenges the "philosophical invulnerability" of the living present in affirming the originary finitude of temporaliza-tion (HQ 139/210). Against the metaphysical, subjective, and hu-manist determination of transcendental subjectivity and its classi-cal attributes of activity, freedom, and volition, the autoaffection of time at the heart of the living present allows Heidegger to describe the originary passivity of *Dasein*, the passive affection of time by itself, as the basis from which all determinations of subjectivity are derivative. As Derrida explains, autoaffection *is* the transcendence of *Dasein*: "The notion of affectivity is at bottom merely the name of the transcendence of *Dasein* towards the Being of beings, and as time, the meaning of the transcendental horizon of the question of Being. Affectivity is in this sense transcendence" (HQ 181/267).

As is well-known, Heidegger argues that humanism, subjectivity, and metaphysics are indissociable (HQ 22/52). It is against any philosophy of life or biologistic or organicist interpretation of the subject that Heidegger defines *Dasein* as a mortal existant rather than as a living being, having the resolute anticipation of its death, the possibility of its impossibility, as its most proper property. But *Dasein* is also a being among others in this finitude and anticipation of death, among a "we" who are also together in a precomprehension of Being, as those for whom Being is a question, who have access to the difference between Being and beings and the capacity to *let beings be* what they are. For Derrida, and again in discussion with Levinas, this letting beings be is the very condition of a certain economy of violence, a nonethical opening of ethics. But if the question of Being is inextricable from a certain humanist propriation, how does this bode for the ecological ethics I am proposing here? As different as are Heidegger and Levinas's thoughts of transcendence, Derrida suggests, and "these two ways of thinking transcendence are as different as you wish. They are as different or as similar as being and the other" (PI 284/298); both follow the same anthropocentric determination of the neighbor in the proximity of my ecology.

Much in Derrida's reading of Heidegger rests on this determination of the proximity of *Dasein* to the question of being and the correlated values of propriety, the proper of man, reappropriation, authenticity, the *Eigen* (proper) and *Eigentlich* (authentic, ownmost) in the *Ereignis* (event, appropriation), and especially the *oikos, oikonomia*. Despite Heidegger's express intentions to the contrary, "it remains that the *proper of man* is inseparable from the question or the truth of Being" (MP 124/148). But what Derrida develops as the trace, as somehow even "older" than the distinction between Being and beings,[15] "exceeds the question *What is?* and contingently makes it possible" (OG 75/110). The trace is the very dislocation of the proper in general, the general expropriation of the proper to man, and it is here that one must seek a different thought of ecology, an-other thought of the transcendence and the ethical passivity of letting, perhaps to name without naming the general ecology of this *oikos*, "the at home in general that welcomes the absolute *arrivant*" (A 34/68).

To mark the necessity of thinking the promise of this expropriation for an ecological ethics, let me recall that justice for Derrida depends on a certain non-self-contemporaneity of the living present. Justice, as he again draws from Levinas, must be thought as "incalculability . . . and [the] singularity of the an-economic ex-position to others" (SM 26/48). Ecology demands that we "join ourselves in this *we*, there where the disparate is turned over to this singular *joining*, without knowledge, without or before the synthetic junction of the conjunction and the disjunction" (SM 35/58), without *knowledge* of the living being of the neighbor dwelling in the proximity of transcendence. In the necessity of maintaining the distances, hiatuses, gaps, and interruptions within this ecological relationality, to avoid a "dedifferentiating discourse of 'life'"[16] perhaps suggested by Derrida's frequent reference to "the living in general," it is perhaps the logic of this generality that must be further explicated. Justice, however, also requires the material inscription of this anachrony at the heart of the living present within a certain place, which one could call "the earth" or "nature." Heidegger for Derrida not only reinscribes transcendental subjectivity into the world bracketed by the *Epokhe*, but radically transforms the interpretation of the word "world" in reinterpreting *Dasein*'s transcendence as being-in-the-world. While for Derrida, humans and other living beings incontestably inhabit the same world at the same time that they incontestably do not, not a single singular living being can be said to inhabit the same world as another: "The difference between one world and another will remain always unbridgeable, because the community of the world is always constructed" (BS2 8/31). And yet, he writes, we can at least suggest that humans, animals, plants, and so on have being alive in common: "We can believe that these living beings have in common the finitude of their life, and therefore, among other features of finitude, their mortality in the place they inhabit" (BS2 10/33). "Whether one calls it the earth (including sky and sea) or else the world as the world of life-death. The common world is the world in which one-lives-one-dies, whether one is a beast or a human sovereign, a world in which both suffer, suffer death, even a thousand deaths" (BS2 264/365).

It is in the finitude that we share across a world that we do not that Derrida famously reads Celan's line "*Die Welt ist fort, ich muss dich tragen*" [the world is gone away, I must carry you] as the very beginning of ethics, indeed of what I'm risking here in thinking general ecology as the disjunction of the living present. The *tragen*, then, must be understood as *both* the necessary and impossible desire of a shared living present—the presumed unity of the world, the "as-if" of a common world assuring the longest sur-vival for all living beings, *and* as a necessary discourse of multiplicity, untranslatability, the nongatherable and expropriation. To think this double logic is necessary to understand what it might mean *to let life on earth live-on* and the sense of passivity in this letting. Derrida importantly reads finitude and mortality not as a power, activity, or agency that would be proper to this or that form of life, but as a radical passivity, as an unpower or nonpower at the heart of power.[17] Within the passivity of the shared suffering before the finitude of mortality lies the origin of compassion, and "it is from this compassion in impotence and not from power that we must start when we want to think the animal and its relation to man" (BS2 244/339). And perhaps everything I have been trying to propose as the expropriation of the proper of the human depends on interrogating this passivity and precisely not as the power to question.

Heidegger himself, Derrida writes, would eventually come to complicate and displace the privilege accorded to the question spanning over thirty years. It is in bringing together the Heideggerean notions of the *Zusage* (acquiescence, response) that Derrida signals toward "a new (post-deconstructive) determination of the responsibility of the 'subject'" (PI 268/282), one we could just as well call "posthuman." The *Zusage*, Derrida explains, can be understood as a promise, affirmation, yes, acquiescing, or consent anterior to all question, to all language, and to all activity.[18] The *Zusage* always comes in the form of a response, but one that must be radically extricated from any humanist, metaphysical, or subjective determination of the subject. On the contrary, the *Zusage* might not be thought entirely apart from a reaction, one within which every living thing is originarily engaged—prior to any active willing on its part, indeed any autonomy—as its originary promising. "The relation to self, in this situation [the *Zusage*], can only be *différance*, that is to say alterity or trace. Not only is the obligation not lessened in this situation, but on the contrary it finds in it its only possibility, which is neither subjective nor human" (PI 261/275–76). One can further understand what I'm proposing as the posthuman promise of the earth in Derrida's reading of Heidegger's concept of *Verlässlichkeit* in "The Origin of the Work of Art,"[19] "reliability" [*fiabilité*], a preoriginary engagement, gift, credit, debt, and duty anterior to any social or natural contract. This engagement is given in what Derrida calls "the silent call of the earth . . . that language without language or correspondence with the earth and the world" (TP 351/402). It issues from a "combat" between the earth and the world, inviting us to think the earth, the world, and Heidegger's *Geviert* or fourfold otherwise—the play of the earth, sky, divinities, and mortals being thought in their "expropriating transpropriation," as he cites it in *Dissemination*, bound together in the ring (*anneau*, *Ring*) (DS 354/430t). This ring is indeed a wedding band, that of a "precontractual or precontracted marriage with the earth" (TP 354/404). The wedding vow of the finite promise of the earth might simply be one that "'lets the earth be the earth.'" (TP 354/405)

III

Through his readings of Nietzsche and Freud, however, Derrida allows us to signal a certain doubling at the heart of the *Zusage*, the promise and affirmation. As Derrida explains, one of the fundamental axioms of anthropocentrism has always opposed the power of the free, active, ethical, and responsible response of the human subject to the passive mechanical reaction and reactivity of nonhuman life. If, however, such a mechanicity were to find itself somehow implicated at the heart of the subject, the whole history of ethical, juridical, and political responsibility, along with the *différance* between reaction and response, would have to be reinscribed "within another thinking of life, of the living, within another relation of the living to their ipseity, to their *autos*, to their own autokinesis and reactional automaticity, to death, to technics, or to the mechanical [*machinique*]" (AA 126/173). The originary acquiescing of the *Zusage*, the affirmation of the "yes" in its address to the other, must mechanically promise to repeat and reiterate itself and is immediately duplicated and contaminated by another "yes." This contamination must be understood both as threat and as chance, as the condition for any life to live-on at all.

However, between the unconditional affirmation of the other and the affirmation of radical evil, there is only a difference of forces; in other words, it is the *différance* between the finite and infinite "yeses" that is originary. What I am attempting to think here is precisely the logic of bringing together the repetition of the mechanical with the organic, living, singular event of ecological relationality, indeed of the machine *as* this spontaneity, "*both as the origin of what flows from the source, and with the automaticity of the machine. For the best and for the worst*" (FK 57/32). The mechanical, in this sense,

at the same time detaches from and reattaches to the family (*heimisch*, homely), to the familiar, to the domestic, to the
proper, to the *oikos* of the ecological and of the economic, to
the *ethos*, to the place of dwelling . . . at the same time expropriates and re-appropriates, de-racinates and re-enracinates,
ex-appropriates according to a logic that we will later have to
formalize. (FK 78/64)

Through this logic Derrida allows us to think an originary technicity at work in the very structure of the living organism. Any living being, he writes, "undoes the opposition between *physis* and
technè. As a self-relation, as activity and reactivity, as differantial
force, as repetition, life is always already inhabited by techniciza-
tion" (NII 244).

To think the living as structured in a difference between forces
is a move common to Nietzsche and Freud. Derrida had examined
this in the logic of a detour between the pleasure principle and the
reality principle in "Freud and the Scene of Writing." The difference
between the two principles, he argues, constitutes the originary possibility of a detour and *différance* as the economy of death within
life. While life protects itself in the repetition of this detour, there
is no original life that would subsequently come to protect itself in
repetition and *différance*. The very concept of a "first time" yields
to an originary dimension of a repetition, "death at the origin of
life which can defend itself against death only through an *economy*
of death, through deferment, repetition, reserve" (WD 253–54/300–
301). The detour through the reality principle allows the pleasure
principle to return to itself within the same domestic economy.
This mastery of the pleasure principle is, perhaps, mastery or sovereignty in general for Derrida. Even Freud's account of the death
drive eventually finds itself reappropriated within its logic. While
the death drive pushes the organism to return to an anterior state
of inorganic matter, the conservation drive protects it from any external forces that would disrupt the immanence of its return-to-self.
Life itself is inscribed in this differantial relay with the transcendence of nature and the material world, "the 'external' force which
disturbs the immanent tendency, and which in a way produces the
entire history of a life which does nothing other than repeat itself
and regress, is what is usually called nature, the system of the earth
and the sun" (PC 354/377). The organism, according to Freud, seeks

to die its own, proper death, just as death, for Heidegger, constitutes *Dasein*'s own-most possibility. Freud's law of *la-vie-la-mort* would be the law of the proper: "It is indeed a question of an *economy* of death, of a law of the proper (*oikos, oikonomia*) which governs the detour and indefatigably seeks the proper event, the prolongation or abbreviation of the detour would be in the service of this properly economic *or ecological law of oneself as proper*" (PC 359/381–82e).

This restricted ecology of the living, reappropriating every material transcendence within the immanence of its own detour, bears for Derrida an almost identical structure to the living present and autoaffective temporalization discussed by both Husserl and Heidegger.[20] But, as I have shown, this immanence and propriation can never fully appropriate the transcendence, alterity, and materiality of the external world, earth, or nature that nonetheless sets the ecology of its mastery into motion. In the law of the *oikos*, "in the guarding of the proper, beyond the opposition life/death, its privilege is also its vulnerability, one can even say its essential impropriety, the expropriation (*Enteignis*) which constitutes it" (PC 359/382). Expropriation thus undoes the logic of mastery and sovereignty from within, or rather, is the relation by which these terms can be deconstructed and deconstruct themselves, whether at play in absolute subjectivity (the living present), human propriety (autoaffective temporalization), or organic immanence (the death drive).

And yet, precisely because thinking together the ecological and the anecological is impossible, is *the* impossible, one might also say that it is a certain drive for mastery, power, propriety, sovereignty, indeed cruelty that is irreducible. Both Nietzsche and Freud, Derrida writes, see cruelty as having no contrary, being tied to the essence of life as the will to power. Thus the apparent difficulty for any progressive politics; one can only negotiate, through the economy of a differantial transaction, differences in cruelty, force, and power, an economy of violence; even the most pacifistic ecological ethics must thus remain somewhat pessimistic. This is certainly how many have read Derrida's remarks on Levinas and "the nonethical opening of ethics" (OG 140/202) or the "economy of violence" (WD 400n21/136n1). But there is also an aneconomic, indirect way of challenging cruelty, sovereignty, or violence, Derrida argues, that must take into account, "in the mediation of the detour, a radical discontinuity, a heterogeneity, a leap into the ethical (thus also the juridical and political)" (WA 273). Beyond the economy of the appropriable and the possible, the "I can," Derrida calls for an unconditional affirmation without sovereignty and cruelty. This affirmation, Derrida writes, comes "from a beyond the beyond, and thus from beyond the economy of the possible. It is attached to a life, certainly, but to a life other than that of the economy of the possible, an im-possible life no doubt, a sur-vival" (WA 276). The economy of life suggested by Freud for Derrida is incapable of justifying a pacifism. "This can only be done on the basis of a *sur-vival* which owes nothing to the alibi of some mytho-theological beyond" (WA 276). It is in the structure of a sur-vivance, then, as the originary expropriation of any identity in death as the condition of ecological relationality, that the terrain for a more gentle and just relationship to the earth's other living beings must be sought.

That "we" are never "ourselves" in the anachrony of ecological relationality is not only due to the irreducibly relational condition of any living singularity. Ethics itself, Derrida's thought seems to suggest, necessitates learning to live with these survivances more justly, more faithfully, perhaps in a promise that can only be in excess of itself, since it must indeed give the world, given to us from even beyond an evolutionary history. "The self [*soi-même*] has that relation to itself only *through* the other, through the promise (for the future, as trace of the future) made to the other as an absolute past, and thus *through* this absolute past, by the grace of the other whose sur-vival—that is, whose mortality—will have always exceeded the 'we' of a common present" (MPD 66/77–78t).

IV

The concept of general ecology I've been attempting to put forward intends to designate the relation between an ethical and ontological transcendence, exposing all the terms therein to a certain sliding as an unconditional affirmation of sur-vivance. It is through Bataille's critique of Hegel that Derrida first develops the concept of general economy, which unfolds as follows. In Hegel's dialectic of mastery or lordship, the master must risk his life in a passage through death, a death that is thus negated, overcome, and reappropriated, conserved, and survived without remainder in the *Aufhebung*, the very logic of the restricted economy of life. The *Aufhebung*, Derrida writes, "is the dying away, the amortization, of death. That is the concept of economy in general in speculative dialectics. Economy: the law of the family, of the family home, of possession. The *Aufhebung*, [is] the economic law of absolute reappropriation of the absolute loss" (G 133/152). In *Positions*, Derrida writes, "If there were a definition of différance, it would be precisely the limit, interruption, destruction of the Hegelian *Relève* [Derrida's translation for *Aufhebung*] *wherever it operates*," following this by posing general economy as "a kind of general strategy of deconstruction" (PS 40–41/55). Indeed, the *Aufhebung* produces the ideality, Form, and propriation of restricted ecology. "The *eidos*, the general form of philosophy . . . produces itself as *oikos*: home, habitation, apartment, room, residence, temple, tomb [*tombeau*], hive, assets [*avoir*], family, race, and so on. If a common seme is given therein, it is the guarding of the proper, of property, propriety, of one's own [*la garde du propre*]" (G 134/152).

For Hegel, Derrida explains, it is necessary for the master to keep the life he exposes to death in the economy and circulation of the *Aufhebung*. For Bataille, however, the risk of death must be absolute and must exceed the dialectic it nonetheless makes possible. Such a death would constitute "so irreversible an expenditure, so radical a negativity—here we would have to say an expenditure and a negativity *without reserve*—that [it] can no longer be determined as negativity in a process or a system" (WD 327/380). General economy engages, thinks, and considers the processes of dialectical economy in relation to their absolute expenditure and loss of sense. But this relation must itself be thought along the interruptive logic of the relation *without* relation I mentioned earlier. To put it otherwise, general ecology exceeds the economy of life it makes possible by supplementing it with an excess, a material remainder or "*restance*" that cannot be reappropriated into the form of its dialectic.

The inappropriable singularity of this alterity or remainder is in fact the *ethical* condition of general ecology, since it structurally interrupts the economy of any good conscience. "The good conscience (*gute Gewissen*) maintains the circle of exchange. . . . A superior calculus without remain(s) [*sans reste*]: what consciousness wants to be" (G 59–60/71t). The remainder for Derrida can never correspond to the ontological question "what is it?" to which one could calculate an appropriate ethical response and thus rest assured in the good conscience of having met one's responsibility. The necessary and impossible task of environmental ethics thus becomes to count and calculate with the aneconomic and the incalculable. General ecology, finally, can be seen as relating whatever one might still choose to call the subject to the without-ground of the origin of its responsibilization, perhaps even relate this responsibility to a certain concept of peace still to be invented, as Derrida puts it, or rather, "another relation to this concept [of peace], as perhaps to any concept, to the non-dialectical enclave of its own transcendence, its 'beyond-within'" (AE 85/152t). . . . "Enclave of transcendence . . . this inclusion of excess, or this transcendence in immanence" (AE 99/173).

Transcendence within immanence, I would argue, the beyond-within, designates the structure of the ethics of ecological relationality—"this interruption of the self by the self, if such a thing is possible" (AE 51/96–97). In the interview "Alterities," however—and again in conversation with Levinas's philosophy—Derrida claims that he is not proposing a praise of this loss or expense, but "a thought of affirmation that does not stop at this loss, or does not dialecticize something like the loss or expense" (ALT 68). Thinking the relation without relation calls for another nondialectical thinking of mediation, one we can think as the very condition of any ecology, living together, symbiosis, or social bond. I permit myself a long citation from this interview here.

> There is a mediation which does not bar the passage to the other, or the entirely other; quite the contrary. The relation to the entirely other (as such) is a relation. It is a relation, of course, without relation to any other relation, it is the relation to someone who, because of the other's alterity and transcendence, makes the relation impossible. This is the paradox. In order to establish a relation with the other, interruption must be possible. The relation has to be a relation of interruption. And interruption in this case does not interrupt the relation to the other, it opens the relation to the other. Everything depends on how one determines the mediation in question. (ALT 68)

Ecological relationality rests on this stricture of a certain interruption or *cut* in relationality, one within which the alterity of the other is respected and "remains absolutely transcendent" (DN 14). The deconstructive thought of ecology might then be thought as a "community that does right by interruption . . . the placing in common of that which is no longer of the order of subjectivities, or of intersubjectivity as a relation—however paradoxical—between presences" (TS 25). "A pluralism of radical separation, a pluralism in which the plurality is not that of a total community, that of the cohesion or coherence of the whole" (AE 96/169), a "sharing out without fusion" (OT 195).

This strange ecological bond across anachrony and alterity allows me to conclude in articulating ecological relationality in terms of the posthuman promise of the earth. Anterior to any "I promise," "you promise," "he, she, or it promises," Derrida writes, "*nous nous promettons*," we promise one another, we promise ourselves. To paraphrase, ecology on this view must be understood as "a 'we' without assured gathering, without intersubjectivity, without community or reciprocity, a strange dissymmetrical 'we' anterior to every social bond"; it requires "dispersion or distraction, the absolute interruption of absolutes, the *ab-solute* or *ab-solved* [*l'ab-solu*] in a certain being-in-the-world that will have preceded everything" (AV 49/40). These interruptions are necessary so that the promise exceeds any horizon of anticipation for its being kept, any calculation of its ends. Without this excess, the promise would come down to a simple programmable prediction of the future. The promise of the earth, as I suggested in my introduction, must be ambiguously read as beyond the *ends* of the human, as it is irreducible to any anthropological subject, egological consciousness, or *Dasein*, but also beyond one's teleological goals and projects. And the promise also gives itself to be thought, now more than ever, beyond the *ends of the world*. Derrida writes, the promise "is not in the world, for the world 'is' (promised) within the promise, according to the promise" (AV 47/39). As I have shown, Derrida's concern for the unnamable in the living present is at work since *The Problem of Genesis*. Even if we persist in using the same words and names in thinking the earth we promise ourselves and one another,

> [These names] will be other and will bear the trace of anachrony within themselves: "we" will no longer be (simply) gods nor humans; the world will no longer (simply) be neither the world (Christian concept) nor the (Greek) *cosmos*; life itself will no longer be what we thought it was, not always but more often than not until now: the simple contrary of death, as philosophers, biologists and zoologists believe they can define it, nor even the being of beings in general, the "there is something rather than nothing." (AV 48/39)

To take up again the stakes of my first two sections, the absolute anteriority of the promise "would be just as foreign to the egological horizon that structures a phenomenology of time (Husserl) as it would to the order or existential horizon of temporal ecstases (Heidegger)" (AV 32/29). The promise would be "older" than the living present, autoaffective temporalization, and death drive it nonetheless makes possible, since temporality itself must be thought on the basis of the promise rather than the contrary. The logic entirely follows what I've been developing here as general ecology; the promise must be essentially excessive and must promise more than it is able to keep. The anachrony of the promise is the aneconomic itself, "time of the unkeepable promise, the time of an expenditure without possible restitution, what in truth never lets itself be returned" (AV 29/27), . . . "originary expenditure of this other time, this time before time" (AV 33/30).

Environmental ethics, then, with both words under erasure, must be thought as an "inheritance of the unkeepable promise" (AV 29/27). But to inherit this promise from a time even before evolutionary history is perhaps itself no longer reducible to a simple question, perhaps even less so one of what to do and of what to do with the earth. To inherit justly would perhaps not only come down to restituting this originary aneconomy, but to maintain the disjunction in the passivity of a certain letting. Derrida writes about this beautifully in *For What Tomorrow*: "One must do everything to appropriate a past even though we know that it remains fundamentally inappropriable" (FW 3/15t). To reaffirm the inheritance of a promise before time not only accepts it, "but relaunch[es] it otherwise and keep[s] it alive" (FW 3/15). Life itself must be thought on the basis of this inheritance, and the double injunction to both passively receive and to affirm, to say yes to, and to choose to make what one inherits live-on otherwise are the entire stakes of the posthuman promise of the earth.

> Two gestures at once: both to leave life in life and to make it live again, to save life and to "let live" in the most poetic sense of this phrase, which has unfortunately been turned into a cliché. To know how to "leave" and to "let" [*laisser*], and to know the meaning of "leaving" and "letting"—that is one of the most beautiful, most hazardous, most necessary things I know of. . . . The experience of a "deconstruction" is never without this, without this love, if you prefer that word. (FW 17/4–5)

March, 2015

Notes

1. Jacques Derrida, *Advances*, trans. Philippe Lynes (Minneapolis: University of Minnesota Press, 2018). For another reading of this text, see my introduction "Auparadvances" therein.

2. Derrida: "Préférez toujours la vie et affirmez sans cesse la survie . . . je vous aime et je vous souris d'où que je sois"; "Jacques Derrida," *Rue Descartes* 2, no. 48 (2005): 6., http://dx.doi.org/10.3917/rdes.048.0006.

3. See Cary Wolfe, *What Is Posthumanism?* (Minneapolis: University of Minnesota Press, 2010), 62.

4. Derrida here recalls a syntagm used years earlier in "Il faut bien manger" (translated as *Eating Well*, in PI).

5. One can see the ethical problematization of the general and the singular through Derrida's reading of Levinas as early as "Violence and Metaphysics," a problem that recurs in *The Politics of Friendship* and *the Gift of Death*.

6. Cf. GD 78/109–10.

7. Cf. AA 111/153.

8. Cf. VP 9/9.

9. Derrida's interest in the presubjective, pre-egological transcendental field in Husserl's philosophy is one shared by Deleuze, both in collaboration with Guattari in *Qu'est-ce que la philosophie?* (Paris: Minuit, 2005); trans. Hugh Tomlinson and Graham Burchill as *What Is Philosophy?* (London: Verso, 1994), and in Deleuze's final essay, "L'immanence: Une Vie," in *Deux régimes de fous: Textes et entretiens 1975–1995* (Paris: Minuit, 2003); trans. Ames Hodges and Mike Taormina as "Immanence: A Life," in *Two Regimes of Madness: Texts and Interviews 1975–1995* (New York: Semiotext(e), 2006).

10. "All living things are the power of auto-affection"; OG 165/236t.

11. As he puts it elsewhere, this self-interruption in phenomenology is another name for *différance*; see Derrida, "Hospitality, Justice, Responsibility," in *Questioning Ethics*, ed. Richard Kearney and Mark Dooley (London and New York: Routledge, 1999), 81.

12. Cf. PM 143/376. Derrida says the same of Heidegger; see especially the sixth session of Derrida's 1964–65 seminar on Heidegger, abbreviated as HQ here.

13. Cf. Derrida, "Avec Levinas: 'Entre lui et moi dans l'affection et la confiance partagée,'" *Le magazine littéraire* 419 (2003): 30, as well as AI 31/43.

14. Cf. David Wood, "Specters of Derrida: On the Way to Econstruction," in *Ecospirit: Religions and Philosophies for the Earth*, ed. Laurel Kearns and Catherine Keller (New York: Fordham University Press, 2007), 285.

15. Cf. OG 23/38.

16. Cf. Wolfe, *Before the Law: Humans and Other Animals in a Biopolitical Frame* (Chicago: University of Chicago Press, 2013), 58.

17. Cf. AA 28/49.

18. Cf. OS 130n5/115n1.

19. Martin Heidegger, "The Origin of the Work of Art," in *Off the Beaten Track* (Cambridge: Cambridge University Press, 2002), 1–56.

20. Cf. PC 359/382. Derrida here indicates his intention, in a footnote, to take up this comparison in *Given Time* (GT). But this analysis does not appear in the published version of the book, which itself constitutes only the first five sessions of a fifteen-session-long seminar given in 1977–78. The actual analysis can be found in the thirteenth session.

5

Un/Limited Ecologies

Vicki Kirby

> To say that one always interweaves roots endlessly, bending them to
> send down roots among the roots, to pass through the same points
> again, to redouble old adherences, to circulate among their differ-
> ences, to coil around themselves or to be enveloped one in the other,
> to say that a text is never anything but a *system of roots*, is undoubt-
> edly to contradict at once the concept of system and the pattern of
> the root.
> —*Jacques Derrida*, Of Grammatology, *101–2/150*

The meaning of "ecology" in these contemporary times is no longer
confined to the intricate interactions between organisms and their
environments, to meanings that refer to the adaptive survival behaviors of life's will to reinvention. The term now carries a political
weight, an interpellative imperative that summons us to acknowledge the apocalyptic force of climate change and the role that humans
have played in the accelerated pace of environmental degradation
and species loss. Most scientists are persuaded that anthropogenic
causes explain "the sixth mass extinction" of both plants and animals. Indeed, the impact of human activity on the environment
over several hundred years or, arguably, even millennia represents
a break from the previous order of such legible significance that the
official naming of the current geological epoch is mooted to become
the "Anthropocene."[1]

Although the urgency of this call to action feels compelling, and
I don't recuse myself from this summons, there remains a nagging
suspicion that this story of human hubris and culpability withholds
a lot more than it can comfortably acknowledge. I want to explore
the underbelly of this confession, this bearing witness to the error of
past and current acts, because it assigns a central place to the human

condition as the agent of change. The task I have in mind aims to explain how a revelation can prove an effective ruse to conceal something unwelcome. However, if the act of trumping or unveiling error also harbors error, indeed, if political life, from whatever perspective, involves these intimate contaminations and inadvertent repetitions, then perhaps we have found a way to consider the political as inherently and always ecological: critique is a messy business that can surreptitiously recuperate and affirm what it claims to reject. My argument follows a well-worn path, intended to show that every individual, whether living creature, subject, object, concept, or even ideological position, is a processual and entangled entity: it is internally fraught, and necessarily so, because the enduring constant of its apparent integrity is always "coming into being." Importantly, the sense that there is complicity in this scene of genetic production is not so much a moral judgment about integrity undone (although the drive to reprimand is irresistible for most of us) as it is a description of life's self-involvement and perverse modes of reproduction: there is no "outside" of this agential dynamic.

The knotted indebtedness and intraweaving of living ecologies is conceded by most of us, even though we routinely exempt political differences that rest on contested concepts, commitments, and behaviors from these same histories of involvement. It seems to go without saying that species of ideas arrive fully formed and independent of each other: they are not environmental adaptations, forged in the cross-infection of problem-solving experiments and motley social perspectives. The need to segregate conceptual life into independent political positions makes perfect sense if moral authority and the political good are to retain their appeal to purity. Further to this, the separation of the material from the ideational is also necessary if humanity's ascendancy over nature's brute physicality, or what is comparatively primordial and deficient, is to be explained and maintained. Although antihumanism and certainly posthumanism have argued that this logic installs an egological self-certainty (cogito) that severs itself from its own history, its ecological roots, it seems fair to say that a recuperated Cartesianism has been hard to shake. My own view is that the fault line of Cartesianism is poorly understood when it is diagnosed and condemned with a perfunctory flourish. An unpalatable fact of Cartesianism is that its questionable logic remains necessary to our sense of self, our belief that we exist as a unique site of reflexive inquiry, comprehension, and intentional action. The very notion of decision, that it is an act that requires forethought and responsibility, presumes this self-possession. Why and how would we undercut our specific ontological significance, our sense of self-control, even if we wanted to?

Cartesianism isn't an accusation that can be remedied with a corrective because both its affirmation and its critique install the cogito as the site of claim and counterclaim. In other words, self-diagnosis and even self-dismantling require the presence of an enduring, coherent, and agential self to manage and transcend, indeed, to survive, this taking apart. Our perception of experiential stability, of ego, discovers and represses the volatile and even aleatory aspects of *be-ing oneself*. Yet despite what is almost an imperceptible concession that our sense of self is a sort of phenomenological flicker effect, it seems fair to say that something important has been conceded, however fleetingly. If the subject is not a preexistent entity *in* a field of social, political, and historical forces, if the social is not outside the subject, then the interiority of the individual is constitutively alien or, more accurately, uncannily familiar. Antihumanism argues that the subject internalizes these seemingly external and impersonal forces; indeed, the genesis of the subject is determined with/in them. So far so good. However, this insistence that we are each made rather than born, a reminder of Beauvoir's famous rallying cry against biological determinism, "One is not born, but rather becomes, a woman,"[2] should make us pause. Beauvoir's intention was to denaturalize any appeal to essence, to givenness, to a static program of prescriptive determination. However, what we need to consider is whether this appeal to a more open set of possibilities means that this process of being made—a mode of production of sorts—can be circumscribed. Does the inside, the *oikos* of a subject that must accommodate the collective identity of human-species being remain intact, autonomous, whole—one species among others? Should otherness—here, the ecology in general—retain its feminized, outsider status as mere support for human identity?

The centrality and implicit superiority of being human recuperate the cogito, without any qualification, by denying that the sociality of the larger environment, the "big picture" in this case, might also be internalized and necessary to what human-species being is and can be. Posthumanism in its more nuanced and informed manifestations, as with poststructuralism,[3] explores the processes through and within which an entity comes into being. However, even these

discourses that include current concerns about the myopia of anthropocentrism and its destructive legacy can celebrate the mea culpa of
this insight in terms of Cartesian certitude. The species' self-interest
of anthropocentrism, for example, is answered by leaving the identity of the human intact and supplementing this one identity with
others that remain other and, inevitably, alien and exterior to it.
It seems that more is better, and the recognition of difference as
an aggregation of entities will counter centrism, admit to ecological indebtedness, and work toward a true cosmocratic horizon. But
can we put aside the lessons of antihumanism, posthumanism,
and poststructuralism in this rush to identify (and indemnify) our
species as the one—indeed, the only one—who can gift this more
inclusive revision? Doesn't the political expediency of capitalizing
identity in this way and equating it with agential choice require
further qualification?

I want to explore the difficulties that attend the question of
agency by revisiting the work of Jacques Derrida, a philosopher who
teaches us about the paradoxical nature of radical interiority. We
are probably familiar with certain maxims that appear throughout
his work: "no outside text," "no outside logocentrism," "no outside
metaphysics," "no outside phallogocentrism." At first glance such
statements appear as injunctions against access, as if we must reconcile ourselves to entrapment and the loss (if we ever had it) of an
outside. For many readers of Derrida this interpretation, which rests
firmly on identifying what belongs where, a form of linguisticism
that Derrida routinely denied,[4] is taken at face value. Consequently,
"text" is made synonymous with symbolic systems, representations, languages of various sorts, and social rules and regulations; in
short, with ways of thinking and making sense of a world. According to this view, because humans are interpreting animals, this exercise in "worlding" inevitably mediates, or distances us, from the
world in its pure immediacy—or, as the linguist Émile Benveniste
describes it, the world as it might appear "under the impassive regard of Sirius."[5] There are two important corollaries to this interpretation that cast language or culture as an enclosure, even a prison
house. The first of these consequences, and perhaps it goes without
saying, is that Derrida draws an indelible line between the human
condition and its others. Nature, now under erasure, is the stuff of
this outside, and it is crossed out because any attempt to know it,
to represent or even experience it, will be construed through the
organizing ciphers of human invention. The conclusion here is that

the cultural translation of nature must conceal an inevitable mistake because human symbolic systems are self-referential: "there is no outside language." A well-known aphorism comes to mind that explains this act of making same. *"Traduttore traditore,"* usually rendered, "translator traitor," is meant to convey that translation is something of a sleight of hand because it effects a distortion or betrayal of the original. The Italian example aptly illustrates this loss because, when spoken quickly, the sounds of the words are sufficiently similar to appear the same, even though the actual meanings bear no resemblance.

The second corollary, which confirms the first, yet with awkward and confusing irony, is that this absolute prohibition effectively and convincingly incites the robust appearance of an outside, an otherwise, an inaccessible. No one, even one who ascribes to this particular reading of deconstruction, can exempt himself from talking *about* nature, the body, matter, and so on. In other words, the character and operations of the indescribable are described relentlessly and in graphic detail. And, surely, this feels right because we experience the presence of nature in our corporeal being and in what feels like direct contact. However, the perception and immediacy of what appears external to the self, this extralinguistic or extrasymbolic reality, is mere semblance according to this interpretation. Presence is forfeited because mediated, such that nature, in truth, can only be a "second nature," a re-presented origin that masquerades as the negative image of culture's substantive grounding and difference from itself.[6]

I want to proceed quite slowly in order to suggest how this particular understanding of deconstruction might open itself to an exploration of the more intricate dimensions of Derrida's neologism, *différance*, a nonconcept whose subtlety and divergence *from* difference are also *interior to* its identifying discriminations—their very possibility, but also their failure to be properly and finally realized. We are in the knotty problematic of quantum separability and inseparability, where the integrity of any departure point—origin, interpreting subject, this present space/time configuration here and now—becomes an "always already not yet" that mangles the coordinates that difference and causality assume. There is now an acceptance within quantum physics that its counterintuitive discoveries about the material world—nature, no less—cannot be explained away as the vagaries of human interpretation or what we learned previously was the enclosed self-capture of ideational symbolic systems that cannot, by definition, speak directly of nature. Quantum "weirdness" suggests that identity of any sort is entangled, local, *and* dispersed, such that an interpretation is materialized in/as the very ontology of the object under investigation. This means that the material world *is* inherently ideational and in such a way that the division between the two (nature versus culture) must suffer a collapse.[7] The scandal in this last description is easily elided, as its assault on our dearest sense of self is so alarming that we tend to repress its broader significance. We might concede, for example, that the agent or author of an experiment is the human subject and that the apparatus or instrument of our inquiries will have material consequences for how we comprehend the object. In other words, cultural constructionism in its various guises has made us comfortable with the assertion that interpretive frames of reference have constitutive or interfering force. However, if we think about the ways in which quantum mechanics complicates and disperses what we mean by the agential, then quantum measurement is poorly understood as a unidirectional process from cause (human authorship) to instrumental effect: interpretation is not a local act that has yet to be applied to a preexisting object/phenomenon. The latter assumption attributes atomic integrity to its analytical terms of reference even as it argues that atomic integrity—namely, indivisibility—is impossible. Contrary to common sense, this means that a starting point, first cause, or origin is not located *in* time and space and *then* brought to bear on an object whose existence endures in a spatiotemporal elsewhere. The predictive success and practical deployment of quantum entanglement on a grand scale can attest that the latter is not the case.[8] It seems that what we thought was *radically* outside—the ineffable, the non- and outside-human interpretive frames of reference—is somehow at the very center of who and how we are and even what we decide. It is this conundrum wherein radical alterity appears as the agential force of radical interiority—the cogito, no less—that exercises this meditation on grammatology as ecology. If radical alterity *is* us, if context is not an external supplement to an identity that prefigures its addition, then to return to our primary theme, what and where is the environment, the ecology? Of course, the same question can now be posed in regard to human identity: what and where is human identity if the *oikos* of its "being at home with itself" includes what can only appear as utterly foreign, corporeally dissimilar and distant? Is it possible that ecology is not external—the ground and environ to and within which our identity is indebted, but instead, something more intimate, familiar, already received and "owned"? Or to put this in a way that recalls Martin Heidegger's contribution to a well-known structuralist aphorism, "language speaks us," are we more possessed than possessing, more authored than author? Heidegger argues that humans are not the inventors of language because our very being (ourselves) is already articulated, or spoken, by language. "Language is not a work of human beings: language speaks. Humans speak only insofar as they co-respond to language."9

But let's begin even more cautiously with what is now well understood as the political legacy of reading difference as negation. Difference finds its essential identity in an "outside," a "not this," an absence. We find that difference is "made to measure" because it is comparatively determined, much like the workings of a figure/ground gestalt. Either side of the line that divides the familiar face versus candlestick example can potentially stand for the positive figure. However, regardless of the way our perception is oriented, we can only see one figure at a time. The "other" side, the side that attracts the description "different from" the image, must assume the status of supporting backdrop. The identity of the environment, or ground, is entirely qualified in this account, as it "exists" in negative relief as no-*thing*. An entire industry of political analysis has engaged and contested this logic that interprets otherness as a lack that unsurprisingly discovers nature, the feminine, the racial other, the disabled, all somehow caught in the passivity of corporeal/material imminence. We have already mentioned Beauvoir's intervention nearly seventy years ago, intended to disrupt the equation of woman with a set of compliant behaviors prescribed by nature. The political violence of the latter's A/-A two-step is that it installs a strict evolutionary determinism whereby a constitutive deficiency is overcome with the advancement, or "figuring forth," of male (culture) over female (nature). This logic is pervasive and structurally compelling, as we see in the triumphalism that again aligns human-species being with culture (complexity, intelligence, self-governance) against nature and its nonhuman others (lack of guile, passivity, need).

Within the conventions of this logocentric reading of difference, time unfolds in a narrative of discrete moments, or episodes, and this forward movement equates with progress—good, better, best. "Evolve" means to make more complex—literally, to roll out of or unfold. Myriad examples of animal and vegetable life, again, discrete and fully formed, are ranked accordingly in a ladder of achievement—"the survival of the fittest." However, Herbert Spencer's now famous phrase was not meant to evoke a race to the finish line where individual capacities of physical fitness, strength, and superiority would explain the inevitable discrimination of winners from losers. The intention behind Spencer's phrase was to support the *involved* machinery of evolution that was implied in Charles Darwin's term "natural selection." Importantly for this argument, "fitness" remains something of a tautology—the motor of workability that is systematicity *as such*. Although this sense of "tautology" captures the intricacy and intimacy of nature's self-involvement or how it reproduces itself, and I use it in a positive way here, the description may sound odd, given its conventional meaning as a fault in logic or an error in style. The etymology of the word preserves this sense of something extraneous or redundant; late Latin *tautologia* is defined as "representation of the same thing in other words," derived from the Greek, *tautologia*, from *tautologos*, "repeating what has been said," from *tauto*, "the same."[10]

Of course, difference is what is *not same*, and yet the habit of a tautology is to execute an apparent differentiation *within* sameness. We have returned to the conundrum of separability with/in inseparability: the *same* problematic that discovers *différance* with/in difference. This scene of iteration, which I am going to call nature, does not evolve by way of an encounter or communication between previously independent and self-enclosed individuals. If we think of evolution in terms of quantum entanglement or Derridean *différance*, then "entities" do not preexist their encounter, their communication, nor does reproduction take place in only one circumscribed moment. For our purposes, differentiation is better conjured as a superposition of possibilities, something we could liken to the riddle of a 3D magic-eye image. At first glance we observe a riot of electric colors, computer-generated squiggly lines, and perhaps a sense of geometric repetition. And then, from within that *same* image, a *different* image leaps out at us in three dimensions. We have all seen them: three dolphins, a love-token, a word or a vase of flowers, and we have all encouraged our friends to find the difference that we have already discovered within the very same image under scrutiny. Although the difference between our first perception and that Ah-ha! discovery of something more seems immense, it is fair to say that it is also no difference at all: the difficulty is that each individual image arises with/in another individual, each genetic inheritance, or what is specific to one image, strangely same, shared, and held in common.

It might be helpful to keep this last example in mind as we focus on the question of ecology, this bundling of forces and appearances with neither beginning nor end—an enfolded genesis wherein every ending is an inventive reiteration, a beginning. In keeping with this, let's imagine that the second image in our magic-eye example is just one of many; in other words, if we continue to look, more and more images will appear. What is fascinating in this example is that the differences that are thrown up are not aggregations of identity where the second is separate from the first and the third is again divided from the second ad infinitum, as if each new perception emerged from a completely different frame of reference. Here, the second is a reconfiguration of the first; indeed, in a very real sense the second *is* the first's *intra-activity* with itself,[11] its *différantial* possibility, because the frame of reference holds an almost infinite potential within its seemingly finite parameters. As Derrida constantly reminds us, if difference is a supplement, then that supplement is internal to our starting point, a sort of genetic transformation or originary epigenesis wherein *différance* inhabits difference.[12] If we think of this *same* frame of reference as the ecology writ large—life itself—then life's constant is an ability to reproduce itself in forms that are morphologically plastic, forms that are akin to many dimensional jigsaw pieces that *necessarily* "fit" because they express life's "own" changing patterns of reflexivity. Importantly, however, this frame of reference needn't operate as an enduring surround, a sort of fixed orthopedic structure within whose prohibitive boundary movement and play are the order of the day. This "system"—and we could call it "language in the general sense," "a general writing," as well as "ecology" or "life"—is certainly subject to structures and rules; however, these structures are also outcomes of the system's own internal movement: they are neither pre-scriptive nor preexistent. In other words, life is always in touch with itself, even with the directions of its missteps and seemingly random experimentations. And if evolution is a form of "self"-adaptation, then life's ability to differentiate itself, to offer up what seem to be separate and independent manifestations, could never be reduced to the linear "rolling out or unfolding of complexity," as the word "evolve" literally suggests. It is sobering to think that even Charles Darwin eschewed the word "evolution" for these same reasons, preferring to describe life's movement as one of adaptation, "the modification of a thing to suit new conditions."

However, to build on this last point, with Derrida, I need to adapt the conventional sense of adaptation to a system whose involved interiority "contradict(s) at once the concept of system." Following Derrida, "the thing" that must change or adjust can have no initial independence, no atomic integrity, no definitive identity that is separate from the causal initiative of an environment that will purportedly change it. Again, even Darwin's understanding of what constitutes a species, the entity upon whose adaptations he grounds his theory, is missing, its origin or circumscription strangely unexplained.[13] Thus, rather than read this scene of adaptation as one that must explain the relationship between creature and environment, my strategy is to cast the environment as a force that speciates, a force that individuates *itself*, a scene of production wherein all differences are reinventions of that same scene. With no outside this general frame of adaptability or "general writing" in Derrida's terms, the local or species specificity might be likened to a "restricted writing" that the general bodies forth. Thus, what appears restricted is actually an articulation or speaking by the general. To return to quantum explanations here, nonlocality (the general) is not the opposite or the refusal of locality, but the concession that a more comprehensive force field can be operational *within and through* any particular instantiation/exemplar.

The sense of fitness, then, is a sort of ontological tautology wherein existence of whatever sort, its very specificity, is exemplary of universal possibilities, just as we saw with the magic-eye example. I am calling this frame of reference a system, indeed, an ecology, because its evolution is not a linear narrative from failure to improvement, a hierarchy of competitive exclusions whose final resolution is the ascendancy of man. The evolutionary biologist Stephen J. Gould discounted images that evoke a ladder or hierarchy, preferring the branching intricacies of a bush or tree. However, the movement from root to branch may still be interpreted as a sequence of things coming into being and then passing away, swallowed up by time as they fall into maladaptive redundancy. Even if we think of a tree of life or what continues to endure in a contemporary, synchronic snapshot, every branch an alternative expression of viability, then even here our habit is to equate the roots of the tree with what came first. What is early seems logically more primitive or simple, whereas the uppermost branches are interpreted as more recent and inherently superior. And yet if we take Gould's statement that there is "no inherent vector of progress" in Darwin, "no cosmic betterment,"[14] then a branch is a root is a branch. What would the notion of initial or final conditions even secure?

In this essay's opening quotation from *Of Grammatology*, Derrida argues that firstness is ubiquitous: the ground or transformative nourishment that will produce what has yet to come has already arrived. When he insists that origin and end remain *rooted* with and in each other in a confusion of deferred limits, an entanglement of identity, the implications of this, if taken seriously, are profound and disorienting. For if we can't equate difference with no-*thing*, with the space or empty void that separates entities, then even death, the final limit and absolute outside, has a robust and enduring inner and afterlife that is difficult to corral in terms of absence.[15] To return to the species question—the unanswered problematic of how a species could be identifiably separate *and* inseparable from its surrounding environment—the biologist Brent D. Mishler helps us to think this conundrum. We could be forgiven for suggesting that Mishler, in "Species Are Not Uniquely Real Biological Entities,"[16] unknowingly discovers Derrida in Darwin. Clearly, both thinkers are enchanted by "the big picture," the "scene" or "system" whose topsy-turvy operations show that identity is made perversely, emerging across myriad strata. Interestingly, Mishler's citation from Darwin on the tree of life has remarkable resonance with Derrida's reference to the entanglement of the root inasmuch as both representations, again in Derrida's words, "contradict at once the concept of system." It appears that where Derrida finds only roots, Darwin finds only branches. Mishler concludes his argument by noting:

> Evolutionary processes are not just operating to produce what we happen to call species; they operate at many nested levels in producing the tree of life "which fills with its dead and broken branches the crust of the earth, and covers the surface with its ever branching and beautiful ramifications."[17]

Darwin finds life in death here, wherein broken and dead branches are internal to processes of viability, and the vertical, progressivist stretch upward, to the heavens, instead becomes a horizontal network of interaction and perverse inversion. Darwin further underlines this sense that there is no outside branching (networks) when he describes the "beautiful ramifications" of this implicated scene. Importantly, the word "ramification" derives from the Latin, *ramus*, for branch, a connection that Darwin would have made quite deliberately, given his education in the classics. In sum, what Darwin emphasizes is that what follows from branching is yet more branching, just as Derrida finds that the root of things is strangely ubiquitous. Perhaps unsurprisingly the etymological dictionary notes that *ramus*, branch, is "akin to Latin, *radix*, root."

But where has this fascination with life's networked superpositions landed us? Have we forfeited our ability to take political action and assume ethical responsibility for anthropogenic climate change if we challenge the humanist subject's self-proclaimed position at the top of the evolutionary tree? And can we reasonably suggest that the agential subject is more than human, taking the comprehensive form of the ecology itself? What do we risk if we embrace this last suggestion—namely, that nature decides, intends, and indeed, chooses how it wants to perceive and represent/reproduce itself? Such awkward questions, regardless of how we answer them, remain wedded to the pragmatic workability of notions such as decision and choice, capacities that routinely describe the intending (human) subject. However, as both biological research and post-humanist/poststructural theories of the subject argue persuasively that the presumed site of conscious reflexivity is more of a retrospective construct,[18] a phantasmatic necessity or, as noted earlier, "a phenomenological flicker effect" with no enduring stability, then surely the question we are really investigating here is that of cause and agency as such.

Jacques Derrida is well-known for asking similar questions, as we see in the collection of essays *Who Comes after the Subject*,[19] to which he contributes. The book's title seems to anticipate the rather odd possibility that there will be a subject after the subject—the human subject surely—and this suggests that perhaps there was even a subject before the human subject. Where and how do we locate ourselves in this space/time mirror-maze wherein what comes next may already have arrived? In his interview with Jean-Luc Nancy, "'Eating Well,' or the Calculation of the Subject," Derrida admits that he tends to avoid the notion of subjectivity altogether:

> because the discourse on the subject, even if it locates difference, inadequation, the dehiscence within autoaffection, etc., continues to link subjectivity with man. Even if it acknowledges that the "animal" is capable of auto-affection (etc.),

this discourse nevertheless does not grant it subjectivity.
(PI 268/283)

But Derrida goes further. "The difference between 'animal' and 'vegetal' also remains problematic. . . . The question also comes back to
the difference between the living and the nonliving" (PI 269/284).

As I interpret this maneuver, it is not a gesture of generosity that
would repair a previous exclusion with its accommodation. The latter strategy is something that Bruno Latour, among others, deploys
in his notion of distributed agency, something that both human
and nonhuman, even things, can enact.[20] But here is the rub. Latour
resolves the dilemma of human exceptionalism by aggregating different entities into participating networks of action, a move that continues to equate agency with authorial efficacy, albeit shared, in this
case. But whereas Latour's democratic concession allows each individual identity (regardless of its supposed difference) an opportunity
to participate in a parliament of interrelations, Derrida's question
about the agential is more preliminary. Derrida wonders "who" determines that the human is, indeed, ontologically individual and different from its purportedly nonhuman others. The question seems
bizarre yet somehow brave and necessary because it troubles the
identitarian politics that Latour's remedial assumes and preserves.

It is the utter mystery of how the identities of the inanimate,
the vegetable, and the animal inhabit each other, how they can express the agential force of decision as "the who" that initiates, that
should give us pause. The very style of this consideration is surely
maddening. However, as the focus of quantum physics is similarly
exercised, such that a seeming lack of conventional logic can have
considerable explanatory reach, we really do need to take stock of
what has become routine in our problem solving. Quite obviously,
Derrida is fascinated by this question of systematicity, *because what
renders an aggregation of entities "systemic" is not their simple
aggregation.* If we concede that relationality or systematicity must
already inhere *within* the entities that are seemingly separate and
independent, then what agential force identifies the human as "the
one" whose choice it is to distribute agency, "the one" among others who can choose to refuse this exceptional status or, just as easily, lay claim to it, and always in the same breath? The messiness of
this riddle returns us to the quandary about our place in the broader
ecology and whether our self-appointment as nature's presumptive
guardian and change agent holds credible weight. In short, what is a

"self-appointment" if the intending subject has nonlocal, ecological distribution?

In the opening comments to this essay I remarked that a confession—in this case, our acknowledged responsibility for climate change and ecological devastation—might operate as a ruse or defense against something that we can't quite put our finger on. I also suggested that Cartesian certitude isn't easily abandoned. Not unrelated, contemporary debates about climate change, animal rights, and ecological degradation are fond of condemning anthropocentrism, as if the error, if it is an error, is easily remedied. However, Derrida's exploration of the political as radical interiority, whether with/in logocentrism, ethnocentrism, or phallogocentrism, is equally relevant for how we engage the catalog of claims and counterclaims that attend anthropocentrism. Similarly, by working with/in the knotty mangle of anthropocentrism's unlikely accommodations, we get some sense of why the attribution of human exceptionalism is anything but straightforward.

The backstory to this assertion is important. It is well-known that Derrida is interested in the logic of narrative persuasion, its hidden economy, and political trade-offs. However, and importantly, he does not regard rhetorical convention and its creative deployment as epiphenomenal to an underlying truth: language is not a frame of reference that prevents or allows (both arguments presume that language is a separate instrument) access to an enduring and different reality that precedes or subtends it. Less understood is the inverse of this claim—namely, that deconstruction is not a more elaborated form of ideology critique where the axiom "no outside text" discovers the enclosure of cultural solipsism. For most readers, these assertions will seem contradictory, leaving us disoriented and struggling to find our bearings. Is language a frame that focuses our attentions and thereby shapes what we see, inevitably distorting or interfering with reality? Is language a lens of reflexive self-capture, a sticky instrument that prohibits access to a substantive world? In other words, what could a thoroughly textual or linguistic reality/truth imply that isn't just another way of evoking cultural constructionism? This is the challenge we have set ourselves, especially in the current context that calls for immediate political action over climate change and an acknowledgment of our central place in ecological destruction. Can we condemn anthropocentrism and human exceptionalism on the one hand—the human is the ultimate culprit—while embracing and reaffirming these centrisms with the other—the human is the only one who can save the day?

What is uncannily familiar about this story of violation against the natural order, this tale of human culpability for which atonement must now be exacted, is its predictability. There is an obvious theological dimension to this representation of an Edenic nature, brought undone by a species whose original sin was to question its rules and rhythms and to alienate itself as a consequence. This style of narrative is leveraged by way of a fantastic leap from purity to corruption: harmony is opposed to discord and conflict, and a naïve and trusting simplicity falls prey to the ruthless opportunism of one species' calculating intrigues. This origin story about purity versus danger has so many iterations in so many different contexts that its truth might seem incontestable, and surely, this same logic remains the explanatory frame to which questions about the ecology are tethered. For these reasons, the provocations in Jacques Derrida's earliest work can prove salutary as they speak directly to this longstanding quandary about our relationship to nature. And yet Derrida will refresh this moral tale in a way that is surprising.

In *Of Grammatology*, Derrida takes us into the jungles of Brazil to revisit a scene of first contact, reported by Claude Lévi-Strauss in *Tristes Tropiques*.[21] Indigenous peoples have been given an important role in this story, which unsurprisingly runs to the very same script that is now under investigation. The Nambikwara appear comparatively puerile, their attentions focused on the trinkets that the anthropologist has brought for them in the hope of buying their cooperation. However, Lévi-Strauss's own interest is very much elsewhere. The backdrop to the drama is the laying of a telegraph line by a band of laborers, and the fascination is the inevitable sense of ambush that these unsuspecting, illiterate people are about to experience as their world is changed forever. There is something prurient about watching this encounter through Lévi-Strauss's eyes, especially when the tribal chief, "the only one among them to have understood what writing was for,"[22] mimes the act of reading from a scrap of paper to impress his followers. For Lévi-Strauss the act is empty yet portentous, because the posturing affectation and self-importance are testaments to their opposite, the pathos of a primitive society's inevitable demise at the hands of an encroaching and superior technology. Although Lévi-Strauss appears forthright in acknowledging his part in this, Derrida reads the anthropologist's mea culpa as self-deception:

One already suspects—and all Lévi-Strauss's writings would confirm it—that the critique of ethnocentrism, a theme so dear to the author of *Tristes Tropiques*, has most often the sole function of constituting the other as a model of original and natural goodness, of accusing and humiliating oneself, of exhibiting its being-unacceptable in an anti-ethnocentric mirror. (OG 114/167–68)

Derrida reads Lévi-Strauss's self-identification as guilty author of the other's demise, as "an *ethnocentrism* thinking itself as anti-ethnocentrism, an ethnocentrism in the consciousness of a liberating progressivism" (OG 120/175–76).

I want to use this reading as a provocation for our current question about anthropocentrism, human exceptionalism, and ecology as grammatology. I will stick to only certain aspects of this story that directly relate to the issue at hand: first, how ethnocentrism, here anthropocentrism, is reaffirmed in the self-deprecating act of admitting guilt and insisting that humans are the change agents of environmental degradation or salvation; and second, how a particular contestation of anthropocentrism/human exceptionalism might effectively reroute these conventional terms of exchange and leave us with something much less predictable to think with.

For Derrida, the identity of "writing," even in the restricted sense of the term, is not clear-cut: "Actually, the peoples said to be 'without writing' lack only a certain type of writing" (OG 83/124). Derrida is alert to the ways in which we "refuse the name of writing to this or that technique of consignment" (OG 83/124–25), and we see this when Lévi-Strauss describes the Nambikwara's efforts as mere "imitation," "without understanding its meaning or its end. They called the act of writing iekariukedjutu, namely: 'drawing lines.'"[23] What is important in Derrida's response is its relevance for our discussion of anthropocentrism, given that the special capacities that explain human exceptionalism are leveraged against an other—nature/the ecology—that is similarly illiterate, uncomprehending: only capable of *imitating* cognition.

Derrida asks why "drawing lines" is an inadequate description of writing:

It is as if one said that such a language has no word designating writing—and that therefore those who practice it do not know how to write—just because they use a word meaning "to scratch," "to engrave," "to scribble," "to scrape," "to incise," "to trace," "to imprint," etc. As if "to write" in its metaphoric kernel, meant something else. Is not ethnocentrism always betrayed by the haste with which it is satisfied by certain translations or certain domestic equivalents? . . . By way of simple analogy with respect to the mechanisms of ethnocentric assimilation/exclusion, let us recall with Renan that, "in the most ancient languages, the words used to designate foreign peoples are drawn from two sources: either words that signify "to stammer," "to mumble," or words that signify "mute." (OG 123/180)

Given this legacy's similarity with our representation of nature as mute or babbling without purpose, waiting to be represented, perhaps we might ask why the myriad patternments of "fitness," the shape-shifting morphologies of exquisite ecological adaptation, are not forms of writing; why are they not rememorations; indeed, why should they not be counted as deliberations? Derrida notes that when we perceive complexity in nature "the notion of *program* is invoked" (OG 84/125) to secure its difference from what we are and what we do. Such protective gestures explain nature's apparent societal complexity, swarming behaviors, and cooperative strategies in terms of "unthinking" algorithms, whereas human exceptionalism is considered exceptional because irreducible to these primitive roots that precede culture. But is a program or algorithm without internal differentiation and inventiveness, given Derrida's notion of "originary *différance*," where writing is imprinting, tracing, grafting, moving?

We could retrieve human exceptionalism as it is routinely understood by conceding that we are *a part* of nature; a unique part. However, grammatology, or *différance*, mangles this understanding of difference as metonymic, the aggregation of identifiable differences that somehow sit together. When the voice of radical exteriority—here, the ecology or environment—speaks *as* human exceptionalism, then its adaptive possibilities remain open, myriad, negotiable, murderous, positive, negative—fraught. Further, the sense of exceptionalism as a quality or capacity among others is instead dispersed and morphologically mutable. We could even say that originary writing manifests as a sort of "originary humanicity,"[24] a specificity of decisive adaptation whose ubiquity *is* ecological change.

However, I suspect that my imagined reader will complain that the life of such an ecology/grammatology could prove indiscriminate in its judgments, even perceiving death and destruction as fascinating opportunities to explore what might be, even what should be. Within this particular scene of ecological vitality (which is in no way alien to being human), the opposition between life and death collapses in the very act of ecological genesis. And with no *absolute* limits—for all limits are grammatologically entangled—every decision is a *systemic* adaptation, every "author" authored by the ecology at large. Importantly, however, this needn't mean that nature or the ecology is a *pre-scriptive* adaptation, as the term "program" might imply. Natural selection is a messy, yet exquisitely intricate and poetic business, because "fitness" is a general manufacture, a general writing. If all of life's protagonists, even those who appear murderously opposed, are ecologically bound with/in each other, and if the outcomes that we seek, however different, are rooted in life's own will to self-understanding and reinvention, then how should we proceed?

Derrida's contribution to questions about ethics is to favor creative invention over partisan political compliance. However, Geoffrey Bennington captures the glum frustration that we might feel when a call to act, especially one that feels urgent, remains open-ended, even its terms unresolved:

> Derrida's regular appeals to the need for invention in the fields of ethics or politics necessarily disappoint: we would obviously like to be told what to invent—at which point we would be released from the responsibility of invention.[25]

I have argued that a decision is inherently ecological because it is "made" *in relation*, not because we are in a relation *with* the ecology—the object of the decision, the thing about which a decision is made—but because the ecology is the subject, "the who" that decides. Bennington is again helpful here:

> if we are to talk intelligibly of decisions and responsibilities, then we must recognize that they take place *through the other*, and that their taking place "in me" tells us something about *the other (already) in me*, such that, following another "axiom" of deconstructive thought, "I" am only in so far as I already harbor (welcome) the other in me.[26]

Although deconstruction offers no program for change, it does shift the terms of political contestation. If anthropocentrism and human exceptionalism are not *one* perspective, if the alien is not outside, then who are we, and how might we learn to be at home with *différance*?

Notes

1. Jan Zalasiewicz et al., "The New World of the Anthropocene," *Environmental Science & Technology* 44, no. 7 (2010): 2228–31.

2. Simone de Beauvoir, *The Second Sex*, trans. H. M. Parshley (New York: Vintage, 1973), 301.

3. By "more nuanced," I mean those thinkers who don't see the "post" in poststructuralism and posthumanism as a break of some sort. The more difficult task is to explore the intricacies of what is involved in "structure" or "being human," rather than to presume that such questions have been resolved.

4. Derrida explains that "the concept of text or of context which guides me embraces and does not exclude the world, reality, history . . . the text is not the book, it is not confined in a volume itself confined to the library. It does not suspend reference. . . . *Différance* is a reference and vice versa"; LI 137/253.

5. Émile Benveniste, *Problems in General Linguistics*, trans. M. E. Meek (Coral Gables, Fla.: University of Miami Press, 1971), 44.

6. Judith Butler's work, especially *Bodies That Matter* (New York: Routledge, 1993), is exemplary of this position.

7. Karen Barad's work, especially *Meeting the Universe Halfway: Quantum Physics and the Entanglement of Matter and Meaning* (Durham, N.C.: Duke University Press, 2007), provides an excellent introduction and detailed elaboration of these themes; see also Joseph Rouse, "Barad's Feminist Naturalism," *Hypatia* 19, no. 1 (2004): 142–61.

8. The predictive capacity of quantum mechanics has enabled the development of atomic clocks that aid GPS navigation and telecommunications, uncrackable encryption codes, supercomputing, radical innovation in microscopes, and laser technologies, to name just a few.

9. Martin Heidegger, *The Piety of Thinking*, trans. James. G. Hart and John C. Maraldo (Bloomington: Indiana University Press, 1976), 25.

10. This and other definitions have been taken from the *Online Etymology Dictionary*, accessed May 19, 2015, http://www.etymonline.com.

11. "Intra-action" is a word coined by Karen Barad to describe this sense that any identity is "already implicated" with/in another; see Barad, *Meeting the Universe Halfway*, 175–78.

12. See Derrida's essay "*Différance*," in MP.

13. For a survey of why the notion of "speciation" has remained fraught, see James Mallett, "Darwin and Species," in *The Cambridge Encyclopedia of Darwin and Evolutionary Thought*, ed. Michael Ruse (Cambridge: Cambridge University Press, 2013), 109–15.

14. S. J. Gould Lecture, "Why Didn't Darwin Use the Word 'Evolution'?," accessed May 19, 2017, https://www.youtube.com/watch?v=v0BhXVLKIz8.

15. Derrida's unpublished seminars, "La vie la mort" (1975), which engage questions of biological life, underline this point; see Jacques Derrida Papers, MS-CO1, Special Collections and Archives, University of California—Irvine Libraries, Irvine.

16. Brent D. Mishler, "Species Are Not Uniquely Real Biological Entities," in *Contemporary Debates in Philosophy of Biology*, ed. Francisco J. Ayala and Robert Arp (New York: Wiley-Blackwell, 2009).

17. Ibid., 120; the Darwin quote is from *On the Origin of Species* (London: John Murray, 1859).

18. Perhaps the best-known example is still Benjamin Libet's "Unconscious Cerebral Initiative and the Role of Conscious Will in Voluntary Action," *Behavioral and Brain Sciences* 8 (1985): 529–66.

19. *Who Comes after the Subject?*, ed. Eduardo Cadava, Peter Connor, and Jean-Luc Nancy (New York: Routledge, 1991); page references given to "'Eating Well,' or the Calculation of the Subject," in Derrida's *Points . . . Interviews, 1974-1994* (PI 255/269).

20. See, for example, Bruno Latour, *Reassembling the Social: An Introduction to Actor-Network-Theory* (Oxford: Oxford University Press, 2005).

21. Claude Lévi-Strauss, *Tristes Tropiques*, trans. John Russell (New York: Atheneum, 1964).

22. Ibid., 288.

23. Lévi-Strauss cited in OG 123/180. As Lévi-Strauss's discussion of the "Writing Lesson" moves between *Tristes Tropiques* and several notebooks, I have used citations in OG for convenience.

24. See Vicki Kirby, "Anthropology Diffracted: Originary Humanicity," in *Quantum Anthropologies: Life at Large* (Durham, N.C.: Duke University Press, 2011).

25. Geoffrey Bennington, "Deconstruction and Ethics," in *Deconstructions: A User's Guide*, ed. Nicholas Royle (Basingstoke: Palgrave, 2000), 77.

26. Ibid., 73–74.

6

Ecology as Event

Michael Marder

As, I am sure, readers will recall, Jacques Derrida paid meticulous attention to titles and their relation to the texts or events they promised. Thanks to the efforts of this collection's editors (Matthias Fritsch, Philippe Lynes, and David Wood), deconstructive thought is proceeding today under the heading *Eco-Deconstruction: Derrida and Environmental Philosophy*. The title is economical—perhaps, excessively so—to the extent that it condenses "ecology" in the first three letters: "eco-." This, no doubt, is a matter of convention. Everywhere "eco-friendly" products, practices, or discourses automatically refer us to ecology. If we pause for a moment, however, we will realize that the abbreviation is not as innocent as it seems. It suppresses "logos," which is the second half of "ecology," and so renders indeterminate the distinction between this term and "economy." Before reviewing, in broad outlines, the relation (or the nonrelation) between the two variations on the "eco-" in deconstruction, I would like to consider, even more briefly, what this hyphenated notion would mean if taken literally.

"Eco-" is the way we have inherited the Greek word *oikos*, meaning "house" or "dwelling." And Derrida's deconstruction has never been of anything but the dwelling. That is one of the things the French thinker took from Heidegger's famous dictum "Language is the house of being. In its home man dwells." Derrida's style of inhabiting this house was by deconstructing it: by exploring its hidden corners and secret passages, exposing whatever has been swept under the rugs, pulling skeletons out of closets, descending to the basement and showing how the pillars that support it make it, at once, possible and impossible, stable and ready to cave in. More literally still, the dwelling is the family abode, with all the conflicts that happen there, with all the rivalries, minor squabbles, and bloody feuds, repressed incestual desires, friendly quarrels and epochal contests between different arrangements or laws to be followed—from Antigone to Hegel. As family property, this sense of dwelling leads us, through an obvious semantic thread, to "economy" or "the law of the household." We might also add to these two general sites of deconstruction Derrida's profound feeling of not being at home, not even in his own language, the feeling that resonates with uncanny notes throughout his body of work.

It is, perhaps, this last variation on the dwelling permeated by the uncanny that comes closest to ecology as an event, breaking through the economic routine of habitation. If economy is the default organization of our dwellings—whether at the level of the psyche, of the household, or of the planet as a whole—then ecology is the rupture that no longer obeys the economic order and that, within its confines, is experienced, precisely, as a disorder, if not chaos, a harbinger of crisis. For all the platitudes about ecology as an object of environmental concern or an amorphous mesh of relations among all beings, it announces itself whenever we feel not at home in our economically arranged homes. This is the historical permutation of ecology for our times—the times of rampant economism, eclipsing all other modes of thinking, and of an unprecedented environmental crisis that provokes ecological thought as a reaction to the emergency of a planetary house on fire. At the current conjunction, not only ecology but also all the "others" of economic "sameness" must be experienced as events because there is no place for them to be registered otherwise within the hegemonic framework superimposed on our dwellings. (These "others" include, above all, art and ethics, standing, respectively, for a noncalculative relation to the world and to another human being.) Unless they are assimilated to and dissolved within the economic sphere, as it happens when artworks are bought and sold at auctions, ethics is based on considerations of utility, and ecology is forced into the mold of ecosystems, governed by relations of metabolic exchange and a circulation of forces, where nothing is lost. The very notion of event as a break or an interruption, a disruption of the same, responds to this assimilation but, at the same time, remains tethered to what it reacts against. Though, in my view, this notion is unavoidable, it ought to be historicized, together with our approach to ecology (or to whatever we comprehend under its name), against the horizon of the current techno-economic permutation of metaphysics.

Deconstruction's Allergy to Ecology

Derrida wrote virtually nothing on ecology, and this omission, it seems to me, is not fortuitous. His silence on the subject speaks volumes: it is every bit as evasive as that which he discovers in Heidegger's avoidance of the word "spirit."[1] In part, the absence of "ecology" from the deconstructive vernacular has to do with the persistence of "logos" in this term, which perhaps all too quickly affirms everything Derrida is so uncomfortable with: speech, the voice, presence, assemblage, gathering into one. . . . Worse still, in "ecology," the markers of metaphysics are combined with the dwelling, the *oikos* belonging to the family and standing for property, the proper, *domus*, one's own domain. The double closure of the dwelling and logos, each promising the fullness of presence while infinitely mirroring and expressing the other, is more than he can bear. It is beside the point that Derrida's concerns with animality, with biodegradability, with the place of the human, and with the very place of place dovetail with the signature themes of ecological thought. The nonappearance of the word itself is not without significance (above all, for anyone who respects the protocols of deconstruction), and neither analogies, comparisons, applications, extrapolations, nor similar hermeneutical gestures will be sufficient to remedy this deficit.

I would go so far as to say that Derrida is allergic to "ecology." Usually alive to the excesses of metaphysics, he becomes hypersensitive (and what is allergy if not hypersensitivity?) and overreacts to this word together with everything it articulates and signifies. Hence, the silence, supplanting the voice, or the speaking of the dwelling. And, hence, also, the intense scrutiny to which Derrida submits another arrangement of the house—namely, its law or nomos.

If my hypothesis is correct, then, besides being the residue of structuralism in deconstruction, economy plays the part of immunotherapy, a strategy for tackling a certain overreaction to ecology. Sometimes, allergies can be cured by introducing small amounts of the allergen into the bloodstream, gradually habituating the body to the irritant. That is precisely what happens with Derrida's fixation on economy: he separates the allergy-inducing dwelling and the circularity of a return from the essentializing routines of logos only to conjugate *oikos* with nomos, which has an air of arbitrariness and exteriority. Ostensibly denaturalized, the abode finally becomes tolerable, especially because, perpetually en route away from or toward it, one barely abides there. The sense of *remaining*, which the economic odyssey foregrounds, does not refer to "staying in place" but to the leftovers, to what remains after all the dizzying spins of exchange cycles and appropriations of family property. To sum up, then, economy authorizes us to speak about dwelling while forgetting what it means to dwell.

I am not suggesting, to be sure, that Derrida blindly adheres to the discourse underwritten by economy; far from it, the event occurs as a breakthrough, an interruption in economic circuitry, as in the case of the gift that "must not circulate" and "must not be exchanged." "If the figure of the circle is essential to economics, the gift must remain *aneconomic*" (GT 7/19). This negative definition of the gift, or of the "gift event" (GT 11/24), maintains economy as the dominant pole that, though disrupted, presides over whatever no longer follows its established procedures. But what if that other pole were not merely aneconomic but ecological? How to discern another sort of positivity underneath various negations of exchange, circulation, memory, and the subject's odyssey without falling into the traps of metaphysics? What would be the implications of dwelling otherwise: at home—whether it stands for one's own body, the psyche, family residence, or the world—which is not secured as one's own property or possession?

The prominence of economy in Derrida's oeuvre is explicable in phenomenological terms, assuming that he starts with the things as they are (in his historical moment, which is also ours, and which has lasted for much of the history of metaphysics). The way things are is structured by economic preoccupations, including not only the much-vaunted drive toward "job creation" and "growth," but, more broadly, by the laws of the psychic, familial, and social dwellings. It is in this general situation that ecology, along with or in the shape of the gift, is experienced as a disruption befitting the epoch of a profound environmental crisis and rooted in the forgetting of what it means to dwell, to abide, to be at home without imposing an ideal mold on it. Ecological thought becomes purely reactive, responding to the uncontrollable effects of the crisis and foreclosing the possibility of a positive, affirmative notion of ecology. Like the gift and the event as such, the logos of the dwelling comes to name a trauma outside the clasp of consciousness, which operates based on the economic precepts pertaining to the psychic dwelling. Now, this state of affairs is not a transhistorical ontological given, but an effect of hegemonic economicism, of *la pensée unique* responsible for the current predicament of our planetary dwelling and arrogant enough to outline the framework for dealing with the crisis.

Through a series of seemingly automatic associations, Derrida tightly interlaces the dwelling and economy, as though the nomos were a priori built into the *oikos* and vice versa. He leaves no space vacant for ecology. Not even a place. Whenever a dwelling of any variety is established, it is segregated from the outside: the fire of the hearth is separated from cosmic fire, subjective interiority is set apart from the world. . . . Economy is "a matter of the law (nomos) of the home (*oikos*); and of the space of separating or associating the fire of the family hearth and the fire of the sacrificial holocaust" (GD 88/122). Moreover—and this near-definition is worth citing at length—"nomos does not only signify the law in general, but also the law of distribution (*nemein*), the law as partition (*moira*), the given or assigned part, participation. Another sort of tautology already implies the economic within the nomic as such. As soon as there is law, there is partition: as soon as there is *nomy*, there is economy" (GT 6/17). The dwelling is the upshot of the law that, in distinguishing it from the nondwelling or from another dwelling, brings it into being. The "yes" of the inside is the result of the "no," directed to the outside. But this affirmation, uttered in an *après-coup* of negation, is ineluctably formal, empty, and abstract. Division, partitioning, and separation draw the admittedly porous boundaries between the house and the outside world (the fire of the hearth and cosmic fire), but they are powerless when it comes to articulating the divided, separated parts among and, especially, within themselves. Nomos is transcendentally erected with every concrete act of constructing house walls or putting a fence around a plot of land. But it does not gift the interiority with any positive content, which can only derive from an ecological, rather than an economic, comportment (for instance, from discernment without segregation, or from articulation without appropriation).

Another Aneconomy—of Life

When Derrida contemplates "a gift outside of any economy," he does not hesitate to equate it—as he does, for example, in his reading of Isaac's sacrifice—with the gift of death "accomplished without any hope of exchange, reward, circulation, or communication" (GD 95–96/132). That death is the singular and singularizing event is a familiar thesis he adopts from Heidegger's philosophy, with the twist of calling it *aneconomic* in comparison with the economy of everyday life. And, just as in Heidegger being-toward-death individuates an authentic life, so Abraham's incalculable renunciation of the singular and irreplaceable reinvigorates continued family existence: "Through the law of the father economy reappropriates the *an*economy of the gift as a gift of life or, what amounts to the same thing, a gift of death" (GD 97/133). With life having been absorbed into an economic routine, death is charged with the task of breaking through the chain of exchanges. The relation of life to death is one of the calculable to the incalculable, of the exchangeable to the irreplaceable. The two, nevertheless, participate in the same movement of the pendulum, which is why the gift of life "amounts to the same thing" as that of death. No longer perceptible in this scheme is another kind of life gathered into an ecological configuration, incalculable and irreplaceable in and of itself, inexpressible in any nomos that would immediately economize it. Such a life is still more alien to our understanding than death with its *mysterium tremendum* of "the *gift that is not a present*" (GD 31/50). Its workings correspond to the literal sense of allergy—*allos ergon*, the strange work immeasurable on any economic scale and incommensurable with life's biological and metaphysical conceptualizations.

Indeed, for Derrida, a fixed address where one dwells signifies no more and no less than a tomb. *Oikos* is evocative of *oikēsis*, which is to say the indwelling of writing in speech, death in life, space in time: "The *a* of *différance*, thus, is not heard; it remains silence, secret and discreet as a tomb: *oikēsis*. And thereby let us anticipate the delineation of a site, the familial residence and tomb of the proper in which is produced, by *différance*, the *economy of death*" (MP 4/4). Stated in these terms, death is exempt from the aneconomic singularity of the gift; instead, its law guards over the property, wherein the subject or her family are entombed. Life, too, bows before this figuration of death, at least insofar as its living strains to preserve itself, to attain a deathlike stability through an incessant rotation in the circle of self-reproduction, coming back home over and over again: "This economy of life restricts itself to conservation, to circulation and self-reproduction as the reproduction of meaning" (WD 255–56/376). All the departures from and returns home are meant to give one a sense of vivaciousness without really breaking the hold of death, without interfering with its economy. Unless, someone says or writes, "I never go away on a trip, I don't *go*, I never put any distance whatsoever between me and my 'house' without thinking—with images, films, drama, and full orchestral soundtrack—that I am going to die before I return" (CP 5/15). And even there and then, despite the noncircular, noneconomic itinerary of consciousness, the remains will find their way home, will be repatriated,[2] keeping the *oikēsis* of economy intact.

The tragedies of Antigone and of Hegel's Spirit will be replayed in every effacement of ecology by economy, drowning the singularity of life in deadly generality. "The tomb is the life of the body as the sign of death. . . . But the tomb also shelters, maintains in reserve, capitalizes on life by marking that life continues elsewhere. The family crypt: *oikēsis*" (MP 82/95). That is where Antigone deposits the remains of her dead brother, doing justice to the dead, and where Creon imprisons Antigone, doing injustice to the living. Nothing under Creon's (and Hegel's) vigilant gazes will be permitted to live except in a negated, relieved, and mediated form. The spiritual law of our planetary dwelling will, for its part, consider pure corporeality to be the tomb of spirit. It will demand that the biological lives of the largest family, encompassing everyone and everything living, be productively destroyed in the service of higher goals. On the condition of having exchanged their immediate vitality for a mediated life, plants, animals, and humans will join one another in a common family crypt. Humanity is preparing a tomb for itself and for those with whom it shares its planetary dwelling, still hoping that life would continue elsewhere—in other corners of the universe, in the immortal sphere of Ideas, or with God.

"The family of the living" is not an idle figure of speech eliciting emotional proximity to human and nonhuman creatures, but an expression stressing the reproductive function of a household (referring back to Aristotle and the faculty of *to threptikon*), where the dead are replaced by the new arrivals of the living. For all the interminable work of mourning that takes place in the family, its economy is predicated on the implicit postulation of substitutability, factored into the very dynamic of family relations (the son becoming a father for his child; his father turning into a grandfather). The nomos of economy converts death in a household from a singular event, the end of the world, each time unique (as Derrida used to put it) into a precondition for repetition that, in fact, idealizes family roles. Not to mention the pure reproducibility and substitutability

of the nonhuman members of our planetary family, grown, cultivated, produced to be cut down or slaughtered, with the view to
replacing them with others of the same kind. Their deaths fuel the
economy that is, itself, *of death*—the economy entirely beholden
to destruction such that every new birth becomes a vehicle for the
demise of whoever is so generated. "The family figures mourning,
the economy of the dead, the law of the *oikos* (*tomb*) . . . ; the house,
the place where death guards itself against itself, forms a theater or
funeral rite" (G 143L/162L). All over the planet, life is reduced to
this very death that "guards itself against itself."

Derrida, of course, resists the idea of an absolute singularity, the
irreplaceable, or the unrepeatable par excellence. The testimonial
condition, for instance, is such that the "singular must be univer-
salizable": "The exemplarity of the 'instant,' that which makes it
an 'instance,' if you like, is that it is singular like any exemplarity,
singular *and* universal" (DF 41/48). There is no inscription that does
not quasitranscendentally reproduce itself as a reinscription, and
there is no law that does not universalize the singular. The nomos of
economy, however, loses track of singularity altogether and spawns
empty universals, iterable forms, for which singular contents are
discardable. The economic form of money is the prototype of this indifference to content, particularly when anything can be exchanged
for it, and it—for anything whatsoever. The actual lives of the members of our planetary family also become nothing but dispensable
contents of malleable genetic forms that no longer require a singular instant, or a unique instantiation, to universalize themselves.
Perversely, "pure singularity" is identified with a corpse—"neither
the empiric individual that death destroys, decomposes, analyzes,
nor the rational universality of the citizen, of the living subject."
(G 143L/163L) That is what constitutes the current irreplaceabil-
ity at the heart of exchange, and that is what our economies trade
in everywhere, all the time: corpses. (Ecology is not free of bodies,
whether living or dead; far from it. But it does not drive a wedge between them and their *eidos*/image/form.)

The stakes in this distinction are high. On the one hand, Derrida's critique of logocentrism tends to impute to logos the totalizing
ramifications reminiscent of nomos. On the other hand, the event of
ecology, consistent with the aneconomy of life, foretokens another
logos, mindful of singularity, which it refrains from consigning to the
voice or says in secret, so much so that one cannot even hear oneself
speak when it is uttered. Although Derrida appreciates the German

provenance of the secret (*Geheimnis*) from home (*Heim*), he still feels the urge to couple it with nomos, ignoring the context wherein "a certain language," "one's so-called natural or mother tongue" (and, therefore, logos) is paramount. The untranslatable statement *tout autre est tout autre*, which he analyzes in *The Gift of Death* (and not only there) "is there before us in its possibility, the *Geheimnis* of language that ties it to the home, to the motherland, to the birth-place, to economy, to the law of the *oikos*, in short to the family and to the family of words derived from *heim*—home, *heimlich*, *unheim-lich*, *Geheimnis*, etc." (GD 88/122). Why the slippage from *oikos* to *oikonomia* and from the uncanny (*Unheimlich*) to the aneconomic, rather than the eco-logical? Is a birthplace, be it the same as the family dwelling, not a singularity in excess of the economic circle, the singularity to which there is no return, which cannot be exchanged, and which articulates the one who is born there with everyone else inhabiting that place? Is the cryptic side of one's home, abstractly encapsulated in the term "the private sphere," necessarily a crypt, a tomb, or is it the manner in which the dwelling is arranged—the way it lives, arranges itself—without my interference? Isn't that the uncanny secret of dwelling-articulation, or, in a word, ecology?

The Art Analogy

Since Derrida does not correlate the aneconomic event with ecology, we must proceed obliquely, seeking pointers from the deconstructive figures of the event and, notably, from art. In "Economimesis," he begins with the Kantian imputation to art of freedom from economic plans and calculations: "Art, strictly speaking, is liberal or free [*freie*], its production must not enter into the economic circle of commerce, of offer and demand; it must not be exchanged" (E 5/61). Further, the freedom of art invests the essence of human freedom in general with meaning. On the hither side of the "natural" logic of ends-and-means, a "free man is not . . . *homo oeconomicus*," whose "essence," *le propre*, is to be "capable of pure, that is, non-exchange-able, productivity" (E 5/61, 8–9/66). Let us pause for a moment here and ask, what sort of an escape from economic rationality is that? Does it, in one fell swoop, cut itself loose from the dwelling and from its law? After all, "the aneconomic" is not a simple negation, but one that may repudiate either the dwelling or the law or the composite "law of the dwelling." In question is the alpha-privative of an-economy and what or whom it actually negates.

Freeing itself from the exigencies of exchange, aesthetic activity does not abandon the dwelling of a "free man." To the contrary, it moves deeper into his *oikos*, where essence itself resides, touching upon the proper that ought not to be conflated with property as the container for the possessive will. At home devoid of nomos in art, we do not appropriate anything, let alone the beautiful, but, thanks to free play, are appropriated by the universal and admitted into it as though it were our most intimate dwelling. Who can share this abode with us? Who is this "we," in the first place? Derrida's response: "This pure productivity of the inexchangeable liberates a sort of immaculate commerce" of "universal communicability between free subjects" in Kant's text. "There is in this a sort of pure economy in which the *oikos*, what belongs essentially to the definition [*le propre*] of man, is reflected in his pure freedom and in his pure productivity" (E 9/67). But is it at all conceivable that subjects who dwell without imposing their law on their home would replicate, at a higher, spiritual, "immaculate" level, the appropriation and alienation of property that move every exchange? Is this another economy or, rather, an ecology we are talking about, seeing that logos is also not a matter of appropriation but a captivating, appropriating arrangement into which we are assembled? And is this "we" exclusively human?

While deconstructing the Kantian separation of art from nature, Derrida chances upon a crease or a fold that will help us respond to the preceding questions. "One must not imitate nature," he writes, "but nature, assigning its rules to genius, folds itself, returns to itself, reflects itself through art. This specular flexion provides . . . the secret resource of *mimesis*—understood not, in the first place, as an imitation of nature by art, but as a flexion of the *physis*, nature's relation to itself" (E 4/59). The immanent fold of nature bespeaks its aneconomic, noncircular return to itself. This flexion is ecological insofar as it claims the genius and its art for the assemblage of *physis* that opens up a dwelling on each side of the crease in contact with the other: nature dwells in artistic genius, and art finds an unacknowledged (secret, once again) home in nature. Ecology itself may be characterized as the aneconomic return of nature to itself through us, humans, as well as though all other living beings who inhabit the planetwide house of this extremely extended family and whose logoi commune on every side of the fold. Prior to any appropriative overtures, we, whether human or nonhuman, are gathered into the secret reserve without depth of exteriority folded upon itself. The capacity to dwell in this fold does not depend on surrounding and securing a piece of territory with a fence or a wall, from which the nomos of *economy* stems. It implies, alternatively, residing in exteriority within a whole that affords its residents freedom from itself.

It will have become clear by now that *art* is not a mere analogy to ecology but its inner *articulation*. By analogy with art, ecology flashed before our eyes an event of freedom from pragmatic calculations and the incalculable appropriation of appropriative subjects to a nonideal assemblage of singularities in beauty, in *physis*, in the planetary dwelling. At the heart of Kantian art, too, there is an analogy between the freedom of genius and that of God, the "'true' *mimesis* . . . between two productive subjects and not between two produced things," the subjects who have an aptitude to logos in common (E 9/68). That is why the analogy within art, like that between art and ecology, is more than an analogy; it is an indicator of logos, which is factored into it and which, in turn, points toward the event of production, leading forth, and appropriation of purely appropriative subjects, be they as omnipotent as God. "Nature is properly [*proprement*] logos toward which one must always return [*remonte*]. Analogy is always language" (E 13/74).

And yet, Derrida economizes on the fold, giving it a short shrift, refusing to spend time and intellectual resources on a study of the alternative mode of dwelling it inaugurates, and handing it back over to economic reason. For no apparent purpose, God freely plays with himself through the poet and "breaks the infinite circle or contractual exchange in order to strike an infinite accord with himself" (E 11/71). As in the "immaculate commerce" of free and equal subjects, the irruption of the aneconomic—aesthetic, ecological—event is co-opted for the purpose of cementing the economy of the same— the Same—whose inner harmony and peace would ensure the stability of the rest. Nonetheless, not every return is circular; the flexion of *physis* in a crease is the aneconomic contact of *physis* with itself across a straight or jagged seam that traverses its fabric. In the logos of the poet, God comes back to himself not only as another but also without himself, and, if we are to believe Plotinus or Leibniz, such odd returns happen in and through the logoi of every creature in the world. If, on the ecological plane, we still wanted to retain something of this divine figurehead, we would have no choice but to admit that God is exteriority, exterior even and especially to itself, that, folded upon itself, unlocks the possibility of dwelling. That would be one way of approaching a dwelling outside the circle or, speaking three-dimensionally, outside the economic *sphere*, where the dwellers are enclosed, feeling secure while being blockaded and suffocated there.

Derrida senses that something momentous transpires in Kantian free play, but he prefers to fold the event into a fading difference *within* economy, thus eliding the ecological fold: "As soon as the infinite gives itself (to be thought), the *opposition* tends to be effaced between restricted and general economy, circulation and expendiary productivity" (E 11/71–72). In other words, Derrida announces that the conceptual frontier dividing the two economies—a kind of Rubicon—has been crossed, deconstructed. This is his signature move that repeats the deconstruction of the opposition between a circle and a line striving to infinity or, more precisely, the Hegelian "good" and "bad" infinities. Whereas Hegel is the thinker of a circular return of the same to itself with the booty of difference, and whereas Levinas valorizes what Hegel would deem to be "bad infinity" in the movement of the "I" to the other without return to ipseity, Derrida's *différance* shifts the spotlight to the break *in* the circle, in the rotations of which something or someone does not return, does not circle back, drops out of circulation, is lost for economy. Schematically, the relation of ecology to economy is that of *différance*—a break in a circle, affording us a breath of fresh air. Except that the circle continues to shut ever tighter, at the risk of its own viability and the existence of our planetary dwelling. What is Derrida's response? Stressing, as always, the noncoincidence of the same with itself, he tends to rediscover the breach not in the other of economy but in another economy, in economy itself, for example, borrowing from Bataille the classification of general and restricted economies—which makes the effacement of the opposition between the two economic modalities all the more tragic because it seals the circle anew.

So, the fold of the ecological dwelling has mutated into a break in the circle, immanence has flipped into transcendence, and the event has turned into a violent disruption instead of the possibility of a continuous habitation. These reversals are happening for a good reason: we are expelled from the dwelling by our own unremitting economic activity and, at best, are glancing at it from afar, from the other side, like Moses who overlooked the Promised Land from Mount Nebo. General economy has occupied the spot reserved for ecology as an evental break in the cycle of exchange. But doesn't such economy make dwelling even more impossible than its restricted variety? General economy is "not a reserve or a withdrawal, not the infinite murmur of a blank speech erasing the traces of classical discourse, but a kind of potlatch of signs that burns, consumes, wastes words in the gay affirmation of death: a sacrifice and a challenge" (WD 274/403). The generality of general economy is condensed into incalculable expenditure, the utopia of consumption without production and of death without a crypt ("not a reserve or a withdrawal"). It disturbs the economic balance of a dwelling where accounts need to be balanced such that the incoming goods or money would be equal to the outgoing merchandise and expenses. Perhaps it anticipates the ecological fold with its lack of interiority, which, in a restricted economy, is but a delay between production and consumption or income and spending. But the path back into the fold of ecology requires significantly more than an exclusive focus on one moment of the economic process at the expense of the other. Expenditure without reserve violates the law of the dwelling and obliterates the intersection at which production and consumption meet in the spatiotemporal interiority of a delay. In so doing, however, it demolishes the dwelling itself because it diminishes to zero the time and space necessary for abiding.

Dwelling, Undecidably

"We must not hasten to decide" (MP 19/20). Undecidability is a sign of the event, which vacillates indeterminately between two determinate poles in an opposition. Between economy and ecology, too, "we must not hasten to decide," Derrida would counsel us, were he able to overcome his allergy to logos. For the time being, this advice pertains to economy and its negation that ought to be thought together, even if the task at hand is unthinkable: "It is evident—and this is the evidenceitself—that the economical and the noneconomical, the same and the entirely other, etc., cannot be thought together. If différance is unthinkable in this way, perhaps we should not hasten to make it evident" (MP 19/20). The difference between the economic and the aneconomic, the economic and the ecological, "is" différance, which means that the decisive advantage of either way of organizing or being organized by the dwelling threatens to obliterate the spacing of time and the timing of space for habitation. The indeterminacy of oikos without the additions of nomos or logos is not an abstract potentiality but the place of différance, the place as différance.

Intuitively, it seems that the event is on the side of the entirely other, of the aneconomical, of ecology. But, factoring *différance* into the equation, the event comes to pass in both modifications of the dwelling and, above all, between them. Within its cycles and, to a large extent, despite them and despite itself, the economy of iterability permits something or someone that does not conform to the idealizing designs of repetition to return in every single rotation. The iterations of nomos recapitulate the unlawful or the extralegal together with the law; the aneconomic derails and constitutes economy with respect both to its nomos and its *oikos*.

At certain moments Derrida is tempted to aggravate the self-deconstruction of economic cycles by thematizing economy's unintended, disruptive, self-interrupting consequences wherein one of the senses of the event resides. He invokes an "economics taking account of the effects of iterability, inasmuch as they are inseparable from the economy of (what must still be called) the Unconscious as well as from a graphematics of undecidables, an economics calling into question the entire philosophical tradition of the *oikos*—of the *propre*: the 'own,' 'ownership,' 'property'—as was the case of the laws that have governed it" (LI 76/144). Deconstructive economics suspends the idea of dwelling tethered to property as much as the law that guarantees its perpetuation—that is to say, it deconstructs both elements of eco-nomy in the name of the undecidable, what cannot be appropriated, legislated, inhabited. By no means does the deconstruction of economics spell out the ruination of the house, leading to absolute homelessness, or the destruction of the law, heralding the worst image of anarchy as sheer undifferentiation. Instead, it strains to imagine how a dwelling would look like at a distance from "the entire philosophical tradition of the *oikos*" and what sort of law would be appropriate to its inappropriability. Or, would this be not a law but another sort of articulation, a logos?

In other instances, Derrida desires simply to nurture the tension between the circle of economy and the noncircular, aneconomic event, situating himself neither too close to nor too far from the routines of the same. Aristotelian moderation, steering clear of both excess and deficiency, and the Freudian uncanny overlap here: "If the figure of the circle is essential to economics, the gift must remain *aneconomic*. Not that it remains foreign to the circle, but it must *keep* a relation of foreignness to the circle, a relation without relation of familiar foreignness. It is perhaps in this sense that the gift is the impossible" (GT 7/19). Although it is not in the circle, does not participate in a circle, the gift event relates to economy, while insisting on being subtracted from its operations. If it could appear on economic balance sheets, the gift would have been the absolute, irremissible debit, which amounts to the same as the absolute credit. (In "absolutes" opposites always coincide.) At any cost, Derrida wants to preserve this "gap between gift and economy" (or, in our terms, between ecology and economy) that is "not present anywhere," "resembles an empty word or a transcendental illusion" (GT 29/46). In the uncanny, "familiar foreignness" of the gap, one cannot dwell. But, provided that the distance it has adumbrated is mended, dwelling becomes impossible, as well. The gift is the impossible *and* a condition of possibility for the cycle of exchange, toward which it keeps its distance.

Bequeathed by Derrida—albeit not as result of his conscious decision, his "will"—the intricate work of translation is cut out for us: the aneconomic, the gift, and pointless aesthetic pleasure ought to be retranslated into the vocabulary of ecology, with an eye to their untranslatable singularity, their semantic and conceptual nonequivalence. The entire problematic of translation itself belongs to the "relation without relation" of the economic and the aneconomic unless it signifies a straightforward exchange between one language and another, an operation of symbolic economy that neglects the untranslatable in its midst. The strong resemblance between the gap separating the two arrangements of the dwelling and a transcendental illusion outlines a pattern whereby the relation of exteriority animates interiority's inner core. The gift "is this exteriority that sets the circle going; it is this exteriority that puts the economy in motion" (GT 30/47). Aesthetic "priceless pleasure" is "the origin of value," "the value of values" (E 18/83). Ecology, it would seem, is the exteriority, which belongs to no one and which dictates the law of laws for the organization of a dwelling. Both on the formal and material levels, however, direct translation runs up against a limit here, the limit crucial to the tension permeating the relation of economy to ecology.

On the one hand, in the unexchangeable gift and aesthetic disinterested pleasure, the negation of the hegemonic system—which is invariably economic—causes something to drop out of its mesh in such a way that, at a distance, the negating aspect comes to determine the negated. Nonidealizable leftovers, the remains that could not be incorporated into the same, are recovered and elevated to the status of material yet quasitranscendental conditions of possibility for economy. From afar, they confirm that there is nothing but economy and that the aneconomic events that define it *are not*, which is to say are not present, are nowhere to be found in the order of present being or even of logical possibility. Hence, "economics . . . is not one domain among others or a domain whose laws have already been recognized" (LI 76/144). It is the domain *as such*, naming and instituting the domesticity of all other domains, by virtue of taking care of the *domus*, the dwelling.

Ecology, on the other hand, does not inherently negate economy but accesses the *oikos* from another side or rather indicates a different direction from which one can access it. It is not a collection of excluded leftovers, sacralized as the disavowed origins of the system that has spat them out, but a positive elaboration of the dwelling, respectful of its singularity and inner articulation or disarticulation. Ecology is not a domain because the mode of dwelling it underwrites is not dominion, predicated on appropriation, but a fold of exteriority. Full-fledged undecidability, moreover, may be experienced between two positivities, as opposed to a dominant term and its supplementary negation, which turns out to be the origin of the origin, in a situation where the dice are loaded and, currently, reloaded in favor of the "underdog." The relation without relation of ecology and economy depends on what they share—the *oikos*—and what sets them apart—nomos and logos—rendering each the same and the other vis-à-vis each other. Dwelling undecidably is shuttling between economy and ecology, interiority and exteriority, the circle and a line tending to infinity, the organized and the organizing house, which only divided will stand.

Spectral Economies, Ghostly Ecologies

In Derrida's later writings, the aneconomic, noncircular return passes under the heading of "haunting." Given the knot in which aneconomy is tied to ecology and the event, an obvious question is whether the ghostly *revenant* is how the nonhuman voice, the logos, of ecology resounds in a planetary dwelling over which human beings have arrogated to themselves full rights, mastery, and control. More precisely, is haunting how economic subjects experience the sudden surges of ecological worries and of the suppressed self-organization of their dwelling?

Just as in the capitalist economy the most physical and concrete aspect of the thing, its use value, recedes to the undifferentiated background of the economic agents' concerns only to haunt the ideal cycles of financial capital in moments of crisis, so the entire ecological problematic comes back, ghostlike, only in periods of environmental calamity. Ecology is probably the most abused use value in capitalist economy that is at odds with the very idea of dwelling while substituting the pure law of value for the arrangement of an *oikos*. Capitalism is, in its core, an economy without a home, and it propagates homelessness—both of a metaphysical variety and of the suffering persons without shelter, *sans-abri*, as one says in French—worldwide.

Portraying Marx himself as a specter, Derrida allies this critic of political economy with use value, both in light of his presumed naïve longing for thinking and being free of abstraction and in light of his repressed status, which provokes his ghostly comebacks. "Marx," Derrida writes, "remains an immigrant *chez nous*. . . . He belongs to a time of disjunction, to that 'time out of joint' in which is inaugurated, laboriously, painfully, tragically, a new thinking of borders, a new experience of the house, the home, and the economy. Between earth and sky" (SM 174/276). Does this "new experience of the house, the home" pertain to economy, however radically different it may be from the capitalist variety? Hasn't the Soviet experiment shown that, absent a sustained rethinking of the dwelling, the horrors of capitalism are likely to persist in another regime of political economy? Communal apartments, in one of which I passed my childhood, and rampant environmental degradation are the legacy of turning a blind eye to the meaning of *home* in "actually existing socialism." The "time of disjunction" would need to be supplemented with disjunctive space—a fissure within our experience of the house—for Marxist haunting to become effective. "Between earth and sky" gestures toward this possibility but falls short of explicitly identifying it.

The house of economy, then, is haunted. And the specter that disturbs it is that of ecology. A haunted house is a place where more is received than the owner desires, putting the sovereignty of the host in doubt. Insofar as it imposes its ideal law onto the dwelling, economy is a make-believe that sovereign control is complete: "*oikonomia*, law of the household" is "where it is precisely the patron of the house—he who receives, who is master in his house, in his household, in his state, in his nation, in his city, in his town, who remains master in his house—who defines the conditions of hospitality or welcome: where consequently there can be no unconditional welcome, no unconditional passage through the door" (H 4/19). Insofar as it receives us before we receive anyone or anything else, insofar as it puts the material articulation of the dwelling at the forefront, ecology refrains from defining the conditions of hospitality and extends to all its "unconditional welcome." Crucially, the receptivity of the house that takes in whoever crosses the threshold precedes the hospitality of the host, so that unrestricted openness haunts a legally shut dwelling, the patron's domain, from within. That is the conclusion of Derrida's "Hostipitality," implying (without saying as much) that economic conditions depend on ecological unconditionality. The very tension between conditional and unconditional hospitality, between invitation and a ghostly visitation, fits within the confines of the quarrel of economy and ecology.

In addition to spelling out the formal paradox of hospitality and its relation to sovereignty, Derrida interrogates its "anthropological dimension" and, in so doing, crosses over to the other dwelling—the ecological—still without naming it as such. An "important question," for him, is "that of the anthropological dimension of hospitality or the right to hospitality: what can be said of, indeed can one speak of, hospitality toward the non-human, the divine, for example, or the animal or vegetable; does one owe hospitality, and is that the right word when it is a question of welcoming—or being made welcome by—the other or the stranger [*l'étranger*] as god, animal or plant, to use those conventional categories?" (H 4/18). Derrida abandons this probing thread too quickly and easily, perhaps because it would lead him toward the unthought (repressed?) noneconomy of economy and ecology. Plants, animals, and even a god welcome us before we have an opportunity to welcome them or to keep them at bay. And, therefore, they share more with the dwelling, which receives its resident, than with the human guest or a host abiding there. The trace of such primordial, ecological hospitality is erased not only in the economic regulation of thresholds but also in the broadening of the category of a stranger beyond its human limits. "God, animal or plant" are not strangers in our "dwellings"; to the contrary, from the ecological perspective, we, humans, are the others in *their* world, much like we are alien to the houses that receive us prior to our appropriation of them. Wondering about the extension of hospitality to nonhuman others is something that happens within the economy of economy, in the vicinity of the rift that cuts it off from ecology.

Whereas ecological events lend themselves to sense as ghostly disturbances in the circuits and dwellings ruled by economy, from the ecological side of things, spectrality is the arbitrary and arbitrarily imposed law of the household. It is not the "other" that haunts; we are the ghosts contributing to the disquiet of the planetary dwelling pushed to the brink of destruction. In this respect, Derrida's words acquire an additional tinge of meaning: "We are on the threshold. We do not know what hospitality is [*Nous ne savons pas ce que c'est que l'hospitalité*]. Not yet" (H 6/22–23). Such lack of knowing is due, to a certain extent, to the nonspecific treatment of the dwelling where welcome is extended. In *oikologia* or *oikonomia*? Conceptual precision is of no help here, as Derrida rightly acknowledges (H 7/26). Inhabiting, mastering, opening or closing windows and doors—all these acts presuppose not the concept but the shifting contexts of the house, either delimited by a legally (nomothetically) stipulated perimeter or articulated without human interference. In the first case, the law of hospitality applies; in the second, hospitality is an ontological (indeed, existential) feature of our planetary dwelling itself and of its nonhuman inhabitants (especially plants) that make this dwelling livable.

It would be too simplistic to stick the label "uncanny" onto the ecological sensation of being foreigners in our "own" houses. Ecology estranges us from the estrangements of economy, and, if it denies us the entitlement to being at home in the world, it does not, thereby, expel us from our homes and turn us into nomads. Rather, the awareness that we are the ghosts haunting our planetary dwelling bestows meaning onto another, nonappropriative experience of *chez soi*, of being at a home to which we have no right whatsoever, legal or otherwise. "Home" is that which receives us, accepts us into itself, before we utter our "yes" or "no" to it, or, for that matter, to anyone or anything else we share it with. If the economy of welcome appears to be inverted, put on its head in this other experience of *chez soi*, this is because it is no longer an economy but an ecology that announces itself there. The hosts are hostages, as Derrida reminds us, relying on the thought of Emmanuel Levinas (H 9/30)—not so much of the guests as of the dwelling they inhabit. To repress a profound trauma of the uncontrollable habitation, the hosts do everything in their power to impose the order and law onto the *oikos* that was not, initially, of their own choosing. Yet, the dwelling remains unpredictable, unstable, eventful: "The most arriving [*arrivantes*] things, and often the worst, come to pass in the bosom of one's own home" (CP 15/23).

A House Divided . . .

As though anticipating the dire predicament of the twenty-first century well in advance, the Gospel of Mark passes a famous verdict: "καὶ ἐὰν οἰκία ἐφ' ἑαυτὴν μερισθῇ, οὐ δυνήσεται ἡ οἰκία ἐκείνη στῆναι" (3:25).[3] The conventional translation of this verse, "A house divided cannot stand," occludes both the conditional nature of the statement and the doubling of the dwelling, *oikia*, in it. What the gospel is really saying is, "And if a house is parted against itself, then such a house will not be able to stand." It is not insignificant that this sentence, too, is divided down the middle, illustrating the parting of the house through a semantic repetition that has the semblance of a tautology. The dwelling persists on both sides while threatening to collapse into an impossibility of continued existence. The same dynamic affects our planetary dwelling, split between an economic organization and an ecological articulation. But, what if, rather than a deficiency, the precarious balance of *oikos*'s nomos and logos were necessary for human life? What if the economic hegemony over the dwelling, menacing with the obliteration of its ecological pillars, undermined this dwelling (and our place within in) much more drastically than its parting in two?

I will not bring up the rich history of the biblical phrase, including, in the first place, Abraham Lincoln's 1858 "A House Divided" speech. The desire for originary ontological, political, or domestic unity wells up from the same metaphysical source. Frustrating this desire, Derrida's deconstruction gives us to understand, as I have already noted, that only divided will the house stand: that the dwelling will admit dwellers into its midst only if it, like two lips, parts against itself and accommodates *différance* or, indeed, accretes around *différance*. The persistence of finite beings is contingent on such splitting of their identity. The home is not at home in itself; from its imperceptible unrest everything begins before beginning.

The homelessness of the house becomes more pronounced as the quarrel of economy and ecology intensifies. Further complicating the sense of uprooting, being adrift, and nonbelonging that marks modernity, the splitting of the *oikos* invalidates the projects of a return to the simplicity of a dwelling (in nature, in one's "homeland," in the family, with those of one's kind, *genos*). Similarly, with the event. Whereas ecology opens up the dwelling by articulating the parts of the *oikos* in a mode different from and deferring the imposition of the law, economy signals the closure of the dwelling through its total determination and invalidation of ecological concerns. The event, however, spans both sides of the unequal equation; it embraces possibility and impossibility, preservation and destruction, openness and closure, ecology and economy. The house divided corresponds to divisions in the event, irreducible to potentiality awaiting actualization.

It is true that ecological dwelling amounts to nonexistence within a purely economic framework. With regard to literature, Derrida asserts that "it does not exist" and explains, "It does not remain at home, *abidingly* [*à demeure*], at least if 'abode [*demeure*]' designates the essential stability of a place; it only remains [*demeure*] *where* and *if* 'to be abidingly [*être à demeure*]' in some 'abiding order [*mise en demeure*]' means something else" (DF 28/29). In other words, within the economy of a stable dwelling, literature does not exist, but if dwelling is thought ecologically, or, more precisely, in a rift between economy and ecology, then it abides as this very rift. This is what the words "means something else" mean. An "abiding order," putting the house in the order of abidance, *mise en demeure* is an economic (or nomic) function, oblivious to the possibility of being put in order, captured or captivated, by logos. But what happens to the dwelling in that other arrangement, freed from nomos? What is the happening, the event, of dwelling governed by another meaning of abidance? And how does the event come to pass, given this semantic and ontological cleft?

From the generality of literature, we are invited to descend to the singularity of a short story by Maurice Blanchot, "The Instant of My Death," around which much of Derrida's *Demeure* revolves. Blanchot's text, in turn, begins and returns to "a large house (the Château, it was called [*une grande maison (le Château, disait-on)*],"[4] which, Derrida suggests, is more than the backdrop for the action: "As if the abode [*demeure*]—its abidance [*sa demeurance*]—were the true central character, at the same time being the scene, the place, and the taking place of the narrative" (DF 77/101). This majestic dwelling, then, accomplishes the incredible feat of combining ecology and economy, the active placing of place and the scene for the transactions conducted there. Ecology puts the dwelling itself on stage, not as a prop but as an actor that announces itself in a logos destitute of a voice. It is the event *of* the place (subjective genitive).

Economy formalizes the dwelling, fills it with goods and services to be rendered, and, at best, denotes what happens in a place (the event of the place, now in objective genitive). A house divided is riven between a house-thing and a house-work that, somehow, come together in Blanchot's narrative, or at least in Derrida's interpretation of it. Stable instability, it "does not *remain* like the permanence of an eternity. It is time itself" (DF 69/89). When economy predominates and the dwelling is converted into nothing but property, time itself is violated, taken away, along with the ecological side of the *oikos*. This, too, is an event, but, lest it be tempered with the gift of time and the self-articulation of the dwelling, it will exhaust the generosity of being and lapse into irremediable nihilism.

Postscript (Solicited by the Other)

It came to me as a surprise that, immediately after a version of this essay had been presented and discussed at the groundbreaking 2015 *Eco-Deconstruction* conference in Nashville, its main argument was met with so much misunderstanding and confusion. To address the controversy on the heels of an already very long text, I will not burden readers with further complexities and technicalities. In fact, I will have to economize, more than ever before, on space, time, and the supplementary explanations to follow. Having said that, I stand by every word written here, precisely because these words have been carefully chosen, weighed, and measured before making it into the text—above all, the word "ecology" and everything that happens (or fails to happen) around it in Derrida's writings.

Even the staunchest critic will agree that the lesson of deconstruction (if, indeed, it ever teaches a lesson or subscribes to pedagogic imperatives) is that *words matter*. Synonymy should be appreciated for the subtle differences between the semantic units that participate in its articulations rather than for their sameness. "Ecology" does not mean the same thing as "environment." When I observe that Derrida is allergic to "ecology," I am not arguing that his thought is irrelevant to environmental concerns. On the contrary, my own work on what has come to be known as "vegetal philosophy" is deeply influenced by and indebted to deconstruction. What Derrida is (doubly) allergic to, instead, are the constituents of ecology: the familial dwelling (*oikos*) and, above all, the trenchantly untranslatable logos that gathers into itself voice, speech, and the very act of gathering, all of them mercilessly ground in deconstruction's interpretative mill ever since the first writings on the phenomenology of Edmund Husserl. Given the way Derrida treats the two ingredients of this composite term, it is safe to assume that nothing in his attitude would change once they were put together, their force redoubled.

Compared to the absence of "ecology"—the word and the thing itself, as the author himself would say—from Derrida's oeuvre, "economy" abounds in his writings. This, no doubt, is in part due to the historical conditions of French philosophical discourse at the time, notably the emergence of poststructuralism from the bowels of structuralism with its emphasis on the economic (Althusser, Foucault, Bataille, to mention but a few names). But I also suspect that, all things being equal (and they never are), economy appeared less objectionable because it conjugated the dwelling, *oikos*, with nomos (law, custom, convention), with its air of indeterminacy, arbitrariness, formalism, and a certain freedom. (For the same reason, Gianni Vattimo's hermeneutical position endorses proceduralism against the claims of substantive justice.)

Once he confines himself to the economic domain, however broadly or generally understood, Derrida has no other choice but to escape from it by way of a negation, clinging onto the aneconomic elements the economic totality is incapable of digesting. While they are not entirely interchangeable, the gift, the event, radical alterity all belong to the aneconomic level. My suggestion is that missing from this approach is the positivity of the event in excess of an oscillation between the same (economy, the proper) and the non-same (aneconomy, the improper), the positivity that is experienced only when another pole—ecology—diminishes the magnetic force of economy. Now, the beauty of the relation between ecology and economy is that it is not an oppositional relation, or one that is not *merely* oppositional. "Ecology" and "economy" share the *oikos* (eco-), even as they offer different approaches to it. This shared dwelling, moreover, does not preexist the ways of accessing it; it is only marked or shaped as something determinate after it has been modified either by nomos or by *oikos*. That is why I write that the event, first making the dwelling what it is or foreclosing it, happens between ecology and economy. Nevertheless, the economic determination of the dwelling, reliant upon the logic of appropriation, extinguishes the possibility of the event, whereas its ecological counterpart nurtures such a possibility, especially to the extent that it inverts this very logic and causes us to be gathered or appropriated into the *oikos*. So, both assertions stand: the event may come to pass between economy and ecology, and it may continue to happen on the groundless terrain of ecology.

And a final word. In the face of a crisis, not least in the face of the global environmental crisis we are living or dying through, the work of thinking at the antipodes to knowing is more effective than all the rushed "applications" of what is already known to the emergency at hand. Before applying deconstruction to environmental ethics, it would be advisable to ask what is to be applied to what, not to mention to inquire into the sense and implications of "application." Perhaps one can only apply the inapplicable. Be this as it may, in the text you have just read, I have not argued that Derrida's writings are inapplicable to ecological concerns. I think, on the contrary, that deconstruction and ecological thought would both benefit from a mutual openness. But for that to occur (at least on the side of the former; the latter, most assuredly, indulges in resistances of its own), Derrida would need to be cured of his allergy to "ecology," in the precise sense of the word, which includes an allergy to logos. That is not a minor task. To follow Derrida on the tracks of an ecological attitude he implicitly refuses to follow is to accept the challenge of this impossible possibility.

Notes

1. Cf. OS.

2. "No, I am exaggerating a little, as usual, I don't *only* think about this return of the body, of this form of my return home. But the thought of this traversal made by my remains [*restes*] traverses the rest"; CP 21/29.

3. Cf. also Matthew 12:25 and Luke 11:17.

4. Maurice Blanchot, "The Instant of My Death," in *Demeure: Fiction and Testimony*, trans. Elizabeth Rottenberg (Stanford: Stanford University Press, 2000), 2–3.

Writing Home:
Eco-Choro-Spectrography

John Llewelyn

Specters of Kant

A few months ago I visited the store of a bespoke Edinburgh tailor with a view to purchasing a replacement for my somewhat threadbare winter coat. The assistant suggested I try on what she referred to as their "deconstructed" model. I laughed. Unsure whether she was joking or serious, I asked her whether she had ever heard the name "Jacques Derrida." She had not. Curious as to why I had laughed, she explained to me what is meant in the garment industry when an item of clothing is classified (her word) as "deconstructed." She pointed to the reinforcing patches of fabric over the shoulders and, turning the coat inside-out, drew my attention to the discontinuous layout of its lining. It's a kind of patchwork, she said. Once again, I couldn't believe my ears. This time, however, I resisted the temptation to ask her if she had ever heard the name "Norman Kemp Smith." Readers of this essay and the collection of which it forms part, however, will know this name as that of the Professor of Philosophy at Edinburgh who translated Kant's *Critique of Pure Reason* into English and published a commentary on it defending what is known as the "patchwork" or "multiplicity" theory of that work's composition. This is the theory according to which Kant's *Meisterstück* [masterpiece] was not composed with his usual rigor but was assembled hurriedly from passages that had not been carefully thought through and some of which were mutually contradictory. The most radical representation of this view on the Continent was advanced in Hans Vaihinger's *Die Transzendentale Deduktion*, a title that betrays that the view in question is one that focuses on the part of the first *Critique* to which Kant gives that title. This view is distinguished from the holistic theory of the composition of the first *Critique* advocated by H. J. Paton of Glasgow and Oxford in an essay published in the *Proceedings of the Aristotelian Society* (30, no. 7 [1929–30]), the date of certain other publications that will play crucial parts in the present commentary.

The integralist interpretation was defended on the Continent by Julius Ebbinghaus and the neo-Kantian school at Marburg. Although I can testify that the strengths and weaknesses of that interpretation and its competitor were not matters under discussion at the seminar conducted by Ebbinghaus that I attended at Marburg in the 1950s, the controversy had been represented at least symbolically by the face-to-face meeting of Ebbinghaus and Kemp Smith at which I was present when I was a student at Edinburgh earlier in the same decade.

The period at Marburg during which neo-Kantism dominated was followed by one during which teachers at the university there included Martin Heidegger and Rudolf Bultmann. While still at Marburg Heidegger made use of the term *"Destruktion"* and some of its cognates, stressing that in his technical use it was not to be taken in the familiar sense of doing away with. It was to be understood as denoting a retrieval [*Wiederholung*] of the history of metaphysics focused on the tendency of proponents of classical readings of that history to limit their concern with being to the ontic being of beings, a preoccupation that led them to neglect ontological being as such.

The emphasis put on the integralist reading of Kant at Marburg was paralleled there by Bultmann's inclination toward an integralist reading of the works of his colleague Heidegger. The emphasis put on unity in that reading of Heidegger, without doing away with the ontological difference of being and beings, is converted into an emphasis upon difference when Derrida recasts *Destruktion* as *déconstruction*. The emphasis put by the German antecedent of the latter on the unifying question of being as such in its difference from the being of beings motivates and is motivated by the importance Heidegger gives in *Kant and the Problem of Metaphysics* (1929) to the stress Kant puts in the first edition of the first *Critique* on the power to con-strue [*in-eins-bilden*], to imagine [*einbilden*], to construct or form or shape a world [*weltbilden*], understanding world as a limiting example or Exemplar of an *oikos* [house, household, dwelling], the unit of all economy, with an exception made for the impossible possibility when for the unity of the *oikos* creeping deconstruction substitutes the degenerate disunity of *chôra*.[1]

In the first place, deconstruction has to do with contrary oppositions—except that this first place, this first *oikos*, may turn out to be a second place or an nth place or a place that leads to the mise en abyme of an infinite trail of reflections of reflections. We shall mention also a last place before we are finished.

If we describe a contrary opposition as an antithesis understood as a *contra*-diction, we should not confuse it with an opposition or antithesis understood as an unrepulsive negation arrived at by simply prefixing a "non-" or adding a "not." To make that confusion would be to assimilate Derridean deconstruction too closely to a version of Hegelian spiritualist *Aufhebung* [sublimation, suspension] or its Marxian materialist inversion. It would isolate the particle "con" from its context in the word "deconstruction," thereby overloading it with the burden of synthesis. Synthesis is what deconstruction moves on, for instance from the work of synthesis of the "and" of the Kantian imagination, the conjunction of the sensible *and* the intelligible from which we move on in this essay.

Keeping in mind that imagination takes hold of sensibility by one hand (the left?) and of intelligibility by its other hand (the right?), one may suspect that the specter of Kant haunts the transition from a bias toward unification to an emphasis on dissolution of an indefinitely increasing multiplicity of seemingly antithetical oppositions, for example (as case) or for Example (as Exemplar or paradigm), sensibility or-and intelligibility, the polar terms of which get used by philosophers to found what appear to be mutually competing metaphysical doctrines based on oppositions such as the one just mentioned or used: mention-use, example-Exemplar, matter-form, interior-exterior, marginal-central, signifier-signified, nature-culture, body-mind–mind, sensation-reason, presence-absence, speaking-writing.

The first of these just listed pairs of supposedly antagonistic polar opposites could be read as an invitation to take sides in a debate between materialism of one kind or another and one kind or other of rationalism or idealism—for instance, one kind or other of conceptualist or-and Realist Platonism or Christianity. The last pair in that list could serve to draw attention to what has become the classical paradigm of deconstruction. So paradigmatically classical has it become that, because deconstruction is wary (but not condemnatory) of classicism, by its own criteria it is itself due for self-deconstruction.

Déconstruction in Derrida's construal of it as doubly hyphenated de-con-struction and as the deposition of opposition is cognate with his *se déconstruire* parsed as middle-voiced rather than as simply passive or reflexive, notwithstanding the historical fact that these latter forms replace the medial form at certain stages of the development of ancient Greek and Sanskrit. This historiological coincidence is perhaps a symptom of a grammatological coincidence of the active and passive voices expressing itself as, let us say provisionally, the voice of a *triton genos*, the voice of a third kind, allowance made for the vagaries and-or rigors of this word "kind" and of the concept of concept commonly assumed in our understanding of that word—and of classification. This is a concept of concept that is assumed in all the metaphysical pairings in our list. We should therefore add it explicitly as a component of, say, the couple "concept-intuition," so still under the watchful gaze of one of the specters of Kant.

Specters of Heidegger

The concept of concept just put in question is one whose use in the given pairings undergoes self-deconstruction in the writings of both Derrida and Heidegger, whether the unword "deconstruction" is spelled in the former's French or in the latter's German with a *k* on loan from *Destruktion*. Self-deconstruction calls into question the metaphysical opposition between the total presence attributed to hearing oneself speak and the detotalizing absence typical of writing. This is the opposition that deconstructs itself as early as Derrida's publications of 1962–72. Already in these places self-deconstruction deconstructs itself on the curious condition that if self-deconstruction deconstructs itself, self-deconstruction survives. In disappearing, apparently, it appears, and in appearing it appears to disappear.

Self-deconstruction takes place in the chiasm of what I follow Derrida in calling "something like the middle voice," meaning minimally by this what escapes the either-or of the active-passive but without becoming a neutral synthesis or a formal or dialectical contradiction (MP 9/9). Deconstruction is not simply the performance of an individual's act of free or constrained will. It happens. *Es geschieht*. It is arriving in the continuous tense heard in the final syllable of the neologism "deconstructance" (*Dekonstruktanz?*). It is arriving, but, like the Messiah, according to some believers, without yet having arrived and without having been actually present. It is always on its way, immer *unterwegs*, but without quite reaching its destination, assuming it has one. Hence if Novalis is right in holding that philosophy seeks to be at home everywhere, homesickness is philosophy's spur,[2] and philosophy as deconstruction is *à lieu*. Deconstruction takes place and takes place away. It does this, however, partly on account of forces over which one has only limited control, a limited agency something like that which Kant is said to have in at least some parts of his first *Critique*, according to proponents of the patchwork theory of its composition, and something like what Heidegger calls *lassen* as this verb operates in the expressions *Gelassenheit*, allowance or letting, and *Seinlassen*, letting be, but with acknowledgment of the reservations Heidegger comes to have over the fundamentality he had once attributed to the ontology implied by the first syllable of the word *Seinlassen*.

The limited passivity, passion, and patience of the force carried by the notion of self-deconstruction are also something like, yet something extremely unlike the modality of the movement of consciousness (*Bewusst-sein*) of which Hegel says in the *Phenomenology of Spirit* that all the phenomenologizing philosopher needs do is watch and describe that movement. It is as though *es denkt in mir*, "it thinks in me," and *ich denke in es*, "I think in it," allowance being made this time for the need to say more about this *es* and this *Es*, as Heidegger acknowledges, followed by Derrida, when this impersonal particle recurs in Heidegger's uses of the phrases *es gibt*, "it gives," "there is," and *es ereignet*, "it takes place."

I use the word "place" in this gloss of *es ereignet* as a cue for a further comment on what we have dubbed the "classic example and Exemplar" of deconstruction. I make the plausible assumption that "place" implies space of some sort, but I grant that plausibility is one of the things that is at stake here, along with a certain "perhaps," the *peut-être* that is not the expression of the possible existence of a thing but the expression of the condition "if there is any," *s'il y en a*, of the haunting of a specter, a ghost, like the one whose spirit is mentioned in the title of Hegel's *Phenomenology*.

Plato argues in his *Phaedrus* that language as spoken is closer to reality than language as written because the former is not dependent as the latter is on spacing. If to translate Plato's word "*chôrismos*," "spacing," we use Derrida's word "*espacement*," we see without hearing the vocalic difference between that French word and the French word written as "*espacement*." This difference, without being some *thing*, an entity, a being, an existent, is something like the difference that obtains between *différence* and *différance*. Both forms with the letter *a* in the final syllable make visible "something like the middle voice," and this is something like the verb-nominality of the German word "*Sein*" and the English word "Being."

Hints of the inheritance from Heidegger drawn on in the notion of deconstruction are evident at several places in the comments Derrida makes in the second volume of *The Beast & the Sovereign* on Heidegger's *The Fundamental Concepts of Metaphysics*, one of the factors that made the years 1929–30 such good ones in the field of publication in philosophy.[3] (Another such factor was Heidegger's inaugural lecture at Freiburg, "What Is Metaphysics?").[4] The subtitle of *The Fundamental Concepts* is *World, Finitude, Solitude*, but when the book was first delivered as a course of lectures it was announced as "World, Finitude, Individual," though with the original term "solitude" unerased. The first and at this point most relevant comment needing to be made about what we have been saying about the directionality, if there is any, of deconstruction is that the topic of the first part of the book proper is what Heidegger calls "profound boredom."

Die tiefe Langeweile [profound boredom] is a "fundamental attunement," a *Grundstimmung*. In the context of the course and the book of 1929–30 a *Stimmung* is a mood, something one finds oneself in, a *sich befinden* or *Befindlichkeit*. A mood comes over us. It imposes itself on us. It is not something one imposes on something else or on oneself. It is not a concept. So although the discussion of fundamental attunement takes place in the first part of *The Fundamental Concepts*, our understanding of that discussion is helped by looking ahead to section 70 of the final chapter. There, possibly for the last time in his writings, Heidegger touches on a certain topic that he has raised briefly elsewhere: the topic of *formale Anzeige* [formal indication].

Heidegger's "formal indication" corresponds with Kant's forms of sensibility in contrast to conceptuality and with Kant's notion of exhibition as contrasted with deduction. It is germane to what we have been saying about deconstruction. One is therefore surprised that, unless I have not read Derrida carefully enough (who has?), no explicit mention of it by him is made, at least in his comments on *The Fundamental Concepts*.

One thing the author of that book says in it is that some of the fundamental concepts referred to there are, strictly speaking, not concepts. It will be recalled—and cited by some readers as evidence for the patchwork theory of the composition of the first *Critique*—

that Kant sometimes refers to space and time as concepts, though in the same book he distinguishes space and time from concepts specifically so called [*Begriffe*], arguing that space and time are originally forms of intuition. They are therefore not conditions of the possibility of concepts or principles, but rather conditions of the possibility of indication. When Kant treats space and time as concepts he is thinking of space and time as objects of a science such as geometry or kinematics. Given that Heidegger had been working on his *Kantbuch* while preparing *The Fundamental Concepts*,[5] it is not surprising that he too makes a double use of the concept of concept in the section in which he considers formal indication. When he wants to distinguish the proper concept of concept he describes it as a scientific concept, where "scientific" stands for the broader term "*wissenschaftlich*."

It is arguable that to indicate is "*anzeichen*" and that Heidegger's word "*Anzeige*" would be more accurately translated by "ostension" or "pointing." Thus translated, the word would be the one used by Husserl to designate what he maintains his method of phenomenology can in principle dispense with. Hence it would be the word Derrida uses in his deconstructive reading of Husserl to call into question the latter's antithetical opposition of indication and expressive meaning (*ausdrückliche Bedeutung*) and his downgrading of the former. Husserl sometimes uses the one, sometimes the other, according to how he sees his phenomenology as the description of a theoretically ideal world or as the description of the world *ici-bas* to which that ideal applies in practice. But this is another of the apparent oppositions that Derridean deconstruction questions, simultaneously betraying a readiness to allow that Husserl himself questions it too, creating the impression that each of these philosophers is brought back to life in order that each may ventriloquize the voice of the other. Thus when Derrida writes that he is going to give as "internal" a reading as he is able to of, say, a Kantian, Husserlian, Heideggerian—or Derridean!—text, he expects to discover that the economy of the homely internality of the *chez soi* [at home] of the economy of the *oikos* is set aquaking by the visitation of some external, diseconomic, unsaving, sublime, *unheimlich* [uncanny], *ungeheuer* [unusual, extraordinary, monstrous] unthing of which it can be said in the sinister left-handed script of *chôra* not that it exists, but, partitatively, that *il y en a*: there is some of it about, as one says of influenza, whose name may go some way to explaining why the notion of influence causes Derrida dis-ease.

In question in deconstruction is the rigor of the specific concepts in question. But in question too in the relevant section of *The
Fundamental Concepts* are the powers of philosophical concepts
quite generally. Husserl's former student Heidegger brings this out
by starting with a reminder of the reference he makes in *Kant and
the Problem of Metaphysics* to Kant's invocation of a "dialectical
illusion" that is "certain, and indeed necessary."[6] More fundamental than that, Heidegger maintains, is another illusion, the illusion
from which ordinary understanding suffers of assuming that "philosophical knowledge of the essence of the world is not and never can
be an awareness of something present at hand." He questions this.[7]
The "something as something" that characterizes a proposition or
what a proposition says something about (its referent or its sense)
is a correct but superficial interpretation of a world's *Walten*, its
prevailing. It is also a symptom of a "natural idleness," a *natürliche
Behäbigkeit*, as Kant writes, reminiscent of what in *Being and Time*
Heidegger names *Verfallen*, "falling." This philosophical counterpart in both Kant and Heidegger of original sin is a laziness that levels off all knowledge to present at hand relationality. It seduces the
human mind from acknowledging that a deeper understanding of
the "as" is possible only if each individual person undergoes a transformation from, on the one hand, taking for granted that knowledge
of worldhood is no more than awareness of the "as" as presence at
hand to, on the other hand, enduring the existentiell experience of
the existentials analyzed in *Being and Time* and the deep boredom
analyzed in *Fundamental Concepts*. This transformation is the endurance of the duration analyzed under the name of deep boredom
[long-whiling] and of "the moment of vision," the *Augenblick* (to
translate Kierkegaard's word) in which each *jemeinig* [in each case
mine] human being in its individuality faces the facticity of death.
But whose death? This is a serious question.

When is a question serious? According to Condillac, as reported
by Derrida in *L'archéologie du frivole: Lire Condillac*, serious discourse is the use of a signifier that designates a signified meaning or
referent and is enabled thereby to convey information.[8] For instance,
the proposition "two plus two make four" is serious, whereas "two
plus two make two plus two" is frivolous or, as Locke would say,
trifling. It lacks archaeology. The bedrock of archaeology is an idea
or an object. But in "two plus two make two plus two," counter to
the allegation, nothing is made.

It is, however, the very same John Locke whom Condillac charges with failure to acknowledge that there is no liaison between ideas unless there is liaison between signs. Derrida takes this complaint a stage further when he endorses Peirce's teaching that a signified idea occupies the *oikos* of a signifier. In the guise of a signifier, what is signified gains eternally iterable new life. Endorsement of this is tantamount to recognition that the alleged antithetical opposition of the signifier and the signified de-signates itself, signs in unsigning, so *se déconstruit*. This slide to a deconstruction of the opposition of the serious and the frivolous is accelerated by the possibility that, in addition to the signified's becoming a signifier, the signifier may become a signified. This complication is one reason Derrida has qualms over the opposition of mention and use and the notion of metalanguage.

Derrida does not make very explicit in his book on Condillac what he means by "deconstruction." The word "deconstruction" occurs on only two widely separated pages of it (AC 63/55, 132/118). *Specters of Marx* is remarkable for a different reason. The word occurs scores of times in it, but the reader would be as hard put to track down a definition of it there as he would be to do so in *L'archéologie du frivole*. The reason for this is implicit in what we have said previously about concepts. Like difference, deconstruction is not a concept or a concept's referent. Nor is it a word. It is more like (if it is like anything) a process of referring, something performed in or by a text. This does not mean that—unlike the father figure in the *Phaedrus* who is said to guide the listener toward presence—an interpreter of the performance and the text cannot guide the reader, if only by pointing the latter in the direction of other texts. For example, this one:

> Doch im Erstarren such' ich nicht mein Heil,
> Das Schaudern ist der Menschheit bestes Teil;
> Wie auch die Welt ihm das Gefühl verteure,
> Ergriffen, fühlt er tief das Ungeheure.
>
> Yet not in torpor could I comfort find;
> Awe is the finest portion of mankind;
> However scarce the world may make this sense
> In awe one feels profoundly the immense.[9]

By awe one is *ergriffen*, gripped. One does not grasp awe in a concept, a full-fisted *Begriff*.

Faust still has much to learn, however, if he believes that what he hopes to find is comfort or his personal salvation. As we shall learn soon, the fear and trembling caused by the thought of the risk that I may fall short of realizing my wholeness is infinitely less distressing than the certain knowledge that I have fallen short of my responsibility to my neighbor or enemy. As Kant learns, the awe he knows by the name of the sublime is not comfortable. It is at odds with quietude of conscience, as Heidegger also knows when he writes of the voice of conscience that it comes from within but also from over and beyond oneself. As survivors in more than one sense of that word, the specters of Levinas also have reason to think.

Specters of Levinas

Reference was made previously to the proximity of the speaker listening to himself that is Plato's paradigm of serious discourse. This is a proximity in which the identity of the person hearing himself speak is somehow separate from the identity of the person speaking. There must be that separation if the person sending the message is in some way truly other and the discourse is serious. There must be that separation in personal identity if seriousness defined in terms of what is signified, as it is defined by Condillac, has to be supplemented by seriousness defined in terms of the signifying response that the human signifier makes to the person addressed as "you" when to the latter's call for help she or he replies, "Here I am. Let me stand in your stead, in your place, your *oikos*." This must be if what Levinas titles "humanism of the other human being" is to be otherwise than being and beyond essence, *epekeina tês ousias* [beyond being], meta-physical and properly ethical—which is to say im-proper insofar as what is proper to me and part of my property cannot be constitutive of my ethicality. My ethicality is constituted by being deconstituted, de-con-structed, when the interiority of the identity of myself as signifying respondent is disrupted by the exteriority announced in the subtitle ("An Essay on Exteriority") of *Totality and Infinity* and this title announces a chiasmic intrusion of infinity into totality.

The out-ness of this in-ness in the dialogue of the soul with itself cannot fail to have been the soul's own out-ness a priori, because this apriority is inseparable from an aposteriority. Its in-ness is inseparable from its out-ness. To say this is not to embrace a logical contradiction, any more than a logical contradiction was embraced when the shop assistant turned that "deconstructed" jacket inside-out and outside-in. Even her concomitant sales talk was capable of contravening the principle of contradiction only if in it there was a juxtaposition of explicit or implicit propositions. But in the case of the interiority and exteriority of the dialogue of the soul with itself, the exteriority is not merely the exteriority of one proposition to another. Any such relation as this is possible only on the supposition of a proto-relation between the person who signs and addresses a message, the *destinateur*, and the one or more than one person addressed, the *destinataire(s)*. The proto-relation is personal and interpersonal. This means that although I may regard the person with whom I speak in soliloquy as my alter ego, I must be capable of regarding this alter ego as my other *tout court* on analogy with another person with whom I am face-to-face in public space, or could have been were it not for contingencies such as that of the other's being spatially remote from me or deceased or not yet born. This is why we tend to think of the *destinateur* and the *destinataire* as one person writing to another. But a little reflection assisted by Derrida reveals that the situation of one person talking to another is in principle no different from the situation of soliloquy. What we ordinarily refer to as the "spoken word" shares this likeness with what we ordinarily refer to as the "written word." Derrida marks this likeness by introducing into his and our lexicon the term *"archi-écriture"* [archi-writing], going as far as to say that this introduction admits a trace of materialism, perhaps one of the sorts of materialism to which reference was made toward the end of the first section in this essay and contrasted there with possibly Platonic sorts of rationalism and idealism.

Our judgment of the philosophical importance of Derrida's neographism will depend on our judgment of the importance of the doctrine it questions: Plato's doctrine that it is in speech that language and thought are originally at home. Could it be that the alleged origin of language and thought is not originary and that speech exemplifies the same graphic features that in the *Phaedrus* and elsewhere Plato regards as secondary or derivative? Although his epithets "secondary" and "derivative" carry the implication of something else that by contrast is first and aboriginal, it is this implication that the notion of "archi-writing" seeks to avoid by combining the sense of the Greek *archi*—that is to say, the sense of commencing, commanding, and leading [*führend*], with the sense of inferiority carried by the notion of writing as understood in the "phenomenological" philosophies, as Derrida describes them, of Plato and the "Platonism" of Rousseau, Condillac, Hegel, Husserl, de Saussure, and so on.

The distancing between sender and recipient most commonly associated with writing as ordinarily understood and contrasted with speaking is most graphically displayed in such correspondence as is effected in the exchanges that take up most of the space of *La carte postale* and the Talmudesque columns of *Glas*.[10] Derrida never tailored a more patchy patchwork than these unbookish construals of the book of the world, this world or another. But what matters philosophically more than the writing traced on the pages of these material texts is the archi-writing these writings exemplify. This is because the analogical or anagraphical pairs among those we listed previously must be supplemented not only, as we saw, by that of concept-intuition, but also by space-time. This is indicated when Derrida asks, "*Différance* as temporisation, *différance* as spacing. How are they to be joined?" (MP 9/9) and answers, by way of archiwriting. This answer raises another question: how can the positing of archi-writing as a quasi-condition of the possibility and impossibility of familiar oppositions be consistent with Derrida's frequent warnings against programming? Isn't Egyptio (pyramidoikol)-Greco-Romano pro-gramming synonymous with Egyptio (pyramidoikol)-Greco-Romano-archigraphy?

When Derrida evinces distrust of programming, he is thinking of justice as defined solely in terms of principles from which one calculates consequences in a world where what is given first importance is happiness or well-being, *eudaimonia* kept safe by a guardian angelic *daimon*. Derrida is listening to the voice of a specter of Immanuel Kant bidding us not to forget that hypothetical imperatives call to be subordinated to imperatives commanding autonomy. Derrida is listening also to the specter of Emmanuel Levinas bidding us not to forget imperatives commanding a heteronomy other than the heteronomy that is subordinated to autonomy by Kant.

So to the question of the contribution deconstruction makes to our thinking about questions of ecology, the answer seems to turn in part on the answer to the question of what contribution Levinas's thinking makes to Derrida's and our deconstrual of deconstruction. That latter contribution is minimal if we limit it to the thinking of the humanism of the other human being. In order that it be more than this minimum, a way must be found to include nonhuman beings among the others to which ethical responsibility is owed. A start on that way could be made, I have argued elsewhere,[11] by assenting to what I call a "blank ecology," meaning by that an ecology that defers consideration of the predicates of things to restrict one's attention temporarily only to existents as such in their existence, allowing ourselves to be struck by the consideration that for any given existent its existence is a good, at least for that existent. To be thus struck is to be party to the realization that each and every existing thing has a foothold in the realm of ethical considerability in Levinas's sense of the ethical as meta-physical, beyond being. Of course, it is we human beings who do the consideration and speak, or don't, as advocates for whom- or whatever has no voice of its own. The responsibility is, as Goethe's Faust says, immense, *ungeheuer*. As Levinas says, we are persecuted by it.

Levinas's conception of being persecuted by our responsibility for the other resounds in Derrida's seemingly contradictory notion of a responsibility that is both welcomed and unwelcome or "unwelcome." But it can hardly be denied that the voice in which Derrida expresses his welcome of unwelcome responsibility takes on a tone peculiar to him when his talk of archi-writing and the archi-trace is succeeded by talk of what we could call "archi-calculation." We are licensed to employ this appellation by his reference in *The Gift of Death* to "a calculation that claims to go beyond calculation, beyond the totality of the calculable as a finite totality of the same" (GD 107/145). In these words we hear a reference to the infinity of the different, the other, and the exteriority explicit or implicit in the title and subtitle of Levinas's chef d'oeuvre.

Levinas maintains that his philosophical writings are "Greek," meaning in part by this that they are not based on any oriental religious scripture. Derrida would maintain this regarding his own *Gift of Death*, notwithstanding that that book's early chapters are triggered by the genealogy of Christianity that forms part of Jan Patočka's *Heretical Essays on the Philosophy of History* and that the exergue borrowed from Nietzsche's *Genealogy of Morals* for his last chapter refers to "that stroke of genius called Christianity." What's more, that chapter proceeds to cite the Gospel according to Matthew, including the part to which Levinas takes exception because he sees it as a threat to the extremity of alterity he wishes to maintain for the dimension of ethicality. The part in question is the one requiring love of one's neighbor and enemy "as oneself." This, he maintains, runs the risk of sullying the love of one's enemy by assimilating *agapê* and its Hebrew approximation *ahava*, which is dispossessive of oneself, gift, sacrifice, to erotic self-possession. But the safeguard against this cannot ignore the fact that a residual risk of this sort is presented by the needs that, even when they are another's, are still needs of an eco-nomic kind that graft themselves on to desires—for instance, those desires analyzed in the parts of *Totality and Infinity* that treat not primarily of the ethical but of enjoyment, of the prudent and of the dwelling—that is to say, of the *oikos* treated by ecology.

Eco-Choro-Spectrography

An ecology is a theory of the *oikos*: the home, the house, the household, the family, the political state, and ultimately the world, this one or another. An ecology, writ large, is a cosmology—for instance, the one outlined in Plato's *Timaeus* and palimpsestically written "on" by Derrida in *Khōra*. Anyone concerned to defend as wide, as just, and as generous as possible a conception of the electorate of a household or worldhold might object that although the blank ecology defined previously may admit of being jointed to the chorography to be de-fined later, it would still be inhospitably restricted.

To this objection it may be said that a blank ecology would be less restricted than the ecology of the human being—this human being here, the autobiographical *je-meinig, dasein*-ish me, or of that human being over there, the other I address as "you." Would the existents that make up the electorate of a blank ecology include specters—for instance, those of Kant, Heidegger, Levinas, Derrida, and Marx? In *Specters of Marx* Derrida asks, "What is the *being-there* of a specter? What is the mode of presence of a specter?" (SM 38/69). The term in italics translates Heidegger's word *Da-sein*, and the response Derrida makes to these questions is a challenge to Heidegger's ontology and the confidence with which Heidegger founds it on conceptions of presence [*Anwesenheit*] and absence regarded as mutually exclusive opposites. Specters, ghosts, the dead haunt those who live on. We are unable to say where their *there* is, where resides the being of their *Da*, and whether or how they exist. Instead of affirming or denying existence of them, Derrida sometimes has recourse to the conditional clause "*s'il y en a,*" "if there is any." Here the verb is not "to exist" but "to be," the singularly perchance to be of the "to be or not to be" of an ontology that has become and has always been becoming a hauntology, as hinted in the experience of mourning the death of a friend either retrospectively or prospectively. Spectrality is part of this forward and backward spection. Defined in light of its derivation from "spectrum," it is, according to the Oxford English Dictionary, the "image of something seen continuing when the eyes are closed or turned away." It is the seriousness in friendship and love that, without being simply opposed to frivolity, makes them matters of life and death.

The paradigm of spection here is that of the unexpected ghost of Hamlet's father, whose exspective eyes are hidden by his visor, though it is this through which he looks to his son to shoulder the responsibility of putting the disjointed world to right. Except that, as I argue in defense of a more generously accommodating ecology, the accusing look may come not only from another human being but from any being on account of the death or destruction of whom or of which it may be mourned. Here Derrida's invocation of Shakespeare recalls Levinas's remark that the whole of philosophy may be regarded as a meditation on Shakespeare.[12] It recalls too Derrida's refusal to oppose logos and philosophy antithetically to mythos and literature, as Derrida maintains John Austin does when he distinguishes speech acts uttered in the real world from those uttered on a stage. Logos-mythos is another pair that should be added to our list of suspects, a pairing to which Derrida says Plato says both "Yes" and "No" insofar as the logical and cosmological analysis conducted in the *Timaeus* has as its topic a story deriving from hearsay that serious philosophers might justifiably consider to be as frivolous as Kemp Smith's scholarly colleague A. E. Taylor considered Plato's *Parmenides* to be.

What is *chôra*? It is the unwhat that facilitates the unsyntheticality and unsynthethicality (*sic*) of the terms in each of the apparently oppositional pairs catalogued in the foregoing pages. Plato authorizes us to compare incomparable *chôra* with a midwife who assists at a birth. A birth is supposed to lead to another birth ("be fruitful and multiply") or to a rebirth. Yet *chôra* itself is not generated or generative. It is not an itself, not a definite article or denoted by one. Yet it is a process, the one described when Derrida writes of *chôra*: "It 'is' nothing other than the sum or the process of what comes to inscribe itself 'on' its subject, but it is not the *subject* or the *present support* of all the interpretations, although, nevertheless, it is not reduced to these." The words "sum or the process of what comes to inscribe itself" suggest a patchwork. Or is the process one that, getting close to a direct contradiction of the law (nomos) of noncontradiction, Derrida calls "the economy of difference" (MP 18/19)? How close to direct contradiction this phrase gets turns on how seriously we take the last syllable of its last word. It turns on what difference it makes whether the word is written as "ence" or as "ance." Taken in the former way, "difference" could be in tension with "economy" because it leaves open the possibility that the terms of the difference relate to one and the same point in space and time that is the condition of the possibility of mutual contradiction. This condition is not met if the last syllable of the word is taken as "ance." In being taken thus, allowance is made for temporal lapse or spatial differentiation (*mouvance*), and that allows in turn for the joint possibility of its being both true and false that, for example, the sun is shining. Hegel the phenomenologist would want us to add the adverbial modification "here and now" to the proposition "the sun is shining." Kant declines to build any essential spatiotemporal condition into the law of noncontradiction on the grounds that this law has to do with propositions represented not by, say, "the sun is shining" with a tensed verb "to be," but by the symbols "p" or "q" used timelessly. The requisite abstractness is lost if the law is taken to imply spatiotemporal conditions. Their function is taken care of according to Kant by the process of schematism of imagination. Overlying or underlying that process according to Derrida is the process of *chôra* in which "one is but the other different and deferred, one differing and differing the other. One is the other in *différance*, one is the *différance* of the other" (MP 18/19–20). This "of" marks perhaps a double (so-called subjective and objective) genitive that articulates the spatiotemporal ferance from one oikogrammatical architecture to another.

But we must pause to reflect on the different ways "take" is taken in the preceding paragraph. There is the (Kantian) take of our taking something so abstractly that they ignore the difference between "difference" and "différance" as seen and this pair as heard. This sensory difference is crucial. It is crucial because, in the wake of Kierkegaard, it commits us to make a decision that is mad. It renders undecidable the question of whether when we encounter "difference" on a page written by Derrida it behaves according to the seemingly safe logic of that word spelled with the ending "ence" or whether it behaves or misbehaves according to the dangerous logic of the word spelled with the ending "ance." We do not know which ending the word has because we cannot settle whether the word is one that is written or one that is heard—that is to say, whether it is voiced, whether it is a determinate fundamental or anarchic *Be-*

stimmung. No matter how Derrida himself or his specter may spell out or shout out this word (*Stimmung* is tempered voice), the word's identity remains undecidable

That is to say, *chôra* (*s'il y en a*) would seem to be the unsovereign paradigm that is so like material substance (*hyle*) that Aristotle equated the latter with Plato's conception of the former. But *chôra* has no identity or likeness. Like incomparable Socrates, the *tympanum* of whose ear is receptive to whatever it is struck by, *chôra* is the recipient of every likeness without being a likeness itself and without being a self-identical self, without being a self, without being. So, to extend the sense and sensibility and responsibility of the metaphor, the retinal tabula of the eye of *chôra* (its, his, or hers, for *chôra* is the recipient of "feminine" recipience), like a watermarked banknote bearing a signed promise of what it will pay to the bearer, accepts acceptance as such. It therefore accepts such autobiographical remarks as those made in the first sections of this essay tracing the ungenerative and ungenerated genealogy of some of the debts Derridean deconstruction owes to Kant, Heidegger, and Levinas and now Plato and Socrates, the most recipient and finest-eared thinker of them all.

The autobiographical is always heterobiographical. It responds to the other addressed as "you" historiographically according to the contingent demands of justice. These demands are never finally decided. They never cease to stretch the ethicopolitical imagination, the faculty of the fusion of faculties that becomes the faculty of the conflict of faculties when, supplemented by the initial letter of the name of Derrida and the unname of deconstruction, it is homophonically called "imadgination" (*sic*) in language that Heidegger called the "house of being" but that now and at the threshold to and from that house has revealed itself to be the house also of more or less than being. So the whole of the ghost story we have been retelling— if it is a whole (*s'il y en a*) rather than a patchwork that detotalizes totality—is indebted not only a posteriori to the three philosophers singled out in parts of that story. That story is indebted also a priori to its readers (*s'il y en a*). It is indebted to everyone who took part in the symposium to which this story attempts to contribute the afterthought that, if deconstruction is chorographic in the way that, borrowing the bastard language of the *Timaeus*, Derrida signals it may be, the house and home and pluriverse that are ultimately the topic of ecology are, as Heidegger's ghost goes on telling us, unhomely, uncanny, *unheimlich*, and as the ghosts of Goethe and Nietzsche go on telling us, not profoundly *langweilig*, but profoundly *ungeheuer*, never to be put to right, but ever fit to make us shudder and, when philosophy makes a serious study of the frivolous in its relation to the serious, fit to make us laugh.

Postscript

With Novalis's statement that the philosopher seeks to be at home everywhere, some will agree on the grounds of their belief that home is ultimately what they call the "hereafter," the "next world," the "afterlife," or the "heaven" they hope they are meant for. In the context of the simultaneously widespread belief in the threat of eternal torment, who would begrudge even Bad King John the prayer with which he ends his last will and testament: to be saved from such incessant pain? Even so, that prayer is self-centered. Furthermore, it is inconsistent with the other-centeredness that distinguishes Christianity as defined by, for example, the Gospel of Matthew cited by Derrida.

But Derrida's chorographic deconstruction of space-time outlined previously opens to Christianity the chance of being saved from this charge of inconsistency. Departing from the predominantly chronological image of temporality adhered to by Kant, absorbing Heidegger's *Destruktion* of that account, and taking to heart Levinas's "reversal" of the priority Heidegger gives to one's own temporality over the time of the other, Derrida's chorographic deconstruction of place in space and time (*oikos*) offers something like a "formal indication" of an eco-eschatography that leads one to wonder whether heaven and hell are not antithetically opposed and whether their "where" (*s'il y en a*), rather than being subsequent to a prepositional "after," is always already adverbially inscribed in a renewing retrieval of love of one's neighbor and enemy wherever they may be and therefore jointed with the immense hellish torment of being hurt by their being hurt and by the hurt of our knowledge that we have failed to live up to the hardly imaginable exemplarity of sermons on the Mount and elsewhere, even when succeeding may mean only, as winter approaches, offering the widow, the orphan, and the stranger the warmth of one's perhaps de-con-structed brand-new coat.

Notes

1. Cf. Jacques Derrida, *Khôra* (Paris: Galilée, 1993); trans. Ian McLeod as "Khōra," in *On the Name* (Stanford: Stanford University Press, 1995).

2. Martin Heidegger, *Die Grundbegriffe der Metaphysik: Welt, Endlichkeit, Einsamkeit* (Frankfurt am Main: Vittorio Klostermann, 1983); trans. William McNeill and Nicholas Walker as *The Fundamental Concepts of Metaphysics: World, Finitude, Solitude* (Bloomington: Indiana University Press, 1995).

3. Ibid.

4. Heidegger, *Wegmarken* (Frankfurt am Main: Vittorio Klostermann, 1976); ed. Willam McNeill as *Pathmarks* (Cambridge: Cambridge University Press, 1998).

5. Heidegger, *Kant und das Problem der Metaphysik* (Frankfurt am Main: Vittorio Klostermann, 1991); trans. Richard Taft as *Kant and the Problem of Metaphysics*, 5th ed., enlarged (Bloomington: Indiana University Press, 1997).

6. Immanuel Kant, *Critique of Pure Reason*, trans. Norman Kemp Smith (London: Macmillan, 1968), A 61–62.

7. Heidegger, *Fundamental Concepts of Metaphysics*, 292.

8. Derrida, *The Archaeology of the Frivolous* (Lincoln: University of Nebraska Press, 1980), trans. John P. Leavey Jr., from *L'archéologie du frivole* (Paris: Galilée, 1990).

9. Johann Wolfgang von Goethe, *Faust*, part 2, ll. 6271–75, trans. Walter Arndt (New York: Norton, 1976).

10. Cf. PC and G.

11. John Llewelyn, *Margins of Religion: Between Kierkegaard and Derrida* (Bloomington: Indiana University Press, 2009), and Llewelyn, *The Rigor of a Certain Inhumanity: Towards a Wider Suffrage* (Bloomington: Indiana University Press, 2012).

12. Emmanuel Levinas, *Time and the Other*, trans. Richard A. Cohen (Pittsburgh: Duquesne University Press, 1987), 72; *Le temps et l'autre* (Paris: Presses universitaires de France, 1979), 60.

Nuclear and Other Biodegradabilities

8

E-Phemera: Of Deconstruction, Biodegradability, and Nuclear War

Michael Naas

/?ie=UTF8&keywords=biodegradable+sun+screen&tag=googhydr20
&index7ymv2vzq10_ehttp://decorativeurns.com/urnmaterial/biod
egradableurns?gclid=CL_8mR58QCFQcOaQodPIgAHAhttps://land-
ing.honest.com/diapertrial?sid=10001&cid=google&mid=search&aid
=Diapers%20%28Exact%29:Biodegradable%20%28t%2http://www
.bpiworld.org/http://www.reverteplastics.com/?gclid=UTF8&keywor
ds=biodegradeable+balloons&tag=googhydr0&index=aps&hvadid=3u
w0d81suk_bhttp://www.greensafestore.com/?utm_campaign=Ecofrie
ndly+Products&utm_source=google&utm_medium=ppc&utm_term
=%2Bbiodegradable+%2Bbags&utm_content=1058377x66629688133
?ie=UTF8&keywords=biodegradable+soap+bar&tag=googhydr20&ind
ex=aps&hvadid=70821257129&hvpos=1t3&hvexid=&hvnJACQUES
DERRIDA/BIODEGRADABLES:SEVENDIARYFRAGMENTShttp://
criticalinquiry.uchicago.edu/uploads/pdf/13436921.pdf:http://=1t2&
jk=biodegradableconfetti&jkId=UTF8&keywords=biodegradable+k+
cups&tag=googhydr20&index_ehttp://greenpaperproducts.com/bio-
degradabletrashbagskitchen101.aspx?gclid=CP_XhYOGunnkhttp://
ecogolfballs.com/Gg/40707/greenbiodegradableplasticdinnerware
.htmlhttp://www.webstaurantstore.com/2923/green-biodegradable
-compostable-plastic-take-out-containers.html.
—*Remains of a biodegradable word search*

"Biodegradables": it's first of all a name, very googleable and attached
to a whole range of industrial products available both online and in
stores, everything from diapers and trash bags to balloons, confetti,
and funeral urns. But it's also the name or title of a 1989 article by
Jacques Derrida, just one among hundreds of other Derrida articles,

of course, but also one of just a handful of Derrida works with a legitimate claim to being absolutely *central* to anything that might fly under the banner of "Eco-Deconstruction." For "Biodegradables" contains some of Derrida's most explicit and developed reflections on ecology and the environment, as well as on the related questions of life, survival, waste, remains, and what will remain—or not— after the end.

And yet this work of 1989 has remained largely unread since its publication over a quarter of a century ago, running the risk of going the way of the very best of biodegradable products—the risk, that is, of decomposing to the point of becoming unrecognizable, unread-able, or at least unread, just part of the vast and ever-growing com-post pile that circulates today under the general rubric of "Theory," its distinctive voice all but drowned out by the incessant chatter or white noise of our contemporary archive, "like the sound of the sea deep within a shell." If I feel compelled, therefore, to return here to this rarely read Derrida work—in the company, I am happy to note, of Michael Peterson's superb contribution to this same volume—it is to try to do my part to help save it from loss, from forgetting, from all those processes that collect the daily debris of our scholarly lives and deposit it, with or without recycling, in some common landfill. It is to help save it so that it might help us today to rethink questions of biodegradability, remains, survival, life, and, perhaps especially, the relationship between ecology and war that is at the center of it, to help save it, then, so that we ourselves might become haunted and provoked by something buried deep within it, some-thing that must not be forgotten today since it concerns nothing less than the whole of life itself.

To use a term that Derrida will at once use and comment on in this text, "Biodegradables" is a rather strange "artefact" in the Der-ridean corpus. It is a long piece (some sixty pages), and yet it is un-known to most readers of Derrida and rarely commented on even by those who know his work well.[1] There are several good reasons for this. First, although this work was, like all of Derrida's works, written in French, it has, to my knowledge, never been published in its original language. It is thus available today only in translation, indeed, again as far as I know, only in Peggy Kamuf's English trans-lation. Second, while the essay was published in an important and widely circulated journal, *Critical Inquiry*, it was not republished or included in any of the collections of Derrida's essays before the 2013 volume *Signature Derrida*, a compilation of all of Derrida's essays originally published in *Critical Inquiry*.[2]

But it is not just the language and venue that seem to have doomed this work, at least for a time, to forgetting. Despite its promising title for questions of ecology or environmentalism, the work appears in a section of that particular issue of *Critical Inquiry* entitled, "On Jacques Derrida's 'Paul de Man's War,'" a title that sounds a lot less promising for such topics and that is, in fact, perfectly descriptive of the work's content. For much of "Biodegradables" is indeed about Paul de Man. It is the seventh and final text in this section of the journal devoted to questions that emerged the previous year following the publication of Paul de Man's recently discovered wartime journalism and Derrida's initial response to that publication, it too first published in *Critical Inquiry*. In addition, then, to a number of long reflections on the notions of biodegradability, survival, the trace, and so on, much of "Biodegradables" consists of Derrida's spirited defense of his earlier essay against the attacks of several scholars who accuse Derrida, in essence, of trying to deny, whitewash, justify, or explain away the anti-Semitic language of some of de Man's wartime essays.

Without entering here into that enormous debate over de Man's wartime writings or into the related controversy over Derrida's defense, if it is a defense, of those writings, it is no doubt in large part these reflections on de Man that have contributed to the obscurity into which Derrida's article has fallen or, rather, the biodegradability to which it has become subject. For this context can easily give the casual reader the impression that "Biodegradables" is simply an *occasional* piece, one written in response to a current debate or controversy (the so-called "de Man Affair") that might have been compelling at the time but is of little more than historical interest today, like a newspaper article or piece of journalism that commands our close attention the day after it is written but is largely ignored thereafter.

A strange artefact, then, an odd hybrid, a rather strange mishmash: "Biodegradables" is one of Derrida's least readily available and least-read works and, as a result, among those that would seem to have the least chance of surviving. But then it would seem that Derrida's choice of form and venue corresponds rather well to the theme of the "biodegradable" itself. Written not in the form of a book or treatise or even as an essay but as "seven diary fragments"—the essay's subtitle—the text seems to invite the criticism of being, precisely, *dated*. Written between Saturday, December 24, 1988, and Saturday, February 4, 1989, the seven long fragments that constitute "Biodegradables" bear the dates of seven consecutive Saturdays, a week of Saturdays, as it were, perhaps even a week of Sabbaths, straddling the last two weeks of 1988 and the first five weeks of 1989. Written each time after the end of the work week, as if to intimate that we not mistake these mere diary entries for some serious work of philosophy, "Biodegradables" seems to invite being read and then put aside, taken seriously for a day or two and then discarded or, if we are ecology-minded, recycled. In short, "Biodegradables" appears to have been written as an essentially *ephemeral* text.

It would appear, then, more than a quarter of a century later, that Derrida chose to weave his most sustained reflections about the biodegradable, his most sustained reflections, we might say today, about sustainability, into a text that was dated and so could today appear much less worthy of being retained or preserved than many of his other writings, which concern important topics in the history of philosophy, oftentimes in its most canonical figures, rather than reflections on the writings, the wartime *journalism*, no less, of a contemporary and friend. If Derrida will ask throughout "Biodegradables" about the status that is to be granted to his friend's wartime journalism, works written decades before de Man's best-known theoretical works, this text seems to invite the very same question about Derrida's own work—this strange artefact that not only designates the biodegradable as its subject but seems, in its very form, its very existence, to pose the question of the biodegradable in relation to questions of writing, the archive, remains, and waste, as well as life and survival. For at issue in this work is the status that is to be granted to biodegradables and to "Biodegradables," to biodegradable or nonbiodegradable remains, to the length and form of their survival—that is, whether they will continue to command our attention and our care, for good or ill, as clearly identifiable things, or whether they will continue to haunt or animate us from some more anonymous, subterranean source. But to get to these questions irradiating from deep within the text, from their nuclear core, as it were, a bit of excavation and analysis js necessary.

Derrida begins his seven diary fragments by saying that he has long been interested in the "biodegradable"—that is, he says, in both the *word* and the *thing*, especially since, in this case, it is difficult to distinguish between them (BSD 813). As he puts it just a page or so later, "I wouldn't know how to qualify or delimit my interest in the question of the 'biodegradable': scientific interest? philosophical? ethico-ecological? political? rhetorical? poetic? prag(ram)matological?" (BSD 814). As for the theme itself, it seems to have been, as is always the case with Derrida, recycled from elsewhere, borrowed from someone else, indeed, in this case, from Paul de Man himself. It is de Man himself, in effect, who raises the question of biodegradability, the question of the decomposition or oblivion of literary works in one of the very same wartime articles at the center of the controversy Derrida is addressing. De Man raises this question in an article of November 11, 1941, published in the Belgian newspaper *Le Soir* on *Solstice de juin*, the recently published book by the French writer Henry de Montherlant. De Man there cites a phrase from Montherlant that seems to him to express well the oblivion—the biodegradability, as it were—to which the literature of those times seemed to be destined.

> In this collection of essays by Montherlant, there is a phrase that all those who have followed literary publication since August 1940 will approve. It is the passage that says: "To the writers who have given too much to current affairs for the last few months, I predict, for that part of their work, the most complete oblivion. When I open the newspapers and journals of today, I hear the indifference of the future rolling over them, just as one hears the sound of the sea when one holds certain seashells up to the ear." One could not have put it any better. And this just and severe sentence applies to all the books and essays in which writers offer us their reflections on war and its consequences, including *Solstice de juin* itself. (MPD 187/176; cited again at BSD 864)

It was de Man himself, therefore, who first raised the question of the decomposition or biodegradability of literary works and who, perhaps unwittingly, also demonstrated the unpredictability of a work's decomposition or survival—Montherlant's but perhaps also his own. In his initial response to the publication of de Man's wartime journalism, first published in *Critical Inquiry* under what we now see to be the recycled title "Like the Sound of the Sea Deep within a Shell: Paul de Man's War," Derrida interrogates, among other things, this passage from de Man, which seems to have wagered so fearlessly on the forgetting of Montherlant's work, as well as, perhaps, his own.[3] Derrida writes of de Man's article in *Le Soir*:

> It begins by quoting, as if in epigraph and in order to autho-rize itself, a remark by Montherlant. Then it turns it against him with an irony whose pitiless lucidity, alas (too much lu-cidity, not enough lucidity, blindly lucid), spares no one, not even de Man almost a half century later. Writing by profes-sion on current affairs, he deals with a current affair in this domain and he announces the oblivion promised those who devote their *literature* to current affairs. Do not these lines, that name "the worst," become unforgettable from then on? It is frightening to think that de Man might have handled so coldly the double-edged blade, while perhaps expecting "the worst." (MPD 187/175–76)

From Montherlant to de Man to Derrida and beyond, who could have predicted, for example, that a phrase that speaks of oblivion and anonymity ("like the sound of the sea deep within a shell") would come to survive and live so many second lives, a phrase that, like a signature or proper name, would have come to resist, it seems, the anonymity and forgetting of which it speaks and to which de Man thought it had been doomed?

From the very beginning of "Biodegradables," then, Derrida at-tempts to think, at once with and beyond de Man, the relationship between, on the one hand, the concept of biodegradability, the kinds of things we typically call "biodegradable" (diapers, balloons, fu-neral urns, and so on) and the word "biodegradable," the term as a piece of language, but then also, of course, by extension, language, or the text more generally. By raising "the question of the remainder in general" (BSD 816), Derrida at once proposes a relationship between the biodegradability of things and the (bio)degradability of culture or of writing. What we get, as a result, is an exercise in philosophical speculation, an experiment as well as a provocation and a wager, one that will teach us a good deal, if we are willing to hold it up to our ear, about biodegradability itself, to be sure, but also about the re-lationship between word and thing more generally, about Derrida's approach to these topics, and, thus, about the nature of deconstruc-tion itself, its themes and its interests, perhaps even its obsessions and its fears.

As for the *thing*, the "biodegradable" thing, it is important to note, Derrida recalls, that we typically do not call a natural thing that decomposes of itself "biodegradable" but only an artificial thing that "lets itself be decomposed by microorganisms" in relatively short order (BSD 813), an industrial product that quickly "returns to organic nature while losing there its artificial identity" (BSD 828). Strictly speaking, then, the biodegradable thing is not the naturally organic thing, the plant or animal body, for example, which decomposes of itself and without technical design. For the same reason, we do not usually call "nonbiodegradable" the natural thing, whether rock or mineral, or rare-earth element, that remains more or less intact after many years, more or less unaffected by microorganisms. The nonbiodegradable thing is thus also an industrial product, but one that biodegrades much less quickly than other things, one that *remains* long after it has been used and discarded, for decades, even centuries, or more—the disposable diaper, the plastic water bottle, but also, and this would seem to be the limit case for Derrida, the nuclear waste that can take between twenty-four thousand years (in the case of plutonium) and four billion years (in the case of uranium 238) to "degrade" or, as we say, to "decay" by half (BSD 836).

"Biodegradable" is thus the name given to a certain category of artificial, industrial, often mass-produced "thing." And yet this thing is unlike other things insofar as "it remains," as Derrida puts it, "a thing that does not remain, an essentially decomposable thing, destined to pass away, to lose its identity as a thing and to become again a non-thing" (BSD 813). The biodegradable thing thus has a curious status among things. While it, like so many other things that are not called "biodegradable," is destined to decay, to be destroyed, to pass away, the biodegradable thing is *designed* to decompose or degrade. It is a thing that is made or composed with a view toward its own decomposition or degradation, with a view toward its becoming something it is not. "Biodegradable" is thus the name given to things destined and designed to become other than the things they are—not just other things but *other than things*, nonthings, something that can no longer be *identified*, precisely, as a thing, certainly not as the thing it once was. (Of course, as we all know today, most so-called biodegradable things decompose and disappear to the naked eye but, upon closer inspection, still retain their identity—as microscopic plastic or Styrofoam particles, for example—and so still pollute the ecosystem into which they were supposed to decompose *without remainder*.)

As for the *word* "biodegradable," Derrida recalls—the year, again, is 1988–89—that this word is "a recent artefact" (BSD 815). Part Greek (*bios*), part Latin (*degradare*), but also, as a word of techno-science, part English or, rather, American, this strange hybrid, this "synthetic composite," would seem to be "more decomposable" than other words, more susceptible "to disappear or to let itself be replaced at the first opportunity" (BSD 815)—a bit like, one might be tempted to think, "deconstruction" itself. Unless, of course, the synthetic, hyperartificial, and artful nature of the word makes it less biodegradable than other words, a word whose survival might, says Derrida in a particularly poignant analogy, "rival that of the masterpieces of our culture and the monuments that we promise to eternity" (BSD 815)—a bit like, one might be tempted to think, "deconstruction" itself.

One can thus already begin to see why Derrida would have been interested for so long, as he says, in the "biodegradable." A thinking of biodegradability would seem to be an ideal vehicle for reiterating or illustrating certain lessons about deconstruction—lessons about the essential iterability of the signifier, about how to find and then recycle things from within the tradition, and then how to reinscribe or repurpose them in a new context. At one point in the text, for example, Derrida writes, after using words like "culture," "agricul-ture," and "nature" in a long passage on the biodegradable, "These words are no good. I keep them only in quotation marks; in fact I keep them just long enough to wear them out and throw them away like useless waste products, but ones that are perhaps very resistant, like the mutism of the quotation marks" (BSD 824).

Though Derrida does not say it in quite these words, it is hard not to think that the biodegradable thing bears a certain similarity to the *deconstructible* thing and that it might raise the question of the deconstructibility of things in a novel way, to say nothing of the question of the biodegradability of deconstruction itself—that is, the question of the survival of deconstruction itself as an *identifi-able* thing.[4] Engaged from the very beginning with questions of the trace, of remains, of cinders and ashes, questions of nature and cul-ture, as well as of life, death, and survival, along with questions of destruction and autodestruction, even autoimmunity, deconstruc-tion will have been from the beginning, it seems, a thinking of the "biodegradable."

If the word "biodegradable" is itself a thing, then it is neither a natural thing that decomposes of itself nor a cultural product that degrades like other biodegradable things. For this word, like every other word and, by extension, every text, cannot be reduced to, even if it can never appear without, its "presumed 'support'" or subjectile (BSD 814), whether paper or plastic, microfilm or computer screen. It's a straightforward but essential point: to think the biodegradability of the word, Derrida suggests that we try to distinguish "the survival of the support (paper, magnetic tape, film, diskette, and so on) from the semantic content" (BSD 815)—that is, the material support of textual meaning from that meaning itself.

If the word or text must be distinguished from its support, then the "biodegradability" of the word must be thought not exactly in terms of the decomposition of some industrial product into organic nature but in terms of the "amnesia of which a culture is made" (BSD 813), a sort of "originary amnesia" that allows a culture to live on and live off what it has forgotten (BSD 837). Derrida thus advances the hypothesis that "biodegradability" might be just the word to describe the self-destruction or self-forgetting of cultural objects, the way in which such objects, after a time of being recalled and identified, simply become part of the white noise of culture itself, of the archive, "like the sound of the sea deep within a shell"—an image that seems even more apt today than it was when "Biodegradables" was written, a time before the Internet, before today's massive digital archives, and before all the technologies that challenge our definition of remains and cause the line between the biodegradable and the nonbiodegradable to tremble.

Assuming, then, that one can say, at least figuratively, that "one publication is more biodegradable, more quickly decomposed than another" (BSD 814), then the question becomes for Derrida whether one can "distinguish here the degrees of degradation, the rhythms, the laws, the aleatory factors, the detours and the disguises, the transmutations, the cycles of recycling" (BSD 813–14). For there are some texts that have "resisted or will resist centuries of erosion and hermeneutic microorganisms" and others, he says, that, "from the very first page," we know we will forget almost immediately after reading them; texts, says Derrida, that seem to carry their own microorganisms, and thus their own destruction, within them (BSD 814). The "de Man Affair" and de Man's work more generally would be, in addition to everything else, a good test case for this biodegradability of the word or of the text more generally. With regard to this entire controversy, which was fought out in journals and newspapers over de Man's wartime journalism (1939–43), Derrida asks, "What will remain of all this in a few years, in ten years, in twenty years? How will the archive be filtered? Which texts will be reread?" (BSD 816). Just a year after the beginning of the controversy, which had provoked the publication of so many responses, Derrida already sees that "people are beginning to forget the articles and the names of so many confused, hurried, and rancorous professor-journalists" (BSD 817). Of course, having agreed, "after some reflection," he says, to respond to the six essays that precede his in *Critical Inquiry*, having "made it [his] duty to respond, to leave nothing without response" (BSD 817), Derrida is well aware of the aporia or paradox of responding to texts that probably merit being left to forgetting or to (bio) degradation. He thus asks himself at one point the strategic question of whether, by responding in this way, he is not in fact accrediting texts that are "weak, ridiculous, violent, indecent, in bad faith" (BSD 837), saving them from what would surely be a more immediate, "spontaneous degradability" (BSD 837). Unless he is in fact saving them for the criticism, derision, ignominy they deserve.[5]

All this then leads to the question of the biodegradability of cultural artefacts more generally—the question of what makes cultural artefacts submit to biodegradability and what makes them resist it. It is here that Derrida notes two aspects or two valences of the biodegradability of cultural artefacts: "To be (bio)degradable means at least two things: on the one hand, the annihilation of identity; on the other hand, the chance to pass into the general milieu of culture, into the 'life' of 'culture' while enriching it with anonymous but nourishing substances" (BSD 837–38). "Biodegradability" thus names, first, the effacement of identity, the erasure of a signature, the becoming nonthing of the cultural thing, but then also, and through this annihilation of identity, a survival—albeit an anonymous survival—in the culture more generally. It is in this latter way that a text is able, once "its 'formal' identity is dissolved," "to nourish the 'living' culture, memory, tradition" (BSD 845). This is what licenses Derrida to affirm that "the worst but also the best that one could wish for a piece of writing is that it be biodegradable. And thus that it not be so" (BSD 845). In other words, one might hope that a piece of writing not biodegrade, that it remain identifiable, that its signature remain legible, but this can happen only at the price of that writing *not* being fully assimilated or integrated into culture. And so one might also hope for just the opposite, for a piece of writing to decompose and become part of the general culture, part of its language, its discourses. For then it would be "on the side of life, assimilated, thanks to bacteria, by a culture that it nourishes, enriches, irrigates, even fecundates but on the condition that it lose its identity, its figure, or its singular signature, its proper name" (BSD 824).

If biodegradability entails a sort of annihilation of the proper name, then nonbiodegradability must be understood in terms of the *survival* of the proper name and its *resistance* to biodegradability. The difference, then, is between two kinds of survival, the one anonymous, its singularity or its event indistinguishable from its context or milieu, a sort of *biodegradable* survival, and the other identifiable and identified, a survival that bears "the singular mark of the event, of the date" (BSD 837–38), the mark of a proper name, a sort of *nonbiodegradable* survival. On the one hand, there would be the survival of some easily digestible meaning, of "everything that lets itself be anonymously assimilated into the tradition of a more or less common memory" and, on the other hand, the survival of the proper name, "of the literality of the formula," "of the singularity of textual events" (BSD 864–65).

It is "the proper name," then, "the proper name function," that "corresponds to this function of nonbiodegradability" and thus to the survival of the work as an identifiable cultural artefact (BSD 824). But that does not mean that the proper name is the index of some unique, resistant *meaning*. On the contrary:

The proper name belongs neither to language nor to the element of conceptual generality. In this regard, every work survives *like and as* a proper name. . . . In the manner of a proper name, the work is singular; it does not function like an ordinary element of natural language in its everyday usage. That is why it lets itself be assimilated less easily by culture to whose institution it nevertheless contributes. Although more fragile, having an absolute vulnerability, as a singular proper name it appears less biodegradable than all the rest of culture that it resists, in which it "rests" and remains, installing there a tradition, its tradition, and inscribing itself there as inassimilable, indeed unreadable, at bottom insignificant. A proper name is insignificant. But there are several ways to be insignificant. More or less interesting. One might as well say that meaning is not the measure of interest—or of wearing away [*usure*]. (BSD 824–25)

In other words, "what resists immediate degradation is . . . no longer on the order of meaning" (BSD 845). What resists biodegradation is not the uniqueness of some sense or meaning but, precisely, an *insignificance*, something that is unable to be repeated or reiterated, summarized and digested, assimilated into culture. For the proper name is, in the end, "proper to nothing and to no one, reappropriable by nothing and by no one, not even by the presumed bearer"; it is not some original meaning but a "singular impropriety that permits it to resist degradation—never forever, but for a long time" (BSD 845).

We must distinguish, then, as Derrida suggests, among different ways of being "insignificant," different ways of being "more or less interesting," different ways of resisting assimilation by culture, different ways to "resist it, contest it, question and criticize it . . . (dare I say deconstruct it?)" (BSD 845). Hence "one of the most necessary gestures of a deconstructive understanding of history" would consist of recuperating and valorizing, bringing out of the common compost of culture, discourses that have been consigned to oblivion by the dominant culture, discourses that have been "repulsed, repressed, devalorized, minoritized, delegitimated, occulted by hegemonic canons, in short, all that which certain forces have attempted to melt down into the anonymous mass of an unrecognizable culture" (BSD 821). Deconstruction thus has to do, at least in part, with bringing minority discourses, overlooked or repressed texts or currents, to the surface in order to highlight their significance—that is to say, their insignificance, their signature. But, just as often, deconstruction also tries to bring to the surface something that still resists within canonical or hegemonic discourses, something within well-known texts, within already recognized "great works" or "masterpieces," something that resists and that cannot simply be reduced to some forgotten or as yet undiscovered meaning.

> What is it in a "great" work, let's say of Plato, Shakespeare, Hugo, Mallarmé, James, Joyce, Kafka, Heidegger, Benjamin, Blanchot, Celan, that resists erosion? What is it that, far from being exhausted in amnesia, increases its reserve to the very extent to which one draws from it, as if expenditure augmented the capital? (BSD 845)

It is thus not the masterpiece's *meaning*, as we might think, that makes it resist—and thereby survive—but its lack of meaning. It is the work's proper name function, its signature, that makes it survive, its resistance to appropriation or translation, to reiteration or rephrasing, its resistance to being reduced to some content *without remainder*. It is in this way that a proper name resists assimilation, but then also a "verse" or an "aphorism," which "can survive a long time, thus resisting the 'biodegrading' erosion of culture, for all sorts of reasons not all of which are to be credited to them or to their author" (BSD 831). It is the aphorism, the idiom, that resists, but then also music, which, says Derrida, is "less (bio)degradable" than discourse because its form cannot be so easily "dissolved in the common element of discursive sense" (BSD 847). When Derrida asks near the end of the English translation of "Biodegradables" whether that which "resists translation [is] more or less (bio)degradable?" (BSD 873), we know that the answer has to be simply, "*Yes.*" On the one hand, it is by being translated, by lending itself to translation, that a text assures its greater dissemination and survival in the general culture. On the other hand, by resisting translation, by insisting on the untranslatability of its idiom, a text assures its signature and thus its survival.

It is essentially meaning, then, meaning stripped of its idiom or its signature (though this can never, of course, be total) that lets itself be assimilated, as opposed, it would seem, to the singularity of textual events, the singularity of the proper name. Of course, the fact that nonassimilation is one of the traits of a work's survivability does not mean that "the best way to escape cultural '(bio)degradability' is to be irreceivable, inassimilable, to exceed meaning. For then one would have to say," as Derrida recognizes, "that absurdities, logical errors, bad readings, the worst ineptitudes, symptoms of confusion or of belatedness are, by that very fact, assured of survival" (BSD 845–46). Derrida has a certain faith, at least on his good days, that "that which has no meaning, purely and simply, is almost immediately '(bio)degradable,'" that "what is 'bad' does not resist" and that, "in order to 'remain' a little while, the meaning has to link up in a certain way with that which exceeds it" (BSD 845–46). While it is impossible to say, then, with any certainty whether it is, generally speaking, rich texts rather than impoverished ones that ultimately survive, since "no calculation will ever be able to master the 'biodegradability' to come of a document" (BSD 836), there are nonetheless certain forms of resistance within works that give Derrida hope for the former.

It is this question of surviving or not, for good or ill, in an identifiable or anonymous fashion, that brings Derrida back to de Man

and to something within his work that cannot be assimilated, something regarding war, I am tempted to think, something that remains
inassimilable and that today still irradiates, a mark, a signature, a
ping, deep within the sound of the sea: "There are remains, something surviving that bears his name. Difficult to decipher, translate,
assimilate" (BSD 861). And Derrida would like to think, it seems,
that there is a similar resistance in his own work:

> If such a remark were not indecent or immodest on my part, af-
> ter the reproaches made against me in this way, I would dare to
> say that, in my view, the works that best resist time are those
> which are simultaneously eloquent and enigmatic, generously
> abundant and inexhaustibly elliptical. It is on this condition
> that they are the least—or if you prefer, the most—"(bio)de-
> gradable." (BSD 856)

The least *and* the most: the best way to survive would seem to entail not being assimilated pure and simple, like the biodegradable,
or to remain unassimilated, pure and simple, like the nonbiodegradable, but "assimilated as inassimilable"—that is, "kept in reserve,
unforgettable because irreceivable, capable of inducing meaning
without being exhausted by meaning, incomprehensibly elliptical,
secret" (BSD 845). It is right at this point where Derrida gives us the
secret of survivability, the secret of the secret, as it were, that we
can perhaps begin to hear something within "Biodegradables" itself,
something secret or at least muted, barely audible, a very strange
artefact that resonates from deep within this strange artefactual
text. Michael Peterson has already identified this secret, and quite
rightly, with the *nuclear*. I too hear this very same pulsation and
read this same signature, but I hear it emanating not just from the
dangers of nuclear waste but, in 1988–89, from the threat of nuclear
war, from the threat of the nuclear arsenal that irradiated then and
still irradiates today deep within the earth but also, if we can make
it out, deep within our discourses and our unconscious.

There are no fewer than six different references to the nuclear in
"Biodegradables," five of them to "nuclear waste," to "the long life
of nuclear waste" (BSD 836; see 815, 818, 863, 845), which seems to
survive as identifiable and nonbiodegradable like the masterpiece
we just saw: "Enigmatic kinship between waste, for example nuclear waste, and the 'masterpiece'" (BSD 845).[6] Nuclear waste, like

the masterpiece, would be the limit case, it seems, for the nonbiode-gradable. Though finite, like all nonbiodegradable things (BSD 854), nuclear waste is, when measured against human life and the human measures that must be taken (or not) to deal with it, as close to im-mortal as a product can be. It is why Derrida will say that, "like biodegradable, nonbiodegradable can be said of the worst and the best" (BSD 815).

In a text entitled, "Biodegradables," what irradiates from its core is indeed nuclear waste, that most nonbiodegradable of nonbiode-gradables. But there is also, even if it is named only once, even if it is barely audible and on the periphery, the reminder of another nuclear threat, buried in hidden mountain facilities and in our collective unconscious:

> Today, our means of archiving are such that we keep almost all published documents, even if we do not keep them in what used to be called living memory and even if libraries are obliged more and more often to destroy a part of their wealth. This is only an appearance: the originals or microfilms are elsewhere, kept safe for a long time, barring nuclear war or "natural" ca-tastrophe. (BSD 829)

This single reference to nuclear *war* and not *waste* helps explain, I think, why Derrida speaks throughout "Biodegradables" not just of biodegradable or nonbiodegradable remains but of the possibility of a destruction without remainder, "the possibility of a radical de-struction without displacement, of a forgetting without remainder." This is, he says, what he calls "ashes [*cendres*]. No trace as such without this possibility, which also lies in wait for the (bio)degrad-able and the non(bio)degradable" (BSD 862). If ashes—cinders—are the possibility of the trace or of remains *in general*, this general law seems to be marked by a very specific historical provenance, that of both the Holocaust and the nuclear bomb:

> Of trace and ashes. All that managed to survive, survival itself, are some names, in the large black archives or on the somber wall plaques in a museum in Jerusalem. Even so they are not all there. Even names can be incinerated. Not repressed or cen-sured, held in reserve in another place, but forever incinerated. (BDS 854)

If Derrida hears the Greek and Latin origins of "biodegradability," as well as its source in English/American technoscientific modernity, he also hears the word with a distinctly twentieth-century European ear: "How not to think of the death camps, the mass graves, the recycling of corpses, the fabrication of 'soap,' for example, from animal fat.... How not to think of ashes in general, the ashes of Auschwitz in particular? Of what I several times called 'the worst' in 'Paul de Man's War'"? (BSD 854). "Paul de Man's War," writes Derrida, as if, alas, we all had our own wars to fight and our own fears to face.

Written just five years after Derrida's most direct confrontation with the question of nuclear arms and nuclear rhetoric, his 1984 "No Apocalypse, Not Now," "Biodegradables" is itself, it seems, haunted by the memories and the threats of war.[7] It is thus, in many ways, still a Cold War text, a work about biodegradability and war, about the nuclear threat that, in 1988 and 1989, was impossible to ignore. Indeed, it is easy to forget, some quarter of a century later, what was happening in the weeks surrounding the writing of "Biodegradables." For in addition to the signing on January 11, 1989, of a 140-nation international treaty to ban chemical weapons, the fallout of technologies first developed during the First World War, and the bombing on December 21, 1989, of Pan Am Flight 103 over Lockerbie, Scotland, one of the first strikes in our newest war, the war on or with "terrorism," there were multiple nuclear weapons tests, both in the USSR (on November 12, 1988, December 28, 1988, and January 22, 1989, at Eastern Kazakh/Semipalitinsk, and on December 4, 1988, at Novaya Zemlya) and in the United States (on February 10, 1989, at the Nevada Test Site). Those nuclear tests form the background or the underground, I would like to think, of "Biodegradables," events that then came to be registered in the writing of this strange artefact.

But what about us today? Does the threat of nuclear war—in addition to the hazards of nuclear waste—still haunt us today, whether in our discourses or our unconscious? As Derrida's ultimate example of "some absolute non(bio)degradable," the image of the nuclear seems to emerge from out of the depths where, like the waste itself, it is "deeply immerged so as to neutralize its physical effects, if not the accumulated anguish that will *always* resonate deep within our unconscious" (BSD 862–63; my emphasis). Derrida seems to be recalling here and, indeed, throughout the essay not only a logic of the unconscious with all its operations of "censorship and repression, condensation and displacement," all those operations that "keep what they cause to disappear" simply "by caus[ing] it to change places" (BSD 862), but also, it seems, another unconscious or another law of the unconscious. According to this other unconscious, there are, it seems, things that *forever* resist and can never be assimilated or that can be assimilated only as radically inassimilable, something like a trauma, perhaps, but, in this case, a trauma of the future, a trauma of what has yet to come.[8] "It's as if," writes Derrida, "something nonbiodegradable had been submerged at the bottom of the sea. It irradiates" (BSD 861).

The sound of the sea deep within a shell thus seems to contain something else at its core, something like a shock or a shell shock deep within it, yet another valence, it seems to me, though one that is even more ominous, of that "indifference of the future" spoken of by Montherlant and echoed by de Man and Derrida. At a time when we are witnessing threats of a new nuclear proliferation (in Iran, for example, but then also in the Middle East more generally), a time when our nuclear stockpiles still remain large enough to destroy life on earth several times over, a time when so-called conventional weapons are leading to the massive destruction of archives and great works of art and culture in Afghanistan, Iraq, Syria, and elsewhere, it seems clear that the question of biodegradability cannot be distinguished from that of war. Indeed, with wars destroying both things *and* cultural objects, destroying them and leaving their remains, it seems that nothing that goes by the name "eco-deconstruction" can today avoid the question of war. There can be no thinking of war, it seems, without a consideration of war's ecological effects, and no ecology without a thinking of war. In a word, any eco-deconstruction must also be today an eco-polemology. For the sound of the sea deep within a shell can also be heard as the sound that remains when everything else has been destroyed, the sound of a world that has gone offline, a world without remains.

At the very end of "Biodegradables," Derrida challenges us, even dares us, to read and then discard his work. The final sentence of "Biodegradables" reads, "What can be the future destiny of a document that would now give one to read, like right here, the sole phrase: 'Forget it, drop it, *all of this is biodegradable*'?" (BSD 873). But this line too is a paradox, the paradox of remembering by being ordered to forget. For there is, as we have seen, something nonbiodegradable within "Biodegradables," something assimilated as unassimilated at its core, something that remains resistant to the order of meaning for both Derrida and for us. The wager of this text, Derrida's wager, it seems, is that this text—like a time capsule— will not only offer a few propositions regarding the biodegradable but will also exemplify or bear witness to what is biodegradable as well as nonbiodegradable within it. Derrida thus asks himself near the end of "Biodegradables" why he wrote this text, and he responds:

> Well, for no reason, just to see, to reflect and see what remains of it, perhaps to take the measure of the "(bio)degradability" of this text here, precisely, beyond its meaning, to test its conditions of translation, publication, and conservation. (BSD 866)

"Just to see," perhaps, what others will hear, a quarter of a century later, what they will hear within or beyond Derrida's own intentions or meaning, when they hold this work up to their ear to try to make out what is worrying it, working it, irradiating within it, promising and threatening from within.

In other words, it is a wager, as always, and the wager is, as always, an engagement or a commitment. As Derrida points out earlier in the text, "One cannot wager publicly on the survival of an archive without thereby giving it an extra chance," for the wager itself can always contribute in some way to the survival (BSD 836). For better or for worse, then, for the betterment or degeneration of the culture or for the environment in general, to the benefit or the detriment of some survival, I too will have made my wager, I too will have placed my bet and, just as surely, exposed my fears.

Notes

1. One exception here would be Peggy Kamuf's fascinating piece for the *Los Angeles Review of Books*, "Remains to Be Seen," written for the tenth anniversary of Derrida's death (October 9, 2014). At issue there is the state of Derrida's legacy and archive, the way in which it has been integrated into our culture and tradition, often at the expense of Derrida's proper name. It is surely not insignificant that this essay too should have appeared in such a biodegradable publication—even if I first found it online at http://lareviewofbooks.org/essay/remains-seen.

2. Jacques Derrida, *Signature Derrida*, ed. Jay Williams (Chicago: University of Chicago Press, 2013), 152–19.

3. Derrida, "Like the Sound of the Sea Deep within a Shell: Paul de Man's War," trans. Peggy Kamuf, in *Memoires for Paul de* Man, rev. ed.

(New York: Columbia University Press. 1989), 155–263; *Mémoires pour Paul de Man* (Paris: Éditions Galilée, 1988), first published in *Critical Inquiry* 14, no. 3 (Spring 1988): 590–652; revised in *Responses: On Paul de Man's Wartime Journalism*, ed. Werner Hamacher, Neil Hertz, and Thomas Keenan (Lincoln: University of Nebraska Press, 1988).

4. This then leads Derrida to reflect on the supposed death, degradation, waning, dying out, of "Deconstruction" itself—with a capital "D." "'Things' don't 'biodegrade' as one might wish or believe. Some were saying that 'Deconstruction' has been in the process, for the last twenty years, of extinguishing itself ('waning,' as I read more than once) like the flame of a pilot light, in sum, the thing being almost all used up" (BSD 819).

5. Of course, these other responses are caught in the very same aporia; those who, writing in the wake of de Man's death in the hopes of "put[ting] him to death this time without remainder" (BSD 861), also end up perpetuating de Man's memory and disseminating his work through their very denunciations.

6. See BSD 815, where the word "biodegradable" is itself compared to a "masterpiece."

7. Derrida, "No Apocalypse, Not Now (Full Speed Ahead, Seven Missiles, Seven Missives)," in P1 387–409/395–418; first published in *Diacritics* 14, no. 2 (Summer 1984): 20–31.

8. Derrida argues that the fear caused by 9/11 was the result not only of what actually had happened but also of what might still happen, "the threat of a *chemical* attack, no doubt, or *bacteriological* attack . . . , but especially the threat of a *nuclear* attack" (PT 97–98/149–50).

Troubling Time/s and Ecologies of Nothingness: Re-turning, Re-membering, and Facing the Incalculable

Karen Barad

> No justice . . . seems possible or thinkable without the principle of some responsibility, beyond all living present, within that which disjoins the living present, before the ghosts of those who are not yet born or who are already dead, be they victims of wars, political or other kinds of violence, nationalist, racist, colonialist, sexist, or other kinds of exterminations, victims of the oppressions of capitalist imperialism or any of the forms of totalitarianism.
> —*Jacques Derrida*, Specters of Marx, *viii; 15–16*

In these troubling times, the urgency to trouble time, to shake it to its core, and to produce collective imaginaries that undo pervasive conceptions of temporality that take progress as inevitable and the past as something that has passed and is no longer with us is something so tangible, so visceral, that it can be felt in our individual and collective bodies. This urgency is both new and not new. With fascism on the rise around the globe and the threat of an accelerated nuclear arms race at hand tied to a perverse sense of the usability of nuclear weapons, the false security of global strategic deterrence based on MAD (the military doctrine of Mutually Assured Destruction) left exposed and undone by madness, compulsiveness, and hubris, the twentieth century is anything but past/passed. The same can surely be said of previous centuries. And if debates on marking the origins of the Anthropocene suggest anything beyond an exacting reading of the layering of sediments used to justify adding a new segment of time to earth's geological clock, it is perhaps that the structure of temporality that timelines (in their linearity) smuggle into the discussion is inadequate to this moment. For if the climate experts in their official report to the International Geological Congress meeting in Cape Town in August 2016 mark the origin of the new epoch to be "defined by the radioactive elements dispersed across the planet by nuclear bomb tests" beginning in 1950,[1] and strong arguments have been made by scientists and nonscientists that offer reasons for using other dates as the "golden spike," the debates have mostly been about laying down the marker at the right time (whether at 1492, 1610, 1945, 1950, or 1963–66), and they have not for the most part questioned whether these times ought to be thought of as falling in a line, as if they were separated from one another by temporal distance.[2] But rather than understand these differing proposals as merely a simple disagreement about origins, perhaps we should take this as evidence that faith in the existence of a singular determinate origin and the unilinear nature of time itself (the fact that only one moment exists at a time) is waning. Is there a sense of temporality that could provide a different way of positioning these markers of history and understand 1492 as living inside 1945, for example, and even vice versa?

Introduction

Clock 1

Time isn't what it used to be. Perhaps it never was. It surely hasn't been itself since the "Doomsday Clock" was set at just minutes to midnight—the untimely hour of time's own demise.[3]

The "Doomsday Clock" of the *Bulletin of the Atomic Scientists*, introduced in 1947, represents scientists' estimation of our proximity to global catastrophe. A device once tuned to the Cold War, the setting of the clock was initially solely synchronized to the prospect of nuclear apocalypse, but in 2007 it was recalibrated to include climate change as an additional significant threat to earthly survival. This is a rather strange clock, a nonmechanical symbolic clock, a bit of sincere theatrics, that is an expression of the scientific community's estimation of global precarity in the present. Doomsday Clock time doesn't simply progress on its own, moving forward without fail, and it isn't synchronized to one particular physical

phenomenon, but rather to global politics and technological progress. A nonlinear device that is reset once each year, the Doomsday
device clocks sociopolitical, technoscientific events, and its measure is marked by the distance from the endpoint—midnight, the
apocalypse—rather than some origin point. Time is synchronized to
a future of No Future. This is time fixated on its own dissolution.
Setting time on edge, it offers both a grim view of our prospects
and a false sense of globalism assuming a homogeneity of times and
spaces, eliding the uneven distribution of nuclear and climate crises'
resources and precarity. Furthermore, it has the anesthetizing effect
of diverting questions of responsibility and of focusing the apocalyptic phantasm of total war, thereby distracting attention from the
realities of war in its ongoingness.

And this includes nuclear war. The first atomic bomb was not
the one dropped on Hiroshima. "The world's first atomic bomb was
detonated on July 16, 1945, in New Mexico—home to 19 American
Indian pueblos, two Apache tribes and some chapters of the Navajo
Nation."[4] That is, it was exploded within range of "the Americans'
own people, Turtle Island's original inhabitants, the Indigenous
Peoples of the southwest."[5] And nuclear war didn't cease when the
bomb dropped on Nagasaki on August 9, 1945. "Nuclear war has
been taking place on this earth in the name of 'nuclear testing' since
the first nuclear explosion at Alamogordo in 1945."[6] Since then
more than 2,000 nuclear bombs have exploded. "The primary
targets . . . have been invariably the sovereign nations of Fourth
World and Indigenous Peoples. Thus history has witnessed the nuclear wars against the Marshall Islands (66 times), French Polyne-
sia (175 times), Australian Aborigines (9 times), Newe Sogobia (the
Western Shoshone Nation) (814 times), Christmas Island (24 times),
Hawaii (Kalama Island, also known as Johnston Island) (12 times),
the Republic of Kazakhstan (467 times), and Uighur (Xinjian Province, China) (36 times)."[7]

In our "postatomic age," time is synchronized to the apocalypse-
to-come, and the present is caught in a pose of holding its breath
in an attempt to forestall the onset of nuclear war, as if it had ever
been a thing of the past. This singular sense of temporality is fixed
and fixated on the event horizon of total annihilation, calibrated to
fear and the elision of the ongoingness of war in our hypermilita-
rized present. Masahide Kato calls this totalizing view, the global-
ized spacetime grid, the "obliteration of the history of undeclared
nuclear war" that has been ongoing since World War II.[8]

Clock 2

Time has been shattered, exploded into bits, dispersed by the wind. Moments caught up in turbulent flows forming eddies, circling back around, returning, reconfiguring what might yet have been.

Hiroshima, August 6, 1945, 8:15 a.m. Clock mechanisms melted by the heat of the blast. The city clocks, clocks in plazas, stores, homes, on wrists, and in pockets, forever synchronized to one particular moment. Two hands etched into eternity—a larger one pointing due east and a smaller one pointing a bit south from west. Two hands seared into the face of time. Time is arrested; ghosts roam the streets. Although time is off its hinges, frozen and disengaged for all time, moments continue to pour down like black rain and settle on charred bodies and buildings; sticking to the air, they are breathed in, ingested, and come to rest in the marrow of bones, lying dormant, like little time bombs ticking inside *hibakushas* (atomic bomb victim survivors, literally explosion-affected people).[9] A pocket watch is all a son has left of his father. Clocks are powerful symbols in Hiroshima. The Hiroshima Peace Clock Tower chimes every day at 8:15 a.m. The Hiroshima Peace Memorial Museum "Peace Watch Tower" has a digital clock synchronized to peace instead of war that is reset back to zero every time there is a nuclear test anywhere in the world. Nuclear geopolitics, an entanglement of histories of violence, condensed into this one moment of spacetime, this one clock, this one now.

Clock 3

Time itself has gone atomic. Time no longer has a face or hands, but it does have a rhythm, a pulse. The atoms barely moving, habituated to temperatures near absolute zero, quantum leaps—dis/continuities—define the continuous march of time.

Atomic clocks are postwar gadgets tuned to resonance and precision. Suggested by Nobel laureate physicist I. I. Rabi in 1945, the first atomic clock, a laboratory instrument that must be kept by a high-tech time keeper, was constructed in 1949. Now there's no telling time without it. Global time, universal time, cosmic time—all keeping rhythm with the smallest bits of matter. The total colonization of spacetime synchronized to the heartbeat of an atom. Globalism is tied not only to the militarization of space but also of time. The latest "atomic clock [is] so precise that it won't lose or gain a

single second in 15 billion years—roughly the age of our universe."
"Who on earth needs such precise clocks?," you might ask. Actually, nothing less than the global economy—the mechanical guts
of capitalism, including GPS, telecommunications, and high-speed
transfer on Internet lines—depends upon it.

Telling Time/s

Each of these different clock times—the Doomsday Clock, the Hiroshima clocks, and atomic clocks—is tied to quantum physics.
Quantum physics gave birth to the atomic age. It is no secret that
it is deeply entangled with the military-industrial complex. Though
different from one another, each of these clock times treats time
as determinate and singular; in essence, each clock has a pointer
pointing at a single position on a clock face and marks one time at a
time. Although each of these clocks is informed by quantum physics, none of them is based on quantum physics' radical rethinking of
the nature of time.

Clock time is what Walter Benjamin poignantly calls "homogenous empty time." Whether calibrated to a projected future, an individual event, or a periodically recurring phenomenon, time is tuned
to a succession of discrete moments, where a moment is understood
to be the thinnest slice of time and where each successive moment
replaces the one before it. This is the time of capitalism, colonialism, and militarism.

But homogeneous empty time is not a universal conception of
time. In his article "Indigenizing the Future," Daniel Wildcat,
drawing on the work of indigenous philosopher Vine Deloria, makes
a critical intervention into modernist conceptions of time and
history:

> It is of critical practical importance that some cultures express
> history as primarily temporal and others express history as fun-
> damentally spatial in character. Once history-as-time is univer-
> salized and human beings are, so to speak, all put on the same
> clock, it is inevitable that in the big picture of human history
> some peoples will be viewed as "on time," "ahead of time," or
> "running late." It makes little difference that the clock hands
> rotate in circles, for they are thought of and acted on as if they
> were wheels moving down a single road called progress.

This road ought to be the ultimate metaphor for Western civilization and modernity, for it is an ideological abstraction. As John Mohawk concisely elaborated in his essay "The Right of Animal Nations to Survive," the metaphysics of progress presents itself as the greatest threat to the future biology of the planet. . . . American Indian or indigenous traditions resist ideas of universal homogeneous world history; there is no single road per se to human improvement. There are many paths, each situated in the actual places, such as prairies, forests, deserts, and so forth, and environments where our tribal societies and cultures emerged. The experiences of time and history are shaped by places.[10]

A multiplicity of paths and histories and the situatedness of time are also aspects of quantum temporality, which is not to suggest that (specific) quantum and (specific) indigenous approaches are identical or commensurate or have the same effect or stakes, but they do share in offering profound disruptions of the conception of homogenous empty time.

In this essay on troubling time/s, my focus is on a novella by Kyoko Hayashi and her (semiautobiographical) account of a Nagasaki *hibakusha*'s journey through time, place, history, and memory in search of a way to justly mourn the victims of atomic bombings. *From Trinity to Trinity* brings us full circle, through the unnamed protagonist's pilgrimage from Nagasaki to Trinity, in ways that powerfully attend to entanglements of colonialism, racism, and militarism that connect these disparate lands. And yet, even while the protagonist discovers a profound kinship with the bomb's first victims—namely, the desert plants and animals—in the end, there is no mention of the effects of the blast on the 19,000 people who lived within a fifty-mile radius of the Trinity Test.[11] One might wonder how Hayashi could have neglected this. Although *From Trinity to Trinity* was published in 2010, it is an astonishing fact that this was four years before the U.S. government would recognize the possibility of human casualties from the Trinity test and announce the beginning of a study of the high incident of cancers among the area's inhabitants and whether this could be traced to radioactive fallout from the blast.[12] That is, it took the U.S. government nearly seventy years to acknowledge that it was even worthwhile to do a study of the possible adverse effects on the people who were exposed to radioactive fallout from the 1945 Trinity test, despite the fact that following the test blast, "American Indians would begin to experience many types of cancers—rare cancers as well as multiple primary cancers."[13] Whether or not Hayashi had any knowledge of the increased rates of cancer on the people who were downwind from the blast, it seems crucial to start here. At the same time, since it will take the rest of this extended essay to give my diffractive reading of quantum physics' radical reworking of time through *From Trinity to Trinity*, I need to postpone the discussion of how (specific) indigenous (and Japanese) conceptions of time matter to this account.[14] For now, then, I turn to the question of how quantum physics understands the nature of time, knowing that it will be crucial to return to these threads and weave them into the entangled tale.

Quantum theory troubles time, in multiple ways, some of which will be explored in this essay. Quantum physics not only deconstructs the strict determinism of Newtonian physics, where the future unfolds predictably from the past, but it also blows away the progressivist notion of time—"homogenous and empty" time—disrupting first-world efforts to harness it as a totalizing system on behalf of universalism and its projects, such as imperialism. Quantum physics opens up radical spaces for exploring the possibilities for change from inside hegemonic systems of domination. Its radical political imaginaries might usefully join forces with indigenous and other subjugated knowledge practices rather than being a tool solely in the hands of the National Security Agency, although there is that, too. But tools are never entirely faithful to their masters.

This essay is about quantum theory's troubling of the nature of time and being, or rather, time-being. At the same time, it is also a story about the troubling times quantum theory has ushered in. That is, inside the nucleus of the story of the troubling of time is the troubled times unleashed by quantum theory's role in the making of the atom bomb and vice versa. These stories inhabit each other—a strange topology that already anticipates the kind of temporal imaginaries suggested by quantum theory.

Much as some folks want to exoticize quantum theory and think of it as living off on some remote island (deemed the "microworld"), safely quarantined from life as we know it (here in the "macroworld" where life is fantasized as solidly Newtonian), this geography is but a marker of an imperialist and colonizing worldview (where "anthropologists" for the object world, otherwise known as physicists, get to speak for the "natives," those radically Other beings that re-

fuse to be good modernist subjects, and at the same time are inanimate and lacking in agency). Quantum theory, despite tales to
the contrary, is not restricted to some alleged microrealm (which
always already assumes that the notion of scale is a given, while
this very notion, together with the nature of space, time, and matter,
is radically rethought). Nor does quantum theory live in the realm
of rarified ideas that now and again has applications for the real
world. Quantum theory is a material practice with direct ties to the
military-industrial complex—its very existence entangled with war,
militarism, racism, colonialism, capitalism, and imperialism. At the
same time, quantum theory disrupts classical Newtonian physics
(together with its most cherished ideas of space, time, matter, and
causality), which has its own problematic legacy in the service of
war, colonialism, and empire building. If Newtonian physics had designs on capturing nothing less than the heavens and the earth under
its rule, quantum physics troubles the very ideas of totality and closure—not only Newtonian attempts, but also its own. Quantum indeterminacy works against such attempts. Quantum indeterminacy
is not a form of unknowingness, nor even a kind of formlessness;
rather, it is a dynamism that entails its own undoings from within.
That is, the dynamism of quantum in/determinacy can be found
within physics, and not only within Derridean deconstruction.

This essay seeks to further the political project of opening up the
seeming totality called "Physics" in order to nurture the cracks and
bring forward its radical possibilities.[15] As such, this essay touches at
once upon both the destructive and deconstructive aspects of quantum theory. Raising questions of history, memory, and politics (all
of which are rooted in and invested in particular conceptions of time
and being), this essay is ultimately about the possibilities of justice-
to-come, the tracing of entanglements of violent histories of colonialism (with its practices of erasure and avoidance) as an integral
part of an embodied practice of re-membering—which is not about
going back to what was, but rather about the material reconfiguring
of spacetimemattering in ways that attempt to do justice to account
for the devastation wrought as well as to produce openings, new possible histories by which time-beings might find ways to endure.

No Small Matter

What is the scale of nuclear forces? When the splitting of an atom,
indeed, its tiny nucleus, destroys cities and remakes the geopolitical

field on a global scale, how can anything like an ontological commitment to a line in the sand between "micro" and "macro" continue to hold sway on our political imaginaries? When incalculable devastation entailing uncountable deaths is unleashed in the harnessing of a force that is so fantastically limited in extension that its job is merely to hold together the nucleus of an atom, a tiny fraction of a speck, a mere wisp of existence, then surely anything like some allegedly preordained geometrical notion of scale must have long ago been blown to smithereens, and the tracing of entanglements might well be a better analytical choice than a nested notion of scale (neighborhood ⊂ city ⊂ state ⊂ nation) with each larger region presuming to encompass the other like Russian dolls. That is, when a force extending a mere millionth of a billionth of a meter in length reaches global proportions, destroys cities in a flash, and reconfigures geopolitical alliances, energy resources, security regimes, and other large-scale features of the planet, this should explode the naturalization of the geometrical notion of nested scales that remains operative when the question arises as to what quantum physics has to do with the (so-called) macroworld.[16]

What is the scale of time? When the cascading energies of the nuclei that were split in an atomic bomb explosion live on in the interior and exterior of collective and individual bodies, how can anything like a fixed, singular, and external notion of time retain its relevance or even its meaning? In a flash, bodies near Ground Zero "become molecular"—nay, particulate, vaporized—while *hibakushas*, in the immediate vicinity and downwind, ingest radioactive isotopes that indefinitely rework body molecules all the while manufacturing future cancers, like little time bombs waiting to go off.[17] What would constitute an event when an atomic bomb that exploded at one moment in time continues to go off? The temporality of radiation exposure is not one of immediacy; or rather, it reworks this notion, which must then rework calculations of how to understand what comes before and after, while thinking generationally. Radioactivity inhabits time-beings and resynchronizes and reconfigures temporalities/ spacetimematterings. Radioactive decay elongates, disperses, and exponentially frays time's coherence. Time is unstable, continually leaking away from itself.

What is the scale of matter? There was a time when matter stood outside of time. Matter fell from grace during the twentieth century. It became mortal. Very soon after that it was murdered, exploded at its core, torn to shreds, blown to smithereens. The smallest of smallest bits, the heart of the atom, was broken apart with a violence that made the earth and the gods quake. In an instant, in a flash of light brighter than a thousand suns, the distance between Heaven and Earth was obliterated—not merely imaginatively crossed by Newton's natural theophilosophy, but physically crossed out by a mushroom cloud reaching into the stratosphere. "I am become death, the destroyer of worlds."[18]

The indeterminacy of space, time, and matter at the core of quantum field theory troubles the scalar distinction between the world of subatomic particles and that of colonialism, war, nuclear physics research, and environmental destruction. Quantum field theory (QFT)—a theory combining quantum physics, special relativity, and classical field theory—produced radical changes in our understanding of the nature of space, time, and matter. QFT also enabled the development of a fundamental theory of the nuclear forces (or fields) proposed by Hideki Yukawa in 1935. After the war, Yukawa was awarded the Noble Prize for his accomplishment; he was the first Japanese physicist so recognized. Physicists working at the forefront of the development of QFT (since the 1930s and continuing after the war) were integrally involved in the production of wartime technologies, including the atom bomb.[19]

In these troubling times, how can we not trouble time? Nothing less than the nature of and possibilities for change and conceptions of history, memory, causality, politics, and justice are conditioned by it. At the very core of QFT are questions of time and being. The indeterminacy of time-being opens up the nature of matter to a dynamism of the play of being and nothingness. Is there something about the nature of this dynamism that might lend some insight into what the practice of the politically committed work of mourning attuned to justice might look like? Or that would make it possible to trace the practices of historical erasure and political a-*void*-ance, to hear the silent cries, the murmuring silence of the void in its materiality and potentiality? What are the conditions of im/possibilities of living-dying in voids produced by technoscientific research and development, projects entangled with the military-industrial complex and other forms of colonial conquest?

The structure of this essay is diffractive rather than progressive. There is not a linear presentation of quantum physics. Instead, I present aspects of quantum physics' rethinking of the nature of time (spacetimemattering) and illustrate these by diffractively reading them through segments of a novella, *From Trinity to Trinity*, by

Kyoko Hayashi, a writer who, at the age of fourteen, lived (continues to live) through the bombing of Nagasaki. Hayashi's story and the story of QFT inhabit each other, and this diffractive reading is itself a performance of this strange topology.

Diffraction as methodology is a matter of reading insights through rather than against each other in an effort to make evident the always already entanglement of specifics ideas in their materiality. The point will not be to make analogies, but rather to explore patterns of difference/différance—differentiating-entangling—that not only sprout from specific material conditions, but are enfolded in the patterning in ways that trouble binaries such as macro/micro, nature/culture, center/periphery, and general/specific that tempt and support analogical analysis.

SpaceTime Diffraction and the Superposition of All Possible Histories: Quantum Physics' Disruption of the Imperialism of Universal Space and Time

> Moving from flight to flight, more of us have come to see, not only that we live in many worlds at the same time, but also that these worlds are, in fact, all in the same place—the place each one of us is here and now. . . . Thus, Two does not necessarily imply separateness for it is never really equated with duality, and One does not necessarily exclude multiplicity for it never expresses itself in one single form, or in uniformity.[20]

Diffraction is a matter of patterning attuned to differences. But not all differences are the same. Classical physics figures diffraction in terms of a comparison between this and that. However, from the perspective of quantum physics, diffraction is allied with the fundamental quantum physics notions of *superposition* and *entanglement*, where difference is a matter of *differences within*, not the "apartheid type of difference."[21]

Waves make diffraction patterns (think of the pattern made by dropping two stones in a still pond, for example) precisely because multiple waves can be in the same place at the same time, and a given wave can be in multiple places at the same time. Particles do neither; by definition, particles are localized entities that take up space: they can be here or there, but not in two places at once.

However, it turns out that particles can produce diffraction patterns, given an apparatus that allows for this possibility. How can this be? According to quantum physics this is because *a given particle* can be in a state of *superposition*. To be in a state of superposition between two positions, for example, is not to be here *or* there, or even simply here *and* there: rather, it is to be *indeterminately* here-there—that is, there is *no fact of the matter* (it is not simply that it is unknown) as to whether it is here or there. As a result of this indeterminacy of position (the precise principle is the position-momentum indeterminacy principle), particles exhibit diffraction patterns under circumstances that make evident the superposition (for instance, a barrier of appropriate dimensions with two openings that allow the passage of a particle will do). Or rather, when they do exhibit a diffraction pattern it is an expression of the fact that they are in *a state of superposition*. Note that while it is tempting to say that a given particle in a state of superposition is in two places at once, this is a simplification that doesn't fully capture the complexities: for one thing a *particle*, by definition, has a determinate position (for example, is either here *or* there); and furthermore, if one were to perform a measurement to directly test the hypothesis that a particle is in two places at once, then it wouldn't be(!), because a particle whose position is detected will behave like a good particle and only ever show up in one place at a time, even though the pattern produced when the position isn't being measured (as in the case of a two-slit experiment) can only be accounted for if it were in two places at once (that is, if "it" behaves like a *wave*, in which case "it" isn't a *particle*).

Diffraction patterns are very common, but not always evident. The special circumstances produced in laboratories function to make particular patterns evident (at the expense of others). But patterns of differences (differencing/différancing) are arguably at the core of what matter is (relational différancing all the way down) and are at the heart of how quantum physics understands the world.[22] Indeed, Nobel laureate physicist Richard Feynman proposed an understanding of quantum physics based solely on the notion of diffraction (that is, superposition). To see this, it is first of all important to note that according to quantum physics, *there is no determinate path* that a particle takes in going from one position to another—that is, no such path exists. But what physicists can do is calculate the *probability* that a given particle that starts out *here* will wind up *there*. The quantum probabilities are calculated by taking account of all possible paths connecting the two points. Feynman derives this result starting with a two-slit diffraction grating (a barrier with two slits) and calculates the total probability that a particle that starts out on one side of the barrier winds up at a particular spot on the other side (in particular, this entails a sum of all possible ways of making it from one side to the other, where each possible way is weighted according to its corresponding probability) (see Figure 9-1A). He then takes the limit of considering a diffraction grating with an infinite number of slits that a particle can go through—representing the possibility of crossing the barrier anywhere along its (infinite) length (that is, all points in a given plane)—and summing over an infinite number of such gratings, thereby summing over all planes (see Figure 9-1B), and hence covering all of space (see Figure 9-1C). The total probability then is related to the superposition of all possible paths (see Figure 9-1D); this superposition of all possible paths manifests as a diffraction pattern. According to this Feynman path integral formulation, a superposition is a sum of *all possible paths—they all coexist and mutually contribute to the overall pattern*, else there wouldn't be a diffraction pattern.

Quantum physics opens up another possibility beyond the relatively familiar phenomenon of spatial diffraction: namely, *temporal diffraction*. This takes a bit of getting used to, even more so than spatial diffraction, but temporal diffraction has in fact been observed experimentally.[23] One way to observe temporal diffraction is to take a disk with one or more slits carved in it, make a hole in the center of the disk, push an axel through it, and rotate the disk on the axel; then, direct a beam of light or particles at the rotating disk (so that the beam is parallel to the axel and the light or particles only pass through when it encounters the open slit in the disk). In this way the beam encounters slits *separated in time* from one another (rather than being separated in space, which is the more usual configuration of spatial diffraction). While spatial diffraction is a manifestation of the position-momentum indeterminacy principle, *temporal diffraction* is a manifestation of another, much less well-known, indeterminacy principle: namely, the time-energy indeterminacy principle. As a result of this indeterminacy principle, a given entity can be in (a state of) *superposition of different times*. This means that a given particle can be in a state of coexisting at multiple times—for example, yesterday, today, and tomorrow. However, temporality is not merely multiple; rather, temporalities are specifically entangled and threaded through one another such that there is no determinate answer to the question "What is time?" There is no determinate time, only a specific *temporal indeterminacy*. The diffraction pattern, in

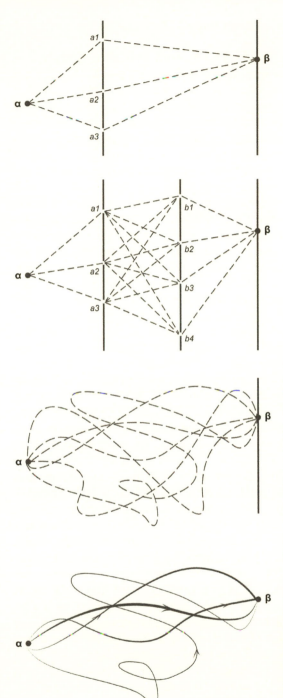

Figure 9.1 In these four diagrams, a indicates the source of particles or the origin point, and β is a point on the screen marking the place where the particle arrives. (A) shows a diffraction grating with multiple slits (a1, a2, and a3). (B) shows multiple diffraction gratings (a & b), each with multiple slits. (C) suggests the limit case in which there are an infinite number of diffraction gratings with an infinite number of slits, which then allows the particle to be anywhere between the source a and the screen β. (D) shows some possible paths, all of which must be included in a Feynman path integral.

this case, is a manifestation of different times bleeding through one another. As in the case of spatial diffraction, this means that it's not that some event involving the entity is taking place at one time or another, but we don't know which; rather, the point is that temporal diffraction is the manifestation of an *ontological indeterminacy of time*: there is no fact of the matter to when it is taking place. (Furthermore, in an important sense, although it's not usually talked about this way, the time-energy indeterminacy principle lies at the very heart of QFT. More on this later.)

In fact, it is possible to do a diffraction experiment in both space and time at once, whereupon *a single particle* will coexist in a superposition of multiple places and times.[24] In this case of spacetime diffraction, a diffraction pattern can be accounted for by taking account *of all possible histories (configurings of spacetime)*, understanding that each such possibility coexists with all others. In particular, then, in its four-dimensional (relativistic spacetime) QFT elaboration, the probability that a particle that starts here-now will wind up there-then entails *taking account of all possible histories*, or rather, spacetimemattering configurings.[25] Crucially, these "possibilities" are *not* to be thought of in the usual way: the diffraction pattern is *not* a manifestation of an uncertainty in our knowledge— it is not that each history is merely possible until we know more, and then ultimately only one will be actualized—the superposition marks ontology indeterminacy (not epistemological uncertainty), and the diffraction pattern indicates that *each history coexists with the others*.

According to quantum physics, then, a diffraction pattern is a manifestation of a superposition. Interestingly, although linearity is a prime target of temporality analyses, superpositions are in fact based on linearity: not a linearity of moments or events evenly distributed *in* time, but a linear combination *of* (different) times. Hence, while contemporary rejections of *linearity* abound, especially in discussions of temporality, this story stays with linearity while opening it up to its radical potential. So, despite the fact that linearity—in particular, linear time—has been fingered as a particularly pernicious idea integral to Enlightenment thinking, the handmaiden to an ideology of progress and associated notions of the unidirectionality of time and temporal successionism, I am arguing that (even) linearity is susceptible to radical reworkings from within. This troubling of the assumed problematics of linearity and the associated quantum reworkings of the classical notion of time operate in concert with, not as a rejection of, an array of recent critical reassessments of temporality that for various reasons trouble the linear conception of time and suggest alternative conceptions of time that include temporal multiplicity and other configurations. (Note that quantum physics' notion of temporal superposition suggests a phenomenon that is far more subtle—that is to say, more complex and far stranger than multiplicity per se.) Needless to say, any suggestion that the notion of the linearity of time is unsalvageable and ought to be replaced with a new, arguably superior, notion of time would be ironic, to say the least, since it would be to fall into the logic of progress and supersessionism. What is needed is an understanding of temporality where the "new" and the "old" might coexist, where one does not triumph by replacing and overcoming the other.[26] Quantum superpositions and, relatedly, quantum entanglements open up possibilities for understanding how the "new" and the "old"—indeed, multiple temporalities—are diffractively threaded through and are inseparable from one another.

From Trinity to Trinity

Time and being are themes at the heart of *From Trinity to Trinity*, a remarkable novella by award-winning author Kyoko Hayashi.[27] Having spent her early childhood years abroad, living in Shanghai, and having returned to Nagasaki at the age of fourteen in March 1945, Hayashi spent the majority of her life chronicling the experiences of *hibakusha* and other victims of colonial violence (paying particular attention to Japanese state aggression against China; also noteworthy is Japan's colonization of Korea, resulting in the little acknowledged fact that between 40,000 and 70,000 Koreans who were conscripted into hard labor by the Japanese died in the bombings of Hiroshima and Nagasaki).[28] Having lived through an event that refuses to end, that decays with time but will forever continue to happen, Hayashi sought to unpack some of the infinite density of one particular spacetime point: Nagasaki, Japan, August 9, 1945, 11:02 a.m.—a moment shot through with many other times, places, and histories.

Kyoko Hayashi's novella *From Trinity to Trinity* traces the spacetime wanderings of an older unnamed woman on a spiritual-political pilgrimage, a journey of re-turning to a land she had never before visited but knew better than the geography of her own body, a wounded land whose history of violation radiates inside her bones.

Making her way to Trinity Site in New Mexico where the first atomic bomb test took place, Hayashi's protagonist "travel hops" from one spacetime point to another, circling back, re-turning and turning our attention to a multiplicity of entangled colonial histories condensed into August 9: she is at once in Nagasaki working alongside classmates in the Mitsubishi arms factory; on a U.S. Air Force base in New Mexico visiting the National Atomic Museum as a lone Japanese visitor among otherwise white tourists who are there to learn about the U.S. "nuclear defense history"; in Nagasaki counting fifty-two empty chairs belonging to classmates who did not return, would never return when school started again; and recounting the history of sixteenth-century Spanish explorers colonizing the land now called "New Mexico" while walking next to Little Boy and Fat Man, sitting like two iron coffins, in the museum at Los Alamos.

Her goal is not one of personal healing per se, but rather a political and spiritual commitment to take responsibility for re-membering the countless people who were robbed of their own deaths by unspeakable violence. Centering the relationship between time and justice, she is committed to the work of mourning as a political embodied labor—a commitment to justice beyond the living present and "concerning those who *are not there*, . . . those who are no longer or who are not yet *present and living*."[29] Nuclear entanglements do not abide by some notion of modalized presents; time is spectral, diffracted. The bomb is still going off when she walks through the ruins of Nagasaki to Ground Zero in the days following August 9, when her gums bleed when she is old, when her son faces with each new day the temporality of the future coming from the past, the prospect of getting leukemia as a *second-generation hibakusha*.

From Trinity to Trinity is a story that embodies questions of history, memory, politics, nationalism, colonialism, race, species, violence, and temporality. Hayashi's point is not to try to make sense out of senselessness, as if a rational story could be made of the madness, or a refreshingly mad story made of the rationalisms, but rather, to take hold of the radical possibility of the *undoing* of August 9. This is a journey across spacetime, nation states, species being, and questions of being/nonbeing.

But it should not be mistaken for a time-travel story, not in the usual sense. This travel-hopping tale is very different from time-travel novels where the protagonist is an autonomous unified subject who continues to live in the time of "his present" while returning to a past that once was, a past that continues to exist and remains accessible to those with sufficient ingenuity and technical know-how, in an attempt to rework some crucial point in a chain of events that will then propagate forward in deterministic fashion in a rewriting of history. Hayashi's travel hopping does not lend itself to such stories. In Hayashi's story, what is at stake is not setting time aright (as if that were possible), but rather the undoing of time, of universal time, of the notion that moments exist one at a time, everywhere the same, and replace one another in succession (much like identical entities passing by in the evenly spaced rhythms of Fordist assembly lines, the new being readied to replace the old); it is also a story of time-being that undoes modernity's unified notion of self and what it means to be human. The travel hopper must risk her sense of self, which never will have been one, or itself. Travel hopping, tracing the entanglements of spacetimemattering, is not the same as writing a linear chronology as a matter of personal or collective history. Travel hopping is the embodied material labor of cutting through/ undoing colonialist thinking in an attempt to come to terms with the unfathomable violences of colonialism in their specific material entanglements. How else might she begin to approach the infinite inhumanity of this weapon of instantaneous mass destruction that in a flash obliterates time?

Tracing Entanglements and the Material Traces of Erasure

Tracing entanglements is no easy task. It takes work.

During the waning decades of the twentieth century, arguably the most murderous century in history, the notion that the past might be open to revision through a "quantum eraser" came to the fore. The quantum-eraser experiment is a variation of a two-slit diffraction experiment, an experiment that Feynman said contains all the mysteries of quantum physics. Against this fantastic claim of the possibility of erasure, I argue that in paying close attention to the material labors entailed, the claim of erasure's possibility fades, while at the same time bringing to the fore a relational-ontology sensibility to questions of time, memory, and history.[30]

The key features of the quantum-eraser experiment are as follows. Recall that the famous two-slit experiment can be used to show that "particles" under appropriate conditions exhibit wave behavior (thereby belying their status as particles)—namely, they produce a diffraction pattern; this pattern is produced only if each individual particle is in a state of superposition that includes the possibility of going through both openings at once, as a good wave does (see Figure 9-2A). On the other hand, if you modify a two-slit apparatus by adding a device to measure which slit a particle goes through, it does in fact go through one slit or the other, like a good particle, contributing to the creation of a pattern characteristic of particles—that is, a scatter pattern, *not* a diffraction pattern (see Figure 9-2B).[31] If the experimenter now adds a device that enables the erasure of the information about which slit a particle goes through *after* it's already gone through the diffraction grating . . . remarkably, a diffraction pattern appears!—indicating that each particle will have gone through both slits at once! (See Figure 9-2C.) This raises the seemingly impossible possibility that one can determine *after* the fact whether the particle will have gone through one slit or the other—like a (well-behaved classical) particle does—or through both slits at the same time—like a wave—*after* it has already passed through the diffraction grating and made a mark on the screen. The claim made by the physicists who proposed and conducted the quantum-eraser experiment is that this is evidence of changing the past. But it's important to slow down and carefully examine the evidence behind this claim because the nature of time and being, or rather, time-being, is itself in question and can't be assumed.

For one thing, something not remarked upon by the experimenters is that what this experiment tells us is not simply that a given particle will have done something different in the past but that the very nature of its being, *its ontology, in the past remains open to future reworkings* (that is, whether it will have been a wave or a particle, which are ontologically different kinds). In particular, I have argued that this experiment offers empirical evidence for a relational ontology that runs counter to a metaphysics of presence. Indeed, I have argued that the quantum-eraser experiment can be understood as offering empirical evidence for a *hauntology*.[32]

The physicists who proposed the quantum-eraser experiment interpret these results as the possibility of "changing the past"; they speak of the diffraction pattern as having been "recovered" (as if the original pattern has returned) and the which-slit information having been "erased." But this interpretation is based upon assumptions, assumptions concerning the nature of being and time, that are being called into question *by this very experiment*.

Crucially, the diffraction pattern is not immediately evident once the information is erased. That is, it is *not* the case that the original

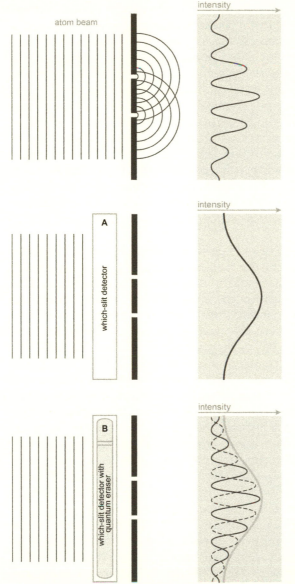

Figure 9.2 This set of diagrams illustrates some variations on a two-slit experiment. The source used in each case is atoms. The graphs to the right show the resulting patterns made after many individual particles pass through the two-slit diffraction grating (one at a time). (A) is an illustration of the usual two-slit experiment. The graph shows the resulting diffraction pattern (characteristic of waves that make a diffraction pattern because they go through both slits at once and combine on the other side of the barrier). (B) is an illustration of a two-slit experiment modified with a which-slit detector that enables a detection of which slit each individual particle passes through. The graph shows a resulting scatter pattern (characteristic of particles), indicating that each particle did in fact go through one slit or the other. (C) is an illustration of a quantum-eraser experiment that entails which-slit detection followed by the erasure of the information of which slit each individual particle passed through. Significantly, the graph shows that inside the scatter pattern there is an extant diffraction pattern that can be found by tracing the entanglements.

diffraction pattern returns. Rather, *a* different *diffraction pattern (not the original one) can be found* within *the scatter pattern* if and only if *the experimenter is clever enough to know* how to trace the existing entanglement. This point is crucially important. *For the labor expended in tracing the entanglements* (including figuring out how to find the extant entanglements and then tracing them) *is a necessary step in making the experiment work.* Remarkably, *this experiment makes evident that entanglements survive the measurement process, and furthermore, that material traces of attempts at erasure can be found in tracing the entanglements.* Indeed, these experiments show that while it is possible to erase particular marks that seem to suggest that the "past" has been changed, it is a fantasy to believe that this constitutes an erasure of all traces of this history. *Erasure is a material practice that leaves its trace in the very world-ing of the world.*

Hence, the quantum-eraser experiment winds up being ironically named, for there is no erasure finally; indeed, the traces of erasure are written into the iterative materializations in their openness. This experiment calls into question the classical Newtonian conceptions of time not only as an unabated continuous flow moving inexorably from past to future, where the past is passed and the future will unfold on the basis of what is the case in the present moment, but also as the assumed existence of a present-past and the very possibility of erasure without trace. I have argued that an interpretation that seems to be in better accord with the empirical evidence than the one offered by the experimenters is that *while the past is never finished and the future is not what will unfold, the world holds the memories of its iterative reconfigurings.* All reconfigurings, including atomic blasts, violent ruptures, and tears in the fabric of being— of spacetimemattering—are *sedimented* into the world in its iterative becoming and *must be taken into account in an objective (that is, responsible and accountable) analysis.*

Our atomic past not only haunts the present but is alive in the thickness of the here and now (a point that will be thickened in the QFT section of this essay). One manifestation of the fact that "now" is shot through with "then" is the Fukushima disaster and its continuing consequences, which are directly entangled with the U.S. bombing of Hiroshima and Nagasaki. In the aftermath of the war, the U.S.-based Atoms for Peace Program was used to convince Japan to develop nuclear energy for peaceful purposes while the United States used the program to shield the buildup of its nuclear arsenal during the Cold War. Hauntings are not immaterial, and they are not mere recollections or reverberations of what was. Hauntings are an integral part of *existing* material conditions. This past—nuclear time, decay time, dead time, atomic clock time, doomsday clock time, a superposition of dispersed times cut together-apart—is literally swirling around with the radioactivity in the ocean. Time itself is nationalized, racialized, out of joint. The entanglements of nuclear energy and nuclear weapons, nationalism, racism, global exchange and lack of exchange of information and energy resources, water systems, earthquakes, plate tectonics, geopolitics, criticality (in atomic and political senses), and more are part of this ongoing material history that is embedded in the question of Japan's future reliance on nuclear energy, where time itself is left open to decay.

History, Memory, and Traces of Erasure: On the Way to Trinity

Soon my eyes caught some big letters on a panel: "Count Down to Nagasaki. . . ."

[Our protagonist is visiting the National Atomic Museum in New Mexico, an unexpected stop on the way to Trinity Site.][33]

I felt time stop in front of the panel.

"Count Down to Nagasaki." While the time towards death in Nagasaki was ticking, what were Kana and I doing in the Ohashi Arms Factory?

. . . at the very moment the bomb left the plane, I was trying to locate the sound of a small roar the factory chief told us he had just heard.

I closed my eyes and bowed my head to the photograph. The ruin of a fire printed underneath the explanation was the city of Nagasaki with Inasayama across the river. <The effect appears to be the same as Hiroshima,> Bockscar pilot Sweeny said in the first report of the attack on Nagasaki. <A majority of the city has been suddenly destroyed. Even though I am watching the actual scene, I still cannot believe it.> Here is the photograph of that destroyed city.

The photo shows a burned field, but under what is seen on that printed paper is teacher T, who died instantly, and classmates A, O, and others.[34]

In this brief passage, where chronology has no p(l)ace, where multiple temporalities present themselves without any one of them
being present, the very coexistence of time-beings disassembling
the allegedly determinate distinction between individual and collective, memory and history, Hayashi offers us a pointed contestation of official museum history: that is, a tale told in chronological
time, a scientized and sanitized account of "objective reality"—the
God's-eye view from above, the view from nowhere. Disrupting this
chronology helps us see through the photograph to what is behind
it: namely, all the various material-discursive apparatuses of production that make up this exhibit—what it contains, what it erases,
which facts matter and how they are collected and framed. What the
official photo shows is an aerial view of a city destroyed, the leveling
of buildings into a structural void. What the museum history invisi-
blizes is *the structure of the void*—the entangled material histories
of death and dying, all the ravages of untold violence, histories of
colonialism, racism, and militarism, and all the attempted erasures
that constitute it.[35] By contrast, what is at stake for Hayashi is a
matter of empirical reality: the reality (literally) on the ground.

We come to see that what the photo shows is *not* the bare facts
of history, but rather a record of erasures: the literal erasure of lives
obliterated like so many buildings, people who had been in the
streets on foot and on bicycles, workers stacking shelves in neighborhood shops, school children working in factories, old people and
children in their homes; but also a particular framing of the event
that makes use of distance to sanitize the suffering and devastation
of lives, while erasing *some* histories of violence and not others. Japanese imperialist aggression is the given background against which
this history is shot, while U.S. imperialism and militarism are outside the frame. Erasures upon erasures.

But erasures are never complete—traces always remain. In her
disjointed time hopping, Hayashi's narrator is bodily tracing these
extant entanglements.

The official photograph freezes time and reifies space. But there
were other photographs taken during the bombing of Hiroshima
and Nagasaki, photographs on the ground, not ones engineered by
humans designed to capture the successes of military operations,
but rather very up-close and personal photos taken by the bomb itself. Shadows of incinerated bodies—human and nonhuman—captured on walls made into photographic plates by the intensity of
the blast.

What lies inside the boundaries of a shadow? Where are its edges? Diffraction unsettles colonialist assumptions of space and time: beginnings and ends, continuity and discontinuity, interior and exterior.

Standing in the museum, Hayashi notes another integral part of the official museum history and its contemporary framing:

> There were no black or Mexican visitors. Not only in this museum but also in Los Alamos and at the <Trinity Site,> all the visitors were white.[36]

Jumping in time but continuing the thought, Hayashi introduces another invisibilized piece of the story, one so covered-over by colonialism's practices of erasure that the very question of the ground on which the exhibit stands seems all but entirely buried. What is the story of this very land that the museum stands on—why here? What is the connection of this land to the obliterated Japanese city shown in the photograph? Standing in the museum, Hayashi traces the entanglements of colonial histories: of late sixteenth-century European colonial conquest of Native American peoples and lands, entangled with the early twentieth-century U.S. colonial annexation of New Mexico in the wake of the U.S. invasion of Mexico half a century earlier, entangled with the Second World War–time designation of native land deemed uninhabited as "Trinity Site," entangled with the testing of the plutonium bomb at Trinity Site, with the same kind of bomb being dropped a month later on Nagasaki, entangled with uranium mining and nuclear-waste burial on indigenous lands in the American Southwest, entangled with the Fukushima disaster, entangled with existing and future cancers of all the *hibakushas* and their offspring, the (human and nonhuman) "no-bodies" who were downwind from the test site and forms of nuclear fallout.[37]

Attempts at erasure always leave material traces: what is erased is preserved in the entanglements, in the diffraction patterns of being/becoming. In tracing the material entanglements extant in practices of erasure, Hayashi's narrator gives us a sense of *how* boundaries of lands and bodies get diffractively materialized and sedimented through one another. The various forms of violence, including all the erasures, are written into the very fabric of the world, into the specific configurings of spacetimemattering, so that it is crucial that she make the pilgrimage to trace the entanglements with her marked and wounded body. Hayashi's narrator bodily traces these entanglements of colonialist histories, violent erasures, and a*void*-ances as an integral part of a sacred practice of re-membering—which is not a going back to what was, but rather a material reconfiguring of spacetimemattering in ways that attempt to do justice to account for the devastation wrought and to produce openings, new possible histories, reconfigurings of spacetimemattering through which time-beings might find a way to endure.

Quantum Field Theory: The Un/Making of Self and the Material Conditions for Living and Dying in the Void

Land occupation, as a mode of empire building, has been and continues to be tied to a logics of the void.[38] Namely, justification for occupying land is often given on the basis of colonialist practices of traveling to "new" lands and "discovering" all matter of "voids": for example, claims of population voids (for instance, lands allegedly unpopulated before the arrival of the settlers), land devoid of property ownership, territorial sovereignty, development, civilization, or inhabitants with specific labor relations to specific parcels of land. The doctrine of *terra nullius* is one such tool of empire building. Whatever the specific nature of the alleged absence, a particular understanding of the notion of the *void* defines the colonialist practices of a*void*ance and erasure.

The void occupied a central place in Newton's natural philosophy. He wavered about the existence of an ether permeating empty space, but unlike many of his contemporaries who were still committed Aristotelians and equated matter with extension, Newton insisted that the void was a spatial frame of reference within and against which motion takes place. Matter is discrete and finite, and the void is continuous and infinite. The void extends indefinitely in all directions, and bits of matter take their position in the void. All in all, the void is quite *literally* universal (measuring the full extent of the universe and beyond) and therefore only very sparsely populated. And since property rests with matter as one of its founding characteristics, the absence of matter is the absence of property and the absence of energy, work, and change. The *void*, in classical physics, is *that which literally doesn't matter*. It is merely that which frames what is absolute. While the so-called voyages of discovery, bringing data (including astronomical and tidal changes) culled from European journeys to non-European sites aided Newton in his efforts to develop a natural philosophy that united heaven and earth,

Newtonian physics helped consolidate and give scientific credence to colonialist endeavors to make claims on lands that were said to be de-void of persons in possession of culture and reason.[39]

If classical physics insists that the void has no matter and no energy, the quantum principle of ontological indeterminacy—in particular the indeterminacy relation between energy and time—calls into question the existence of such a zero-energy/zero-matter state or rather makes it into a question with no decidable answer. Not a settled matter, or rather, no matter. And if the energy of the vacuum is not determinately zero, it isn't determinately empty (since energy and matter are equivalent: $E=mc^2$).

That is, according to QFT, the vacuum can't be determinately nothing because the indeterminacy principle allows for fluctuations *of* the quantum vacuum. How can we understand "vacuum fluctuations"? If the physicist's conception of a field can be likened to a drumhead, with a zero-energy state being akin to a perfectly still drumhead and a field with a finite energy being a drumhead in one of its (quantized) vibrational modes (like the 3D analog of harmonics of a string), then while the classical vacuum state would be perfectly still, without any vibrations, a quantum vacuum state, although it has zero-energy, is *not* determinately still as a result of the energy-time indeterminacy principle. *Vacuum fluctuations are the indeterminate vibrations of the vacuum or zero-energy state. Indeed, the vacuum is far from empty, for it is filled with all possible indeterminate yearnings of time-being; or in this drum analogy, the vacuum is filled with the indeterminate murmurings of all possible sounds: it is a speaking silence.* What stories of creation and annihilation is the void telling? How might we approach the possibility of listening?

Putting this point in the complementary language of particles rather than fields, we can understand vacuum fluctuations in terms of the existence of virtual particles: *virtual particles are quanta of the vacuum fluctuations.* That is, *virtual particles are quantized indeterminacies-in-action. Virtuality* is *the indeterminacy of being/ nonbeing, a ghostly non/existence.* The void is a spectral realm; not even nothing can be free of ghosts. Virtual particles do not traffic in a metaphysics of presence. They do not exist in space and time. They are ghostly non/existences that teeter on the edge of the infinitely thin blade between being and nonbeing. They speak of indeterminacy. Or rather, no determinate words are spoken by the vacuum, only a speaking silence that is neither silence nor speech, but the conditions of im/possibility for non/existence. There are an infinite number of possibilities, but not everything is possible. The vacuum isn't empty, but neither is there any-thing in it. Hence, we can see that indeterminacy is key not only to the existence of matter but also to its nonexistence—that is, to the nature of the void.[40]

In fact, this indeterminacy is responsible not only for the void not being nothing (while not being something), but it may in fact be the source of all that is—a womb that births existence. Particles (together with their corresponding antiparticles, in pairs) can be created out of the vacuum by putting the right amount of energy into the vacuum, thereby giving a virtual particle (-antiparticle pair) enough energy to emerge from the vacuum; similarly, particles (together with their corresponding antiparticles, in pairs) can go back into the vacuum, emitting the excess energy.[41] Hence, birth and death are not the sole prerogative of the animate world. "Inanimate" beings also have finite lives. "Particles can be born and particles can die," explains one physicist. In fact, "it is a matter of birth, life, and death that requires the development of a new subject in physics, that of quantum field theory. . . . Quantum field theory is a response to the ephemeral nature of life."

The void is a lively tension that troubles the opposition between living and dying (without collapsing their important material differences); the void is a dynamism of indeterminacy, a threading through of living with dying and dying with living, a desiring orientation toward being/becoming that cannot a-void matters of life and death. *The vacuum is far from empty; rather, it is flush with yearning, with innumerable possibilities/ imaginings of what was, could be, might yet have been, all coexisting.* Don't for a minute think that there are no material effects of yearning and imagining. Virtual particles are experimenting with the im/possibilities of non/being, but that doesn't mean they aren't real; on the contrary. Consider this headline: "It's Confirmed: Matter Is Merely Vacuum Fluctuations."[42] The article explains that most of the mass of an atom, its nucleus made of protons and neutrons (which constitute the bulk of an atom) is due not to its constituent particles (the quarks), which only account for a mere 1 percent of its mass, but rather to the contributions from virtual particles.[43] *The void can hardly be thought of as that which doesn't matter!*

QFT not only reworks the classical understanding of the void, but also of matter in its inseparability from the void. Consider the classical physics view of an electron, one of the simplest particles—a point particle—a particle so small as to be of zero dimensions. Not only is it without extension, it is without an interior, completely devoid of structure. And yet it causes a great deal of trouble, both for classical and quantum physics.[44]

According to QFT, as a result of time-being indeterminacy, the electron does not exist as an isolated particle but is always already inseparable from the wild activities of the vacuum. That is, the electron is always (already) intra-acting with the virtual particles of the vacuum in every imaginable way. Let's take just a very small peek "into" the electron and the infinite number of wild things going on.

Electrons are charged particles, which means they are susceptible to, or we might even say inclined toward, touching and being touched. Indeed, touching, according to physics, is but an electromagnetic intra-action between charged particles. (The reason the desk feels solid, or the cat's coat feels soft, or we can even hold coffee cups and one another's hands, is an effect of electromagnetic repulsion. All we really ever feel is the electromagnetic force, not the other whose touch we seek.) The electromagnetic force experienced between two charged particles depends on the relative nature of their charges: opposites attract, and like charges repel one another.

Now, since a charged particle emits an electromagnetic field and charged particles positioned in electromagnetic fields feel an electromagnetic force on them, the electron being charged both emits and intra-acts with its own field. This self-touching intra-action—a constitutive part of what an electron is—turns out to be a source of unending anxiety in the physics community. Commenting specifically on the electron's self-energy intra-action, the physicist Richard Feynman expressed *horror* at the electron's monstrous nature and its perverse ways of engaging with the world: "Instead of going directly from one point to another, the electron goes along for a while and suddenly emits a [virtual] photon [which is the carrier of the electromagnetic field]; then (horrors!) it absorbs its own photon. Perhaps there's something 'immoral' about that, but the electron does it!" This self-energy/self-touching term has also been labeled a "perversion of the theory" because its value is infinite, which is an unacceptable answer to any question about the nature of the electron (such as, what is its mass or charge?). Apparently, touching oneself, or being touched by or in touch with oneself—the ambiguity may itself be the key to the trouble—is not simply troubling but a *moral* violation, the very source of all the trouble.

But it's worse (better) that that! For this simple self-energy intra-action is not a process that happens in isolation, either. All kinds of more involved things can and do occur in its intra-action with this frothy brew of nothingness. In fact, there is a virtual exploration of every possibility, an infinite set of possible ways of self-touching through touching others in all possible ways. So there is an infinity of infinities.[45]

In fact, Feynman proposed a "renormalization" procedure that attempts to reel in the electron's queerness, its unruliness. According to this procedure the "bare" electron (which is mathematically infinite) is "dressed" with the infinite contributions of the virtual particles of the vacuum such that, in the end, the physical electron is finite. (I'm using technical language here!) That is, what renormalization entails is the subtraction of two infinities to get something finite. This renormalization procedure necessarily entails taking into account the infinite possible intra-actions with all virtual particles in all possible ways—that is, all possible histories.

Hence, according to QFT, *even the smallest bits of matter are an enormous multitude!* Each "individual" is made up of all possible histories of virtual intra-actions with all others; or rather, according to QFT, there is no such thing as a discrete individual with its own roster of properties. In fact, *the "other"—the constitutively excluded—is always already within: the very notion of the "self" is a troubling of the interior/exterior distinction.* Matter in the indeterminacy of its being un/does identity and unsettles the very foundations of non/being. Together with Derrida, we might then say, "Identity . . . can only affirm itself as identity to itself by opening itself to the hospitality of a difference from itself or of a difference with itself. Condition of the self, such a difference *from* and *with* itself would then be its very thing . . . the stranger at home" (A 10/28). What is being called into question here is the very nature of the "self"; all "selves" are not themselves but rather the iterative intra-activity of all matter of time-beings. *The self is dispersed/diffracted through being and time.* In an undoing of the inside/outside distinction, it is undecidable whether there is an implosion of otherness or a dispersion of self throughout spacetimemattering.[46]

Hence, matter is an enfolding, an involution: it can't help touching itself, and in this self-touching it comes into contact with the infinite alterity that it is. Ontological indeterminacy, an unending dynamism of the opening up of possibilities, is at the core of mattering. How strange that indeterminacy, in its infinite undoing of closure, is the condition for the possibility of all structures in their dynamically reconfiguring stabilities (and instabilities).

According to QFT, there is no a-void-ing the fact that the void is far from empty. Indeed, nothingness is an infinite plentitude, not a thing, but a dynamic of iterative re-opening that cannot be disentangled from (what) matter(s).

Re-Turning and Re-Membering as Counterhegemonic Practices: A Counterpolitics to Colonialism's Avoidances and Erasures

Ironically, the land that was denounced as <a wilderness in which no white people's culture could prosper> became cultivated by the invaders' bloody battles and desires.[47]

Every *hibakusha* knows their survival carries within it the wailing and silence of the dead.[48]

The climax of the novella is the narrator's trip to Trinity Site, the place where the first plutonium bomb was detonated on July 16, 1945, at 5:29 a.m. It is here, at the end of her journey, the very place where it all began, standing in the midst of a desert, inside a fenced area with nothing inside it save a monument to nothingness—to Ground Zero—that the fullness of these embodied tracings of all the various colonial entanglements comes full circle.

Hayashi is committed to being a chronicler of August 9.[49] Given that she deliberately writes against the grain of chronology, perhaps Hayashi's commitment to tracing the material entanglements condensed into the spacetime point of August 9 might be more aptly captured by the more unconventional title "travel-hopping scribe" of August 9.[50] *From Trinity to Trinity* is not a time-travel novel but a time-diffraction tale, an embodied pilgrimage committed to tracing the material entanglements: a risky journey of placing one's body in touch with the matter/materiality of specific colonialist histories—an embodied accounting of some of the sum of all possible histories (Feynman's path integral approach), or "super-many-times" (as in Japanese Nobel laureate physicist Sin-Itiro Tomonaga's QFT approach), an iterative circling back around (as in Japanese Nobel laureate physicist Hideki Yukawa's *maru* or circle approach)—touching the infinite alterity that constitutes a point.[51] What is the structure of the infinity of a point labeled (on some calendars as) August 9? Re-turning to a point to face the incalculable.

Re-turning is a troubling matter, a matter of troubling. Loop diagrams in QFT are calculational devices representing processes in
which there is a re-turn to—a touching of—the self. Loops are the
ones that cause the most trouble for ruling conceptions of space,
time, matter, causality, and nothingness.

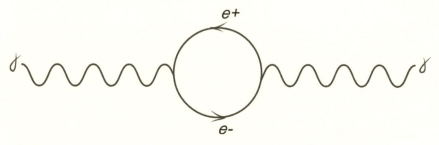

Figure 9.3 This "Feynman diagram" is one expression of the non/
emptiness of the void. It represents the void performing a vacuum fluc-
tuation (just one of an infinite number of fluctuations of the void in
its specific structuration). This one shows the virtual creation and annihilation of an electron-positron pair (which are jointly created and annihilated, and where the positron is an anti-electron—that is, its anti-matter
partner). This can also be understood as a photon self-energy diagram.
Wavy lines represent photons (quanta of the electromagnetic field, or light-
particles, a particular case of which might be a gamma ray or high-energy
radiation relevant to nuclear decay), while solid lines are electrons (and
positrons): e- represents a virtual electron traveling forward in time and e+
a virtual positron moving backward in time). The loop diagram is (itself)
infinite and needs to be renormalized; it represents but one an infinite set
of possible spacetimemattering-histories. That is, there are an infinity of
infinities that constitute each finitude. The diagram displays fluctuations
of the nothingness: virtual creation-annihilation, birth-death, with all the
potential that that holds.[a]

a. This diagram sets up an interesting set of reverberations with a diagram we
might draw of the pilgrimage of Hayashi's travel-hopping narrator who re-turns to
the spacetime point Trinity July 16, 1945, from another crucial spacetime point
Nagasaki April 9, 1945. Braving a re-turn to the void, the narrator risks this self-
energy intra-action, this undoing of self, and is thereby transformed from victim to
survivor, in concert with all entangled beings ("human" and "nonhuman") who are
hibakusha. In particular, this diagram is part of a self-intra-action diagram where
touching oneself involves touching Others. The renormalized self is a collectivity, not an individual, in an undoing not only of self/other but human/nonhuman.
Hence, as Hayashi points out, revenge doesn't make any sense. All this is made
possible by the fact that nothingness (the wounded desert, the devastated cityscape)
is not empty.

Being an August 9 travel-hopping scribe is different from being a historian. For one thing, travel hopping involves making the journey in spacetime, tracing the multiple histories with one's body, putting the self at risk as part of a committed response-ability to those who have died and those not yet born. It entails re-cognizing material kinship with this exploded/imploded moment in time.

> I am going to Trinity [she tells her friend] . . . the truth is, even today I still want to break away from August 9. . . . I have always wished I was not related to August 9.
> . . . Katsura [my son] is a second-generation *hibakusha* . . . he disliked being an inmate on death row without a prison term. . . . He wanted to live away from August 9.
> . . . Trinity is the starting point of my August 9. It is also the final destination of *hibakusha*. From Trinity to Trinity—.
> If I make that journey, I can hold August 9 within my life circle. If I can never be free from the event, I should end my relationship by swallowing it.[52]

What does it mean to swallow an event? Perhaps this is an evocation of the *ouroboros*, the mythical symbol of the serpent biting its tail, representing "creation out of destruction, Life out of Death."[53] Or perhaps it means to ingest the event like radiation: to take it into your gut, to feel it leach into your bones, mutate your innards, and reset your cellular clocks. Perhaps it is about the im/possibility of metabolizing the trauma, transforming the self from victim to survivor. Perhaps it is a way of un/doing the self, of touching oneself through touching all others, taking in multitudes of Others that make up the very matter of one's being in order to materially transform the self and one's material sense of self.[54] Perhaps it is about the willingness to put oneself at risk, to place one's body on this wounded land, to be in touch with it, to have a felt sense of its textures, to come to terms with a shared sense of vulnerability and invisibility, to feel the ways that this land, this void, which marks the colonizers' continuing practices of *avoid*ance, always already inhabits the core, the nucleus of your being.

> I walked to <Ground Zero>. . . .
> From this point in July, fifty years ago, the flash of light of the atomic bomb ran all directions in the desert. I heard, on the day of the experiment, it had been raining hard since morning,

unusual in New Mexico. The experiment was carried out in
the heavy rain. The flash of light boiled the downpour and,
with that white froth, ruined the fields, burned the helpless
mountains, and shot up to the sky. And then silence. Without
time to defend and fight back, the wilderness was forced into
silence.[55]

Let us pause before this silence, before rushing on, this silence
threaded through with all matter of murmurings, so many cries that
might yet have been but never were.

From the bottom of the ground, from the exposed red faces of
faraway mountains, from the brown wasteland, the waves of
silence came lapping and made me shudder. How hot it must
have been—
    Until now as I stand at the <Trinity Site>, I have thought it
was we humans who were the first atomic bomb victims on
Earth. I was wrong. Here are my senior *hibakusha*. They are
here but cannot cry or yell.
    Tears filled my eyes.[56]

Here at Ground Zero, time-being was shaken to its core: matter was
split off from itself—traumatized. Violence tears holes in the very
fabric of the world in its sedimenting iterative intra-activity. Woundedness is not reserved for human beings alone (and an account of
woundedness must at the same time bring into the story the many
thousands of people who were downwind from the blast).[57]
    Landscapes are not stages, containers, or mere environments for
human and nonhuman actors. Landscape is not merely visually
akin to a body; it is the skin of the earth.[58] Land is not property or
territory; it is a time-being marked by its own wounds and vitality,
a layered material geo-neuro-bio-graphy of bones and bodies, ashes
and earth, where death and life meet. Etymological entanglements
already hint at a troubling of assumed boundaries between allegedly
different kinds: Earth, *humus* (from the Latin), is part of the etymology of *human*, and similarly, *Adam* (Hebrew: [hu]man[kind]) derives
from *adamah* (Hebrew: ground, land, earth), giving lie to assertions
of firm distinctions between human and nonhuman, suggesting
a relationship of kin rather than kind—a cutting together-apart.[59]
Time-beings do not merely inhabit, but rather are *of* the landtime-

scape—the spacetimemattering of the world in its sedimenting enfoldings of iterative intra-activity. Memory is not merely a subjective capacity of the human mind; rather, "human" and "mind" are part of the landtimescape—spacetimemattering—of the world. Memory is written into the worlding of the world in its specificity, the ineliminable trace of the sedimenting historicity of its iterative reconfiguring.

> We are heading for Los Alamos. It is a steep mountainous road. One side of the road is a cliff, and below it I can see the mesas we saw on our way to Santa Fe. The wind that blows in the canyon seems to carry grasses and bushes away, so there is no green on the cliffs of the mesas. Stones and dirt also get blown away, so the cliffs have small holes like the ones an insect bites in cabbage leaves. Seen from far away, these holes are round as if they were hit by an adult's fist. Holes of the same size are scattered on the surface of the cliffs, and gray stones are partly visible here and there. The wind blew away the dirt that covered buried stones, and when the stones fell, holes remained. So stones lay under the mesas. These stones that fell off the cliffs are the dead of the mesas. I remember my classroom when the second semester started after the defeat of the War. In my grade, fifty-two students died. So we had one less class when we formed new classes.[60]

Hayashi understands land, in this case, this marked void, this silenced land, as the ground for respectful just and nonviolent mourning, for re-membering. Re-membering is a bodily activity of re-turning. She must place her body on this wounded ground to hear its murmuring silences and muted cries, to re-member and reconfigure the spacetimemattering of all *hibakusha* in their material entanglements.

> I have always been aware of being a *hibakusha*. But as soon as I started walking through the small passage within the fenced area led by a guide, my always-present awareness of being a victim disappeared from my mind. It was as if I became a fourteen-year-old again. I may have been walking toward an unknown <Ground Zero> as though I were someone from the <the time> before August 9, but it was when I stood in front of the memorial that I was truly exposed to the atomic bomb.

Looking back, I did not shed a tear on August 9. As I ran with the pack of people whose hands, feet, faces no longer looked human, no tears came to me. . . .

For the first time here at Trinity, however, I might be crying with human tears that I did not shed on August 9. Standing on the land that speaks no words, I shivered, feeling its pain. Until today, I have lived with merciless pains that hurt my mind and body. But it could have been the pain of the skin that grew from August 9. Here in this desert I had momentarily forgotten my life as a *hibakusha*.[61]

It is here, in the midst of the nothingness, the place where living and dying meet, where time-being is exposed as indeterminately multiple and filled with all matter of desiring im/possibilities, that the travel-hopping scribe can finally lay to rest her fifty-two classmates who were denied their own deaths. Long ago she had taken on this response-ability for the fifty-two and carried them around with her all these years. It is in putting herself at risk, in risking her sense of self, this work of embodied re-membering, that she can finally release her tears and let them rain down on the ground.

When I told you I was coming to Trinity, you asked me if I was an atomic bomb maniac. I wonder with what I can possibly fill the fifty-two spaces that were once lived by fifty-two school-mates in my grade. I want to embrace the emptied spaces but my hand reaches towards nothing.[62]

In re-turning to the nothingness, she brings one void in its particularity (Nagasaki) to another (Trinity), not to renormalize these infinite violences, avoidances, and erasures, but to bring to bear the clouds of im/possibilities that surround these entangled events.[63] What does it mean to confront the nothingness, to touch its fullness? This is a question that cannot be answered in the abstract, not once and for all, but must be asked over and over again with one's body.

This question, which must be lived, re-turns us to a question that has been held in suspension: for whom is Ground Zero empty? Clearly, this land is far from empty: on the contrary, it is teeming with all matter of im/possibilities—material conditions of living and dying. Living and dying in this void are a multitude of beings excluded from the designation of "human". Not only those beings living right at Ground Zero at the time of the Trinity test, including rattlesnakes, insects, plants, rocks, and soil, but also all those time-beings downwind from the test site, including those who often don't get counted as (fully) human, together with the bones and ghosts of their deceased ancestors and their future offspring. That is, what resides in the void are all those who endure despite layer upon layer of colonial and racialized violence, all those whom Man counts as Other, variously marked as subhuman, nonhuman, inhuman, or not even acknowledged as worthy of a mark or being named.

In fact, this parcel of land, designated as "the wilderness of New Mexico," on and around the Trinity site, is "home to nineteen American Indian pueblos, two Apache tribes, and some chapters of the Navajo Nation." The fact that there were 19,000 people living within a fifty-mile radius of the secret test is something that was not only ignored by the U.S. government until 2014, but unfortunately, is not mentioned by Hayashi, though they surely belong among her kin.[64]

For Hayashi, it is precisely the question of re-membering and just mourning that defines being human, not to thereby define its nature as some particular singularity, rooting the story in the soil of human exceptionalism, but rather to bring it back around to questions of the nature of the "human" (in its differential constitution). What makes us human is not our alleged distinctiveness from—the nonhuman, the inhuman, the subhuman, the more-than-human, those who do not matter—but rather our relationship with and responsibility to the dead, to the ghosts of the past and the future.[65] The pilgrimage of Hayashi's unnamed protagonist is a work of mourning, a concerted ongoing labor, never finished or complete; where mourning is not about making memorials, but rather about ontologically reconfiguring a past that never was on behalf of possibilities for a better future, not as performed by a willful liberal humanist subject, but in the tracings of entanglements of multiple time-beings through which the unnamed protagonist is herself constituted. It is in bodily bringing together the different structures of nothingness—tracing their entanglements—that the world can mourn and that the unnamed come to matter and are recognized as part of the ongoing reworlding of the world.

Hayashi's political-ethical commitment to the activism of re-membering the *hibakusha* has been a life practice of tracing the entangled violences of colonialism, racism, nationalism dispersed across spacetime. Crucial to this ongoing labor of mourning is the work of re-turning—turning it over and over again[66]—decomposition, composting, turning over the humus, undoing the notion of the human founded on the poisoned soil of human exceptionalism.[67] Not to privilege all other beings over the human, in some perverse reversal, but to begin to come to terms with the infinite depths of our inhumanity, and out of the resulting devastation, to nourish the infinitely rich ground of possibilities for living and dying otherwise.

Notes

This essay was originally published in issue 92 of the journal *New Formations*, published by Lawrence and Wishart, London, 2017.

1. I am deeply grateful to my very patient editors, Matthias Fritsch, Phil Lynes, and David Wood, for their warm encouragement and for being willing to work with me on questions of timing. All diagrams in the essay have been carefully and skillfully crafted by Elaine Gan; I am very appreciative of her willingness to draw and redraw these diagrams for me. Thanks are due to Cleo Woelfle-Erskine, Vivian Underhill, Lani Hanna, and Noya Kansky for their careful reading and helpful feedback. Any errors are my responsibility.

Damian Carrington, "The Anthropocene Epoch: Scientists Declare Dawn of Human-Influenced Age," *Guardian* (2016), https://www.the guardian.com/environment/2016/aug/29/declare-anthropocene-epoch -experts-urge-geological-congress-human-impact-earth. Under consideration for a specific date marking the new epoch: July 16, 1945, the date of the Trinity test.

2. Particularly incisive critiques include those by Zoe Todd, "Relationships," *Theorizing the Contemporary, Cultural Anthropology* (2016), https://culanth.org/fieldsights/799-relationships; Neel Ahuja, "The Anthropocene Debate: On the limits of Colonial Geology," blog (2016), https://ahuja.sites.ucsc.edu/2016/09/09/the-anthropocene-debate -on-the-limits-of-colonial-geology/); and Dana Luciano, "The Inhuman Anthropocene," *LA Review of Books* (2016), http://avidly.lareviewofbooks .org/2015/03/22/the-inhuman-anthropocene/.

3. The Doomsday Clock of the *Bulletin of Atomic Scientists*, which is now calibrated to take account of climate change as well as nuclear disaster. Introduced in 1947, the distance of the clock hand to midnight measures how closely a panel of scientists believes we are to global disaster.

4. Tanya H. Lee, "H-Bomb Guinea Pigs! Natives Suffering Decades after New Mexico Tests," *Indian Country Media Network* (2014), http:// indiancountrytodaymedianetwork.com/2014/03/05/guinea-pigs-indigenous -people-suffering-decades-after-new-mexico-h-bomb-testing-153856. Here's the more extended text: "Much has been made of the dropping of the first atomic bomb on two now-infamous cities, Hiroshima and Nagasaki, and the health-nightmare aftermath. But only now [2014] is the spotlight

being put onto those who had the actual first atomic bomb dropped in their vicinity—it was the Americans' own people, Turtle Island's original inhabitants, the Indigenous Peoples of the southwest. The world's first atomic bomb was detonated on July 16, 1945, in New Mexico—home to 19 American Indian pueblos, two Apache tribes and some chapters of the Navajo Nation. Manhattan Project scientists exploded the device containing six kilograms of plutonium 239 on a 100-foot tower at the Trinity Site in the Jornada del Muerto (Journey of Death) Valley at what is now the U.S. Army's White Sands Missile Range. The blast was the equivalent of 21 kilotons of TNT. At the time an estimated 19,000 people lived within a 50-mile radius."

5. Ibid.

6. Masahide Kato, "Nuclear Globalism: Traversing Rockets, Satellites, and Nuclear War via the Strategic Gaze," *Alternatives* 18 (1993): 348.

7. Ibid.

8. Ibid, 339.

9. The Japanese word *"hibakusha"* designates surviving victims of the 1945 atomic bombings of Hiroshima and Nagasaki (Wikipedia), https://en.wikipedia.org/wiki/Hibakusha, accessed July 9, 2016.

10. Daniel R. Wildcat, "Indigenizing the Future: Why We Must Think Spatially in the Twenty-First Century, *American Studies* 46, nos. 3 and 4 (2005): 417–40; *Indigenous Studies* 1 (2005–6): 433–34.

11. Lee's essay cited in note 4.

12. Dan Frosch, "Decades after Nuclear Test, U.S. Studies Cancer Fallout: Examination Will Probe Radiation Exposure Near 1945 Trinity Blast in New Mexico," *Wall Street Journal*, September 15, 2014.

13. Quote from essay by Lee (cf. note 4).

14. This essay is an excerpt of a current book project, Karen Barad, *Infinity, Nothingness, and Justice-to-Come*, which contains a more extended discussion of the issues, including an investigation into how specific indigenous temporalities (of American Southwest tribes) and specific Japanese conceptions of temporality matter to this story.

15. See Barad, *Meeting the Universe Halfway: Quantum Physics and the Entanglement of Matter and Meaning* (Durham, N.C.: Duke University Press, 2007).

16. Entanglements call into question the geometrical notions of scale and proximity; topology, with its focus on issues of connectivity and boundary, becomes a more apt analytical tool. It's not that scale doesn't matter; the point is that it isn't simply given and what appears far apart might actually be as close as the object in question; indeed, it may be an inseparable part of it. See the concept of *spacetimemattering* in Barad, *Meeting the Universe Halfway*.

17. "In this flash the body becomes molecules"—Bill Johnston in the film on Japan and the Atomic Bomb class he cotaught with Eiko (4:58 min.).

18. This line from the *Bhagavad Gita* was famously quoted by physicist J. Robert Oppenheimer (his translation from the Sanskrit) in the wake of the first atomic bomb explosion.

19. This is only to suggest the barest hints of a rich history, which can't be told here. For a more in-depth account as it relates to the story being told here, see Barad, *Infinity, Nothingness, and Justice-to-Come*. One crucially important reference is Silvan S. Schweber, *QED and the Men Who Made It: Dyson, Feynman, Schwinger, and Tomonaga* (Princeton: Princeton University Press, 1994).

20. Trinh T. Minh-ha, *Elsewhere, Within Here: Immigration, Refugeeism and the Boundary Event* (London: Routledge, 2010), 56.

21. Minh-ha, "Not You/Like You: Post-Colonial Women and the Interlocking Question of Identity and Difference," *Inscriptions*, special issues *Feminism and the Critique of Colonial Discourse*, 3–4 (1988), http://cul turalstudies.ucsc.edu/PUBS/Inscriptions/vol_3-4/minh-ha.html.

22. And not just some so-called microworld, as if there were a line in the sand between "micro" and "macro," as if scale were already given. As Bohr was fond of pointing out, if Planck's constant (the measure of discreteness or lack of continuity of the physical world) had been larger, then we wouldn't have talked ourselves into a metaphysics of individualism to begin with. In a performative relational ontology, it's differentiating entanglings all the way down.

23. See, for example, Marcos Moshinsky, "Diffraction in Time," *Physical Review* 88, no. 3 (1952): 625–31, and Časlav Brukner and Anton Zeilinger, "Diffraction of Matter Waves in Space and in Time," *Physical Review A* 56, no. 5 (1997): 3804–24.

24. See Brukner and Zeilinger, "Diffraction of Matter Waves in Space and in Time."

25. There were several different variations of the sum over all possible histories' approach to QFT that were proposed: Feynman's path-integral approach, Tomonaga's super-many-times approach, and Yukawa's *maru* or circle approach—each inspired by Dirac's many-times formulation of relativistic quantum mechanics. For more details on these approaches and how they figure in the story of QFT and the atom bomb, see Barad, *Infinity, Nothingness, and Justice-to-Come*.

26. On the irony of the "new" in "new materialisms" and capitalism's push to discard the old in favor of the new (as discussed in Barad, *Meeting the Universe Halfway*), see Barad, "Nothing Is New / There Is Nothing That Is Not New," invited keynote address for the "What's New about New Materialism?" Conference, University of California Berkeley, May 5, 2012.

27. Kyoko Hayashi, *From Trinity to Trinity* (Barrytown, N.Y.: Station Hill, 2010). The translation from Japanese to English and the substantial Introduction and Afterword are by dancer and choreographer Eiko Otake, who has recently done some amazing artist-activist work on Fukushima. I

engage with this latter work in Barad, "Ecologies of Nothingness: Haunted Spacetimescapes, Dances of Devastation and Endurance" (unpublished paper).

28. See especially Lianying Shan, "Implicating Colonial Memory and the Atomic Bombing: Hayashi Kyoko's Short Stories," *Southeast Review of Asian Studies* 27 (2005), http://www.uky.edu/Centers/Asia/SECAAS/ Seras/2005/Shan.htm. For Korean atomic bomb casualty statistics, see, for example, "South Korea A-Bomb Victims Angered by Obama's Hiroshima Visit," NDTV, May 26, 2016, http://www.ndtv.com/world-news/south -korea-a-bomb-victims-angered-by-obamas-hiroshima-visit-1412418.

29. SM xix/15.

30. I offer only a very abbreviated discussion of the quantum-eraser experiment here. For a detailed description and analysis, see Barad, *Meeting the Universe Halfway*. I also try to highlight some of its implications in my article in Barad, "Quantum Entanglements and Hauntological Relations of Inheritance: Dis/continuities, SpaceTime Enfoldings, and Justice-to-Come," *Derrida Today* 3, no. 2 (2010): 240–68.

31. Wave-particle duality is discussed at length in chap. 3 of Barad, *Meeting the Universe Halfway*. On the quantum-eraser experiment, see especially ibid., chap. 7.

32. See Barad, "Quantum Entanglements."

33. The National Atomic Museum was rebuilt at another location under a new name (National Museum of Nuclear Science and History) in 2009.

34. Hayashi, *From Trinity to Trinity*, 16–17.

35. On the point that the void has structure (!), see the section to follow on QFT.

36. Hayashi, *From Trinity to Trinity*, 20.

37. I have taken the liberty of supplementing Hayashi's tracings to include some of the other most evident entanglements.

38. This section includes excerpts from Barad, *What Is the Measure of Nothingness? Infinity, Virtuality, Justice/Was Ist das Maß des Nichts? Unendlichkeit, Virtualität, Gerechtigkeit* (Berlin: Hatje Cantz Verlag, 2012), and Barad, "On Touching—The Inhuman That Therefore I Am," in *Power of Material/Politics of Materiality*, ed. Susanne Witzgall and Kirsten Stakemeier (Berlin: Diaphanes, 2015), 1:153–64 (originally published in *Differences* 23, no. 3 [2012], but with very unfortunate typographical errors).

39. This is far too rapid a trot through a thick set of histories, but I'm afraid it will have to suffice for now. For far more developed and detailed accounts, see Karen O'Brien, "'These Nations Newton Made His Own': Poetry, Knowledge, and British Imperial Globalization," in *The Postcolonial Enlightenment: Eighteenth-Century Colonialism and Postcolonial Theory*, ed. Daniel Carey and Lynn Festa (New York: Oxford University Press, 2009), 290. See also especially Margaret C. Jacobs and Larry Stewart,

Practical Matters (Cambridge, Mass.: Harvard University Press, 2004), and Sylvia Wynter, "Unsettling the Coloniality of Being/Power/Truth/Freedom," *CR: The New Centennial Review* 3, no. 3 (2003): 257–337.

40. In reading this paragraph, in particular, it is well to remember my specific use of the slash, as in im/possibility: to evoke the enactment of an agential cut that cuts together-apart (one move), differentiating-entangling. Hence, as in Derrida's notion of *impossibility*, impossibility and possibility are not mere opposites.

41. It is through this means that physicists create new particles using accelerators, by putting energy into the vacuum. (See, for example, the discovery of the Higgs particle at CERN in July 2012.) The existence of antiparticles was postulated by Paul Dirac in 1928 in an essay in which he put forward a relativistic theory of quantum mechanics. The first antiparticle to be discovered was a positron (an antielectron) in 1932. Antiparticles have the same mass but opposite charge as the corresponding particle (e.g., while electrons have a negative charge, positrons have the same mass as an electron but a charge of opposite sign), and they travel backwards in time. More on this in Barad, *Infinity, Nothingness, and Justice-to-Come*.

42. Stephen Battersby, *New Scientist*, November 20, 2008, http://www.newscientist.com/article/dn16095-its-confirmedmatter-is-merelyvacuum-fluctuations.html, accessed February 2012.

43. As we'll soon see, all particles, including quarks (which are the constituent particles of protons and neutrons making up the nucleus of an atom), are inseparable from and constituted by the virtual fluctuations of the vacuum.

44. From the point of view of classical physics, either the electron is unstable or its mass is infinite—not good choices, but physicists thought this puzzle might be solved by providing a quantum physics understanding of matter. But the quantum account of matter presented its own set of difficulties. The difficulties, whether from a classical or quantum-physics vantage point, stem from the particle's so-called self-energy: in particular, because it is a charged particle, it emits an electromagnetic field, and in calculating its mass one must take account of its interactions with itself (i.e., its infinite self-energy).

45. For more details see Barad, "TransMaterialities: Trans/Matter/Realities and Queer Political Imaginings," *GLQ: A Journal of Lesbian and Gay Studies* 21, nos. 2 and 3 (2015): 387–422.

46. This is true of moments of time as well as bits of matter (being), each of which is indeterminately infinitely large and infinitesimally small, where each bit is specifically constituted through an infinity of intra-actions with all others.

47. Hayashi, *From Trinity to Trinity*, 24.

48. Ibid., xi.

49. As described by Eiko Otake, the book's translator, in Hayashi, *From Trinity to Trinity*, xii.

50. This title is of course inspired by Hayashi's own term, "travel hopping" (which in any case sets up wonderful resonances and dissonances with the overused and poorly understood term "quantum leaping," which has been (mis)appropriated by capitalist markets to sell all kinds of consumer products).

51. These seldom mentioned alternative approaches by Japanese physicists seemed important to include here, even if I don't have time/space to discuss them in this essay. Much more needs to be said, and I consider these approaches in some details in a forthcoming book. The various approaches to QFT—whether that proposed by Feynman (superposition of all possible histories), Tomonaga (super-many-times), or Yukawa (*maru*)—refer to abstract studies in theoretical physics that queried fundamental notions like universality, singularity, materiality, nothingness, and alternative histories. Each approach is shot through with efforts to understand nuclear forces and build atomic bombs. In each case there is evidence of a breech of alleged divisions between social, political, and natural forces, including those distinctions said to exist between practical physics, technological prowess, and highly abstract physical theories; between pure science and militarisms, capitalisms, nationalisms, colonialisms, racisms; and between politics and physics. For a more detailed discussion see Barad, *Infinity, Nothingness, and Justice-to-Come*.

52. Hayashi, *From Trinity to Trinity*, 9, 11.

53. "The ouroboros has several meanings interwoven into it. Foremost is the symbolism of the serpent biting, devouring, or eating its own tail. This symbolizes the cyclic Nature of the Universe: creation out of destruction, Life out of Death. The ouroboros eats its own tail to sustain its life, in an eternal cycle of renewal"; copied all over the web; original source not clear; see, for example, http://www.tokenrock.com/explain-ouroboros-70 .html.

"Life is a circle, into which O'Keeffe offered her bones. Is it reincarnation?"; Hayashi, *From Trinity to Trinity*, 28.

54. Not only her fifty-two classmates, all of whom she's been carrying around all these years, but also her other fellow *hibakushas* here in New Mexico, including the land, the rattlesnakes, the wind.

55. There is a factual error here: while it is true that it did rain that morning and the rain was quite unusual, the test was delayed until the rain stopped. This error is noted and addressed by the translator, Eiko Otake. I take up this point further in Barad, *Infinity, Nothingness, and Justice-to-Come*.

56. Hayashi, *From Trinity to Trinity*, 49–50.

57. See notes 4 and 13.

58. Hayashi's references to iconic American painter Georgia O'Keeffe are frequent and clearly significant. O'Keeffe is famous for painting landscapes not as mere objects, but as bodies, bodies with their own explicit eroticism. The vibratory bodily sensuality of the land is uniquely vividly

expressed in O'Keeffe's nonrepresentational realist paintings of the New Mexico desert. Hayashi specifically mentions the fact that O'Keeffe's bones are scattered on the mountain peak. "We were to drive on that land. My destination was Trinity, but personally I was also looking forward to seeing and feeling the land where O'Keeffe's life and death were united"; Hayashi, *From Trinity to Trinity*, 13.

At the same time, it is important to note that some of O'Keeffe's paintings have been objected for their cultural appropriation. For example, Pueblo neighbors of the Georgia O'Keeffe Museum "have expressed strong opinions against public exhibition of Katsinam, including katsina tithu, in sculptures and paintings," which O'Keeffe began to paint after seeing them in Pueblo ceremonies and dances performed in 1929; see, for example, Martha Schwendener, "The Spirit of Cultural Objects: A Review of 'Georgia O'Keeffe in New Mexico,' at the Montclair Art Museum," *The New York Times*, January 4, 2013.

The very question of different understandings of *landscape*—particularly, the important differences between American cultural conceptions and those of indigenous and Japanese cultures—is warranted in this discussion and requires further elaboration.

59. See Barad, "Diffracting Diffraction: Cutting Together-Apart," in *Parallax* 20, no. 3 (2014): 168–87, for more details on the agential realist notion of "cutting together-apart" (that is, differentiating-entangling).

60. Hayashi, *From Trinity to Trinity*, 29.

61. Ibid., 50–51.

62. Ibid., 33.

63. It was crucial to Hayashi's efforts to come to terms with humanity's inhumanity that she be in touch with all matter of inhumanness, including that which courses through all being. The reference to clouds here is simultaneously to clouds of virtual particles and rain clouds.

64. See note 4.

65. Which is not to suggest that this way of marking the human is yet another opportunity for human exceptionalism, since all time-beings mourn.

66. See Barad, "Diffracting Diffraction."

67. With gratitude to Donna Haraway, Maria Puig de la Bellacasa, and Kristina Lyons, among others, for the rich soil of this fertile material imagery.

Responsibility and the Non(bio)degradable

Michael Peterson

> . . . one transforms while exhuming.
> —*Jacques Derrida, "Biodegradables"*

"Biodegradable" typically designates products that are disposable—in a good way. We don't complain that they are flimsy. Rather, we know that they will "return to nature." This designation only makes sense when opposed to the "nonbiodegradable": products that are durable—in a bad way. The nonbiodegradable will not "go" anywhere. It retains its artificial identity long after it is no longer useful. These words are recent additions to our language. Their artificiality announces itself in their combination of Greek (*bios*) and Latin (*desgradus*). In his "Biodegradables: Seven Diary Fragments"—a text that I will privilege over the next few pages—Jacques Derrida reminds us that the word "biodegradable" seems to be something other: "It does not belong to the organic compost of a single natural language, this strange thing may be seen to float on the surface of culture like the wastes whose survival rivals that of the masterpieces of our culture and the monuments that we promise to eternity" (BSD 815). Here, we might pause and notice that "eco-deconstruction" echoes the Greco-Latin otherness and recent artificiality of "the biodegradable," although what this means for the survival of our texts is not yet clear.[1] Language, waste, masterpiece, culture, monuments, and the promise mark intersecting moments of the question of survival. What seems clear to Derrida at this point is that "like biodegradable, nonbiodegradable can be said of the worst and the best" (BSD 815).

Nuclear waste appears at several points in Derrida's text as a representative or even a sort of ambassador for the nonbiodegradable.

Derrida suggests nuclear waste as a candidate for that which could be thought as some "absolute" nonbiodegradable "in what is called the literal or strict sense" (BSD 863). It is crucial to note, however, that this waste's apparent nonbiodegradability cannot be thought merely as the continued existence of inert matter. Indeed, it is precisely the radioactivity of this waste that forces us, in order to act responsibly, to bury it and keep it buried. A difficulty soon arises, however, when Derrida argues that it is precisely a "singular impropriety" that allows something to resist degradation. Nuclear waste's intergenerational activity precludes the possibility of any essential appropriation while at the same time demanding that its handlers take it up as *their* responsibility. How, then, are we to be responsible for something whose very logic is prohibitory?

When we look closely at American nuclear-waste policy and attempt to read policy protocols and advisories philosophically, we avail ourselves of testaments from tasks forces and expert panels trying to think through precisely these problems. Derrida offers us the opportunity to engage in such reading while resisting the temptation to understand practical problems as merely *technical* problems. The claim that Derrida and whatever is meant when we refer to a kind of legacy of deconstruction can or even must orient us toward the environmental or ecological in all its material irradiance is less surprising when we take seriously the thought that the remainder—ashes, archive, trace, *restant*, waste—simmers beneath the surface of the discourse Derrida helped to inaugurate, erupting and interrupting again and again.

In the pages that follow, I take one particularly troubling passage from a report authored by the Human Interference Task Force— assembled to assess the likelihood of human interference with the underground waste-isolation facility that houses the United States' mid- to high-level nuclear waste and recommend preventative strategies—as representative of a certain view of responsibility. The Human Interference Task Force states, "Future societies with knowledge of the existence and location of the [nuclear waste] repository, its contents, and the risks of interference, bear the full responsibility for any of their actions that can reasonably be expected to adversely affect the performance of the repository."[2]

Thought in such a way, our responsible handling of the waste becomes intimately linked to the possibility of communicating adequate knowledge—whatever this designation might come to mean— to future generations. In fact, the very possibility of responsibility as it is being understood here hinges on the claim that, if certain conditions—in this case, the transmission of a bare minimum of knowledge—are met, a given society can become *fully* responsible for something, and, therefore, another society—even if this society is or is said to be the source of the inheritance in question—ceases to be responsible. Derrida flags the connection between textual transmission and the "ethico-ecological" most clearly in the following fragmentary remark: "enigmatic kinship between waste, for example nuclear waste, and the masterpiece."[3]

To approach these questions responsibly, a brief survey of the practical issue at hand—the long-term disposal of radioactive or nuclear waste—is appropriate.[4] Our use of nuclear fission as a source of energy for civilian and defensive purposes has left us with highly radioactive waste that will remain dangerous to organic life forms for thousands of years. No process capable of speeding up this waste's decay exists today. Throwing the waste in the ocean risks contaminating the majority of the earth, and attempting to launch it into space risks poisoning whatever is left of our atmosphere. The only apparent solution is to bury the waste deep underground in a geologically stable area until such a time that it no longer poses a threat. This is the project currently underway in Carlsbad, New Mexico, at the Waste Isolation Pilot Plant (WIPP). The question confronting those tasked with maintaining the WIPP's titular isolation involves determining how best to ensure that the waste is not accidentally or irresponsibly handled for the "period of concern": 10, 000 years.[5] The challenge is in finding a way to maintain the ongoing isolation of a dangerous material and, at the same time, maintain the warnings required for this unimaginably long period of time.

Creating a warning system that could potentially last for up to 10,000 years is not a simple task, and many different suggestions have been made.[6] From genetically modified blue glow-in-the-dark cacti—to suggest radioactivity and intentional human activity—to an enormous "landscape of thorns"—meant to communicate danger and inhospitableness—the message we send to the future ought to make future peoples understand that the land in question contains something artificial, deliberately isolated, and dangerous. Of course, our own more recent archeological efforts make it seem unlikely that any sort of monumentizing or marking will discourage investigation. There must, then, also be a system in place to communicate more complicated information: that the site is radioactive, what radioactivity means, and the dangers inherent in compromising the site's isolation. A truly multidisciplinary team of linguists, archeologists, futurists, science fiction writers, historians, physicists, economists, and other specialists was assembled in 1991 to discuss the problems of and possible solutions to such an undertaking. This team's findings are available to the public at large through the U.S. Department of Energy website, and I would strongly encourage those interested to review the document.[7]

The practical problems that arise from this project are enormous: how to ensure that our language is comprehensible to people whose identities and sociocultural context are necessarily unknowable to us? How to ensure that the information remains materially legible? How to preempt the varying levels of political and technological progression or regression that will occur in the next 10,000 years? The difficulty of properly handling nuclear waste lies at the intersection of writings on semiotics, sovereignty, materialism, and intergenerational ethics. The problem is that the half-life—the amount of time that must pass for one half of the original mass of radioactive material to decay—of nuclear waste is too long. It *will* survive, and when we deploy an ethical schema that hinges on transmission of information to the future, so must our warning.

The general strategy at work here follows a logic according to which our actions in the present—creating this hazardous waste and interring it—are justified because we will ensure that we have properly isolated the waste and that future societies are able to continue properly isolating our waste as well as, presumably, any additional waste that will undergo disposal in the future. That no such strategy to *in fact* transmit the information required to ensure continued isolation existed at the time that we began to produce nuclear waste or exists today is of secondary concern. Future peoples have been entrusted with stewardship over our waste as well as with developing the techniques required to make this stewardship possible. We are apparently justified in doing so because the resources our society accumulates through the production of nuclear waste are the conditions for future peoples to inherit this waste responsibly. This strategy is not one that we can charge with not caring about or counting future people as rational-ethical agents—in fact, the justification of our actions in the present *depends* on understanding future people in this way. Because future societies *will* inherit the earth as we have left it for them, *and* because future societies *can* inherit the earth, we, the present generation, *can*, in turn, bequeath it to them explicitly and as debt.

To take this further, it would appear that our nonbiodegradable waste- disposal strategy insists that this explicit bequeathal is the very height of responsibility. Indeed, although two teams assembled to assess the possibility of inadvertent human intrusion into these sites recommended that the EPA consider adopting a "no marker strategy," "because markers might draw attention to the WIPP" and so compromise its isolation, this suggestion was not, it would appear, seriously considered.[8] Having no markers at all would undermine the very basis upon which the decision to bury this waste was made. The logic of the EPA and its attendant task forces *needs* future generations to take responsibility for this waste so that our responsibility can come to an end.

To allow future generations to take up this risk, we must not only build sites capable of a degree of nonbiodegradability equal to the waste stored within them, but we must also communicate a message to future societies that resists erosion. The triple-survival of our waste, our site, and our message are deeply tied to the question of intergenerational responsibility. As other authors in this collection make clear, Derrida's own engagement with environmental or ecological problems is often anything but direct. However, any careful reader of Derrida should, I would think, be prepared to admit that he was a careful thinker of textual survival. The question of nuclear waste reveals that this engagement with the survival of texts is at once a thinking through of *waste*. And so here it is worth thinking alongside Derrida as we consider the problem of the possibility of communicating a warning to future societies. The well-known *mot d'ordre* of "Signature Event Context" iterability will help us in this endeavor as we consider for ourselves the enigmatic kinship between nuclear waste and the masterpiece. Iterability can, I hope to show, be understood in terms of the pair "biodegradable/nonbiodegradable."

We may begin to get a sense of what Derrida means when he uses the concept of biodegradability by contrasting it with its apparent opposite: nonbiodegradability. Derrida writes, "Everything that is 'biodegradable' lets itself be decomposed or returns to organic nature while losing there its artificial identity" (BSD 828). This preliminary understanding of biodegradability contrasts the artificial and the natural in the sense that returning to nature simply *is* losing one's artificial identity and vice versa. On the other hand, we have the *"perle"* or "pearl," a translation that brings to mind the expression "pearls of wisdom" but perhaps is more in harmony with our idiomatic English "gems."[9] Derrida describes the *perle* as similar to "a text, a verse, an aphorism, a *bonmot* [sic]" in that both are hard to digest" and "passed on" (BSD 832). The *perle*'s endurance guarantees that it will pass from the hands of its creators. The non-biodegradable survives precisely because it resists decomposition, appropriation, or assimilation.

The masterpieces we inherit from our own cultural history are so inherited on a variety of contradictory conditions. They "belong" to *our* history, but without historic idiosyncrasy. Their meaning remains discoverable or comprehensible without settling into *a* meaning. They remain reinterpretable—and this formulation is both an avowal and disavowal of fixed identity. The danger of our waste lies in its continuing to act. Its isolation is required into the future because it is not inert. And so, just as "Biodegradables" has outlived Jacques Derrida, our nuclear waste will outlive us. The information that we would transmit to the future must outlive us as well if we are to make the reception of this information a condition for the responsible disposal of our waste. The monuments we would make to and for our waste must become pearls or masterpieces in their own right. What, then, of biodegradability?

To understand the sense in which writing "decomposes," we will take a quick detour to a well-known passage from Derrida's 1972 essay "Signature Event Context":

> Every sign, linguistic or non-linguistic, spoken or written . . . , in a large or small unit, can be *cited*, put between quotation marks; in so doing it can break with every given context, engendering an infinity of new contexts in a manner which is absolutely illimitable.[10]

Signs are taken up by the addressee in the context of that addressee, not in the context in which they were written or spoken. This simple point allows us to understand what Derrida means when he writes that every sign can be—that is, necessarily has the possibility of being—taken up "outside" of its original context. As long as signs in general are understood differentially—which is to say that the meaning of a given sign is not imminent to the sign itself—then we can see that the new context in which a sign is encountered will, inevitably, alter the meaning of the sign. Furthermore, this inevitable alteration of meaning is not a defect. That a sign can acquire meaning in a radically new context is the condition for the transmission of meaning in general. If a sign could not be in-formed and taken up anew in a context other than its own, the sign would remain radically idiosyncratic and unable to communicate. That a sign can be read in a new context is a necessary condition for it functioning as a sign at all. Hence, the first point we should keep in mind when considering the project of ensuring adequate understanding of the warning for future peoples: we cannot know in advance the context in which the warning will be read and so can never, in principle, guarantee that the warning will be taken up in the way we need it to be.

Second, however, we must note that Derrida continues with the claim that a sign, in being encountered, will engender "an infinity of new contexts in a manner which is absolutely illimitable." The sign itself becomes a part of the context in which it is encountered and, in this way, alters the context in which it is encountered. So the fact of being read or encountered makes a difference to the context in which something is read or encountered. Derrida is pointing us to the fundamental reciprocity between sign and context: the meaning of a context is determined by the encountered sign, and the meaning of the sign is determined by the context in which that sign is encountered. The language of biodegradability that appears in the essay "Biodegradables" seventeen years after "Signature Event Context" can be understood as an elaboration of the sense in which a sign engenders new contexts. What is biodegradable "loses its artificial identity"—the identity that is made *for it* through its being made (or uttered)—and "returns to organic nature"—which is to say that it actually takes part in making the world in which it is encountered meaningful. One of the reasons that we find the "bio" of "biodegradable" put in parentheses in Derrida's text is to remind us that we should understand this concept not *only* in relation to the scientific-natural world of biologists. Rather, (bio)degradability refers us to a given utterance or text's simultaneous constituting role with regard to context and the absolutely unfixed or decentered structure of its own meaning. Texts, events, meanings, and efforts are all, in this sense, (bio)degradable.[11]

The warning, then, will not only change in meaning as its surrounding context (inevitably) changes, but will also itself engender new contexts. At this point, we can see that the relationship between the (bio)degradable and the non(bio)degradable is not a simply binary. Rather, we should see the relationship between (bio)degradability and non(bio)degradability as one of necessary coimplication.

That a sign is taken up in new contexts (and so undergoes shifts in meanings and engenders changes in context) *is* its survival. Only once it has stopped being taken up or encountered at all can it be said to no longer live—its identity would be completely dissolved. Likewise, the non(bio)degradable must continuously be taken up as meaningful (which would entail a displacement of what we would be inclined to call "its original meaning") in order to endure. So the pair "(bio)degradable-non(bio)degradable" don't oppose each other any more than the pair "life-death," but simply name two different features of survival in general. The enigmatic connection between nuclear waste and the masterpiece is found in the fact that both the waste and the masterpiece remain, decay, and, *in so doing*, live. Nuclear waste lives not *in spite* of its half-life, its slow decay, but precisely *because* of this decay. This is equally true of the warning itself: the message will survive only in being encountered, transformed, exhumed, and dissolved.

Of course, we must also admit that both the warning and the waste will not simply appear for future people like a message in a bottle washing up on a shore. It is conceivable that over a period of 10,000 years there will be a break in continuity, due to war, climate change, disease, but, barring these sorts of situation, we can try to solve the problem of effective communication by appealing to the fact that our descendants are overlapping, and not discrete, units. That there will be overlapping and successive generations who can be taught about the waste and its dangers allows us to think about the waste as an inheritance. Future generations *receive* the radwaste from us *and* from the immediately preceding generation *and* from every generation in between. However, we must insist that continuity is not the same as stability. In fact, we find in iterability—the repetition that can break with any given context and, in doing so, engender new contexts—a logic that allows us to understand the sense in which continuity promises instability. This is to say that there is a break, if we want to continue to use this language, or a rupture or disjointure at the very site of inheritance. Not only can we think that over 10,000 years the probability of an explicit historical break in sociopolitical continuity approaches inevitability—and we have good reason to be suspicious of any political regime that would claim for this kind of uninterrupted regency—we must add that the condition of interruption is the structure in and through which recontextualization, and so inheritance in the most general sense—reading—takes place. So we must be careful not to simply alter or clarify the ethical guideline under critique. That is, it is not enough to say that future people are fully responsible so long as there is continuity, for this continuity does nothing at all to guarantee sufficient transmission of adequate knowledge of the waste.

More to the point, and on the question of the guarantee, if the EPA takes the version of responsibility cited at the beginning of this essay as the basis for the justification of the continued production of nuclear waste, we are forced to admit that this condition fails. Future generations cannot inherit from us a message equal to the task of endowing them with full responsibility for our waste. The logic of language and inheritance expounded by the writers of *Reducing the Likelihood of Future Human Activities That Could Affect Geologic High-level Waste Repositories* is, simply put, not adequate to the phenomena.

It remains true, however, that our remains will be taken up, and, necessarily, not by us. There will be a future, and this future contains within it the promise of (to quote Derrida in *Rogues*) "an unforeseeable coming of the other" (R 84/123)—an unforeseeable coming of the *unknowable* or *absolute* other. Our responsibility can be found in the encounter with—the response to—the mark itself: namely, that we take up the mark and interpret it in its manifold of possible readings. If our responsibility toward future peoples means recognizing that they too will approach the warning and the waste as something to be interpreted *by them* and *in their unknowable future context*—that is, in recognition of their capacity to act responsibly—we cannot decide in advance how they are to interpret whatever it is we will pass on to them. The decision—how a warning is to be read, how waste is to be handled, and who our inheritors will be—cannot be made in advance. There is a bringing together of the ontological and the normative in this move that is characteristic of texts that fall more or less closely within the parameters of what we can call "deconstruction more generally." We *must* decide: there is waste, there is remainder. We *cannot* decide in advance: the future and our inheritors are to come. We *must* decide again and again. We *must* respond and ask or pray that the future recognize us and our remains. This is survival.

The claim that we must not act as though we could decide for future societies in advance of their coming might, under a certain light, provide us with reason to reconsider the no-marker strategy. We might look to reduce the violence we would be doing to future people by taking the possibility—or, as we may now want to say, necessity—of (mis)interpreting our warnings off the table. Indeed, the decision to leave any kind of warning at all commits us at least to a decision concerning what kinds of future Earthlings we hope to warn—anything that cannot read, in the broadest possible sense, or develop a Geiger counter would simply fall outside the purview of our warning. To avoid *that* exclusion, or *that* decision, we might find justification for leaving as little behind as possible. The very idea, though, that during our time on Earth we could take only memories and leave only footprints belies the superficial ontological "honesty" of leaving no markers. The waste is itself a marker of our having been here. Our footprints—radioactive as well as carbon—are also memories and so demand a response.

Because the future is unknowable in principle—all we know is that it is to come (*à venir*)—and yet, at the same time, our responsibility for this waste is indelibly linked with future peoples' ability to adequately take up our remains, we must admit that no specific action or principle can *fulfill* our responsibility toward future peoples. We cannot even appeal to a regulative idea such as "act in such a way that maximizes the context-sensitive decision-making of future peoples" because we would then be positing this idea as binding across time. Rather, through our understanding that a mark's survival is in its being taken up and transformed, we understand that the only way to preserve our warning is to constantly revisit our decisions, our method, and our obligations. The error, then, is made in thinking that our responsibilities have been or could ever be met once and for all. Rather than an indefinite deferral of all decisions, though, there is—and this "is" reveals to us the strength of Derrida's analysis in its being both normative and ontological—a sort of unlimited decision that is always being made. Indeed, the strongest condemnation we could muster against the strategy currently in place for dealing with our non(bio)degradables is that it forces a certain tactic of reading on future people *in order to* make them our redeemers. In another context, we read Derrida affirming that "the only attitude (the only politics—judicial, medical, pedagogical, and so forth) I would *absolutely* condemn is one which, directly or indirectly, cuts off the possibility of an essentially interminable questioning, that is, an effective and thus transforming questioning" (PI 239/252). Here we find reason to question the continued production of nuclear waste. The burden we place on future societies in understanding them as the redeemers of our action commits these future societies to logics and material inheritances that we have decided for them. The survival of our wastes, their (bio)degrad-ability/non(bio)degradability, erodes our ability to ever be finished with them. In *Rogues* Derrida writes, *"plus d'États voyous"*—more and no more rogue states. We, this we that was, is, and will have been responsible for our nuclear inheritance, can take up a general-ized version of such a formulation in our present context and read *"plus d'une fois pour toute"*—both "more than one once and for all" and "no more once and for all." (No) more once and for all, even with nuclear waste.

Notes

1. What this grafting means for eco-deconstruction is thought along and through other avenues in Michael Marder's contribution to this very volume.

2. Human Interference Task Force, *Reducing the Likelihood of Future Human Activities That Could Affect Geologic High-Level Waste Reposi-tories*, Tech. no. 6799619 (Columbus, Ohio: Battelle Memorial Inst., Office of Nuclear Waste Isolation, 1984), 8; NTIS no. DE84013725, *Energy Cita-tions Database*, https://www.osti.gov/scitech/biblio/6799619, accessed November 27, 2012).

3. To borrow just one of many terms that Derrida suggests might fall under the title "biodegradable"; see BSD 845, 814.

4. For a more detailed and brilliantly accessible overview of deep geo-logic waste disposal, see, in particular, Chapter 1 of Peter C. van Wyck's *Signs of Danger: Waste, Trauma, and Nuclear Threat* (Minneapolis: Uni-versity of Minnesota Press, 2005).

5. Despite the fact that the U.S. Department of Energy admits that "the waste will remain dangerous for longer than 10,000 years" and that "government experts agree that 'there is no doubt that the repository will leak over the course of the next 10,000 years'"; quoted in Kristin Schrader-Frechette, "Ethical Dilemmas and Radioactive Waste: A Survey of the Issues," *Environmental Ethics* 13 (1991): 328. For example, the half-life of Uranium-236 is just under 24 million years. Plutonium-239 has a half-life of 24,000 years—much more manageable than Uranium-236, but still more than double the 10,000-year period for which we understand our-selves to be responsible.

6. I am indebted to Alan Bellows of http://www.damninteresting.com for bringing these attempts—and the problems surrounding nuclear waste disposal in general—to my attention through his article "This Is Not a Place of Honor," http://www.damninteresting.com/this-place-is-not-a -place-of-honor, accessed October 20, 2012.

7. See Stephen C. Hora, Detlof von Winterfeldt, and Kathleen M. Trauth, *Expert Judgement on Inadvertent Human Intrusion into the*

Waste Isolation Pilot Plant, SAND90-3063 (Albuquerque, N.M.: Sandia National Laboratories, United States Department of Energy, 1991), http:// prod.sandia.gov/techlib/access-control.cgi/1992/921382.pdf; Human Interference Task Force, 1984. Again, Peter C. van Wyck's *Signs of Danger* is an invaluable resource for those seeking to learn more about the technical and semiotic difficulties such a project presents

8. Hora, von Winterfeldt, and Trauth, *Expert Judgement*, ES-1.

9. I would like to thank Michael Naas for reminding me that when we open our Larousse, we see that the *définition familier* for *perle* is "a mistake [*erreur*], misunderstanding [*méprise*] or gross blunder [*maladresse*] grossière observable in someone's speech or writing."

10. Jacques Derrida, "Signature Event Context," trans. Samuel Weber and Jeffrey Mehlman, in *Limited Inc.*, ed. Gerald Graff (Evanston, Ill.: Northwestern University Press, 1988), 12.

11. I would like to stress here that bringing the concept of biodegradability into relation with writing and survival more generally amounts to only one reason for the insertion of parentheses. Indeed, these parenthetical marks work to simultaneously underline and bury the valences of *bios* and that storied concept's relation to writing. At stake is nothing less than the relationship between survival and decay—a relation whose elaboration I will, on the one hand, defer to some day in the future (it would be necessary to engage in a more sustained and careful reading of, among other texts, "Plato's Pharmacy," *Archive Fever*, *The Beast & the Sovereign*, *The Death Penalty Seminars*, and many of Derrida's other as of yet unpublished seminars and, on the other hand, continue to develop here in whatever partial way I can).

Extinguishing Ability: How We Became Postextinction Persons

Claire Colebrook

I would not be the first person to note that Derrida seemed to pay inadequate attention to *literal extinction*. First, one might consider the ecological criticism articulated by David Wood, Timothy Clark, Tom Cohen, and others: by emphasizing the radical openness of futurity Derrida considered conditions of possibility, but did not consider the nature or climate from which such conditions emerged. For David Wood an expanded deconstruction is not only compatible with but is crucial for environmentalism. Wood's focus on time and futurity in Derrida expands the conditions of temporality (debt, mourning, anticipation) to consider what lies well beyond the human. It is true that within Derrida's work *différance* is radically outside chronological time and any parsed conception of past and future, but Derrida tended to emphasize the ways in which *différance* opened futurity; if the past haunted and disturbed the present in unmasterable ways, this seemed to generate a fragile promise, rather than a dire prediction. Just as Derrida considered the ways in which the Marxist archive might, in a feverish manner, dislodge an actualized Marxism, so David Wood poses the question of being haunted by a past or an archive that is inhuman and that includes all the erased and vanquished ghosts that even in their absence animate the present. (At its most literal and material we might consider the fossil fuels that provide the light and energy for the present):

> Derrida's sense of a "democracy-to-come," one that would break free from the constraints of the nation-state, one that would pursue justice beyond the rules laid down in advance, one that would take seriously the need to represent all the

earth's stakeholders, could surely embrace this difficult but necessary ideal of a parliament of the living. It may be that its decisions would only be enforced once it becomes clear that, to paraphrase Benjamin Franklin, if we do not hang together, we will surely hang separately. But even the thought of such a body begins to animate virtual voices in the human head, voices of creatures whose spectral presences haunt us in so many ways—species that have died out, flocks and herds we breed to eat, animal companions we live with, even the sports teams we name after animal totems. How long can we refuse to acknowledge all these ghosts, just as we balk at acknowledging the source and character of our own animating energies?

. . . environmentalism finds itself in an often problematic and aporetic space of posthumanistic displacement with which deconstruction is particularly well equipped to offer guidance. But . . . equally . . . environmental concerns can embolden deconstruction to embrace at least what I have called a "strategic materialism," or the essential interruptibility of any and every idealization . . . a deconstructive embrace of materialism, would be this: we tend to think of matter and spirit, or matter and mind as somehow opposed, and hence as unable to be thought of in the same space. But we can, I believe, get beyond this reductive understanding of opposition without falling into the arms of dialectical synthesis. The model of the Mobius strip allows for the idea that radical opposition can be combined with deep ontological continuity: at every point two sides but one surface. This gives us a beautiful way of representing our relation to nature—opposed in some sense, and yet at the same time continuous.[1]

However, one might want to ask (as Timothy Clark has done) whether a focus on the conditions of thinking and experience does not generate an openness of futurity that is possible only by precluding thought of ecological conditions and depletions:

A geographical and geological contingency, the finitude of the earth, now compels us to trace the anthropocentric enclosure of inherited modes of thinking and practice. The enlightenment project to render all the elements of nature part of a calculable technics is made to face its own dysfunction in the agency of what had previously been excluded from reckoning, or, more precisely, in that which had always been included-as-excluded. The condition of closure renders anachronistic inherited economic, political practices and modes of judgement without acceptable alternatives appearing in their place. The epoch whose intellectual closure is now visible, the "flat earth" epoch so to speak, inaugurates the need to think a bounded space in which the consequences of actions may mutate to come back unexpectedly from the other side of the planet.[2]

Derrida's arrow of time and promising went in one direction only and tended toward mobility and ability; the promissory "perhaps" of the future was always one of openness to what could not yet be calculated, determined, decided, or imagined. If he deploys such terms as "necessary impossibility," the use of the term "impossible" is a way of keeping possibility open: if there can never be the pure fulfillment of democracy or justice—if such terms are at once necessary and impossible—it is only their full realization that is impossible, and this therefore precludes terms from falling into fixity, inertia, and static completion. Justice, like the text, is never *closed*. Such mobility, promise, and ability were possible by way of what I would refer to here as a "nonliteral" or "nonnatural extinction": whatever the present may appear to offer in terms of dire outcomes and almost certain disaster, one could nevertheless think of a future to come not emerging from calculation. This is especially so in Derrida's seeming critique of Heideggerian potentiality: even if there is an "archive fever" or death drive that will disturb any authentic potentiality that is supposedly one's own, the emphasis is on the promissory and the "to come"—so much so that what is put out of play is a closed death, a death that does not repeat. Death is always tied to a future, and the future can never end, once and for all. That is to say: on the one hand, death is not *literal* (in the sense of actual), for anything that dies in the narrow sense haunts the present and opens it to possibilities it may have thought were impossible. (Marxism may live on in unheard-of ways.) On the other hand, death is truly literal: it bears the structure of a trace, for anything that "is" has an identity by way of sustaining itself through time, marking out what is the same only by way of relations that are never present in any one moment. The *literal* or the force of the letter, of being the same *through difference*, is what takes any "now" always beyond itself.

What makes such a future nonliteral is Derrida's general conception of writing: in order for anything to be sustained through time
it must carry over from the past and into the future certain repeatable traces. But if a trace can be reiterated—and *must* be reiterated
in order for something to be—then it cannot be tied or contained
by any present. The condition for living on is a certain exposure to
death. Far from Derrida's theory of textuality reducing everything
to writing, his identification of the condition of writing (tracing and
inscription) as being the enabling feature of anything at all opens
anything both to becoming and death. Inscription or techne in its
broadest sense is "death in life, as condition of life" (AI 39/62). A
rigid calculus that accepted the world as it is and from there spelled
out the future as an extension of the present would only be possible
by erasing its own conditions of possibility. For any experience of
the present, as present, to be possible one needs to trace or mark a
potentiality that is experienced in the now as what could be recalled
into the future. If this is so then there can be no exhaustive closure
of the present: whatever I might claim to know or predict in the present that I experience, or in the now that is before me, there is always
something *in the present as present* that exceeds any predictive or
calculative power. The present can be known and lived only if it is
stabilized and differentiated from a past and future, but that stability is only possible if the present exceeds itself: what is lived *as* the
now is possible by way of a trace of the past and an anticipation of
the future, but it is that very excess—the mark that traces the present—that also exceeds every present. This future of mobility, promise, the incalculable, and the "perhaps" is a future of *thought*; or,
more accurately, what makes thinking and experience possible—the
maintenance of the same through time, along with anticipation and
protention—is the trace or synthesis. Derrida always acknowledged
that the tracing syntheses that made thought possible would also
disturb thought. Even so, such disturbances were always exorbitant
and hyperbolic, opening thought beyond itself. But what if one were
to think of *différance* beyond the implications it may have for presence and the now? If there is such a thing as *nature*—a world bearing
its own syntheses, order, and temporality, then it may well go into
decline, chaos and nonbeing regardless of the concepts and anticipa-
tions that we may have of it. And, more radically still, if there is no
nature—as both Tim Morton and Karen Barad have claimed—then
there is a radical nonlinear, nonpresent, inhuman, queer existence
beyond thought that does not partake in the experiential structure

of presence (of anticipation, retention, intentionality, and so on). By "queer" neither Barad, nor Morton, nor I refer to the nonnormative, which is a disturbance, veering, or destabilization of stability; rather than there being a norm, normality, or presence that is disturbed, there is force with no end or direction other than the chance encounter with other forces. Thought in the Derridean sense—such as concepts of justice, democracy, friendship, forgiveness—are articulated as idealities that are always "to come" partly because they can only be *thought* as other than any transitory and actualized instance of what claims to be justice, democracy, and so on. In this sense, one might say that the literal—the forces that inscribe these concepts— is queer, transitory, or (to use Derrida's terminology) untamed and anarchic. But the *promise* of those concepts—because of the conceptual intention of ideality—is to survive and live on in the absence of the original context.

Now it might seem that Derridean deconstruction, in its claim for a promissory future, is *only* talking about the literal: there is a certain inscriptive force that makes the present possible. In human terms, one might say that to live in this world, *in time*, one must retain perceptions from the past into the present and anticipate those retentions into the future. The lived world is, then, already inscriptive. All the features that enable writing in the narrow sense—a system of traces that are distributed and dispersed while also *bearing a mobility and repeatability beyond any present*—mark *any* present (such as prelinguistic or animal experience). Whatever happens to the planet in a literal sense (the planet that will go on after humans), is quite different from the ways in which the earth is lived, and it is that lived earth or *world* that bears an ideality that at once subsists beyond any actual present and yet might also be extinguished when the *living* of that world ceases to be. On the one hand, then, Derrida will claim that every death of a lived world (every death of a subject) is the end of *the* world. And this suggests that whatever happens to the planet as such it will bear a thousand tiny extinctions every day. The planet may literally live on, and the world may literally not end, *but* insofar as every lived world relies upon the traces and the syntheses of a consciousness that is nothing more than its inscription of the world, the world is always ending. Here we might need to think of a double, contrary, and difficult difference of the sense of literal: if one thinks about the literal in the sense of writing, trace, inscription, or even synthesis, then one might say that the literal— as the condition of anything at all, including thought—always bears an afterlife. One might even say that what we know of as the earth, the world, the environment, and so on is composed from forces and relations that are not exhausted by the earth *as such*. However, one might also think of the word "literal" *not* in terms of writing, inscription, syntheses, and the potentiality for iteration, but as that which is actually a contraction or diminution of such forces. The literal, in the sense of actuality, perhaps cannot be thought without potentiality (and so when one thinks of writing, after Derrida, the text is always more than itself, always exorbitant). This was, indeed, the force of deconstruction: to see any letter, trace, inscription, or force as uncontainable, untamable, and monstrous. But perhaps we need to think about the capacity of the literal *not only* as repetition, but also as coming to a halt, as losing its feverish and untamed movement. To speak of "literal extinction" would refer both to the ways the movement of the letter disturbs, disrupts, and opens (where the letter would extinguish the rigid, the inert, and the selfsame and where inscription is a potentiality and process and not the text "itself" but its power of repeatability), *and* the "literal" as also opposed to the figural, or what Paul de Man invoked by thinking of a material sublime: a stony, inert, lifeless, inhuman matter that is other than all the figures through which it is phenomenalized.[3] Literally, the earth remains present; but the conditions that allow us to think this presence—conditions of inscription or literal conditions (the brute materiality through which ideality is made possible)—are erased when any factical mind ceases to live. Or, if the literal (or iterability) goes beyond thought and life, then the "to come," "the perhaps" and futurity may not be the most astute ways of thinking about inscription. For Derrida, even though deconstruction opens the thought of syntheses, traces, and inscription beyond the human, and even though the *letter* of deconstruction is at odds with the phenomenalization of the world, it was the letter's role in generating, opening, and extinguishing the phenomenal that was Derrida's point of focus. There is a figurative or ideal extinction of the world every time a mind, with all its traces, syntheses, protentions, and retentions, ceases to be. There is nothing but an ongoing series of extinctions, precisely because every world is the effect of singular inscriptive processes that effect individuated intentional subjects.

On the other hand, the notion of a complete erasure or loss would —by the same logic of conditions of inscription—also be impossible. If there can be a *world*, not just the literal planet, but a temporally synthesized and sustained lived horizon of sense, then there must also be a system of traces through which that ideality is constituted. The world of mathematics is made possible by a constituted system of traces; ideally or in its meaning, the truth of mathematics should persist and subsist beyond its concrete or literal inscriptions. And one might say that even if all the archival traces of mathematics were to pass away, mathematics could not *literally* be extinguished; its sense would remain true regardless of inscriptive remainders. Here, again, is a tension in the literal: the archive—the letters that have composed the history and world of man—could be destroyed; and yet Derrida leaves open the double sense of a remainder. Could there be erasure of a once-constituted sense, or does truth possess an indelibility that no archival destruction could vanquish? Even if there were new texts regarding the history, origin, and emergence of mathematical inscription—imagine a new manuscript emerging that reconfigured the history of number, logic, or geometry—that could (one might argue) neither erase nor transform the ideal and constituted sense of truth, "pure truth or the pretension to pure truth is missed in its meaning as soon as one attempts . . . .to account for it from within a determined historical totality" (WD 160/237). Even though there is a pretension to pure truth that *means or intends* that which is not reducible to a structure, a truth emerges and is witnessed by way of a structure: inscription or the literal dimension is crucial to the history of sense, and even if one were to lose or modify certain textual dimensions, constituted sense would remain. However, as Derrida notes in *Spurs*, how does one decide whether a mark or trace is an accident of the archive (ephemeral and easily capable of appearing and disappearing without a trace) or is crucial to the text's world, bound up with its very sense: is Nietzsche's "I have forgotten my umbrella" an inscription that composes the world of Nietzsche, with its singular life and desire to speak and mean, or could such phrases come into being and die almost as if they were biodegradable—having no sense or force in relation to the meaningful whole?

The logic of extinction, of literal extinction, bears a curious relation to ability that one might define by way of Derrida's concept of biodegradables (BSD). Inscriptive conditions—the fact that sense, truth, ideality, and the world are constituted through traces and syntheses—open up two sides of a biodegradable "logic." A true, meaningful, and profoundly rational claim would, ideally, allow its inscriptive conditions to "biodegrade": the eternal sense of Euclid or even Milton would become so much a part of the world that no concrete *literal* version need survive. A culture would be maximally *able* if it did not require the support of an archive, if it could sustain and nurture sense and truth without material supports. At the same time, one would also hope that the most trivial, toxic, accidental, pernicious, offensive, and stupid texts would be bound so tightly to their textual supports that they might biodegrade without trace. One might hope that hate speech, misinformation, revenge porn, propaganda, and all other cultural pollutants would not survive their brief appearance and moment of distortion. One ideal of human knowledge and culture would be a complete erasure of inscription; the truth and full meaning of a text would require that it pass into ideality and comprehension, no longer requiring an archive. Such a humanity might also be maximally able and ecological: so attuned to its world that there would be no waste, piling up of techne, no systems that could not contribute to the full and clean functioning of the whole. Is this not one way in which a certain utopia of climate change has been thought: that the thought of "our" destructiveness might motivate us to clean up the system, removing waste, excess, distortion, and injustice and allowing for a humanity no longer violently at odds with itself and its world, a humanity that becomes transparent to itself, cleansed of ideology and self-pollutants?

But, as deconstruction tirelessly demonstrated, the ongoing existence of truth and sense not only requires a material inscription; such a materiality has a force that continues to operate in the absence of reading. There is something necessarily disabling and counterecological about ability and ecology. Human abilities—such as speech and writing—are (to follow Bernard Stiegler, after Derrida) pharmacological: by extending, enhancing, and sustaining thought and memory they also require a certain submission and systematization.[4] Civilization's "abilities," such as industrialized agriculture, manufacturing, militarization, and political systems, are at once constitutive of human culture and, while limiting the range of freedom, have an autoimmune quality. It is evident today, in the era of the Anthropocene, that the very agricultural practices that allowed humans to draw more from their milieu in a sustainable manner and protect themselves from the contingencies of nature now render what was taken to be nature less hospitable and less sustainable. Ecology, or the capacity for human bodies to become aspects of a dynamic system of earth-species relations, is precisely what threatens human and inhuman life today. There can only be a human archive if the human body couples itself with various systems (inscriptive, technological, agricultural, scientific, moral, political, familial), but these same tendencies toward order are also generative of *disorder*. Political systems have become so entrenched as to preclude genuine political action; food manufacture has generated abundance (but at the expense of famine, eating disorders, and resource depletion); media technologies have extended human knowledge systems but at the expense of attention, press freedom, and the environment.

It is possible to tie Derrida's early work on the relation between truth and inscription to a broader notion of ability and disability. One might say that the condition for ability is transcendental disability, but this has both a weak and a strong sense. The weak sense is fully articulated in Derrida's conception of writing as *pharmakon*. On the one hand, the ideality of what presents itself as true for all time requires the fact of an inscription that could always be annihilated; and yet, on the other hand, that same inscription opens any seemingly timeless truth to a future that it cannot command. But there is a stronger sense of transcendental disability: our knowledge systems that enable us to diagnose the present as an era of climate change not only emerged from the technologies (industrialization, colonization) that generated climate change; man as *homo faber* or *homo sapiens*, the man of technoscience, the man of striking abilities and capacities—the man who tamed nature so that human life would be sustainable and not subject to the contingencies of the earth—*this* man of striking abilities is not only possible because of his own incapacity (coupling himself to the forces of the earth); he has incapacitated so many others (the colonized, the enslaved, the nonhumans, and those of his own species he has deemed to be less than human).

There is also a highly specific sense of transcendental incapacity that is announced in the era of the Anthropocene. This other sense of transcendental incapacity is both destructive and promissory. Man can now read his place within deep time. Our claim to know the present within history relies upon geological traces that present themselves as signs of a time past and a time future. This claim to know the present, to master all other contexts and scales, relies upon an inscriptive scene that in its very capacity to indicate a knowledge that surpasses human calculation and mastery also opens the possibility that it might offer another sense. No one could say, from within this present, that the geological strata might not—one day—be read as the opening of a new epoch in which a species took hold of the earth and generated new possibilities that might enhance

life beyond anything we can now imagine. However much geolo-
gists and geoengineers might want to erase the inscription of the
Anthropocene and state that there is now one incontrovertible fact
and one future, their very condition for saying so is a scene of inscription that would (perhaps) open other futures. Our ability to predict or to read the signs of the present to indicate a future is coupled
with disability: not only is there a certain not-knowing at the heart
of any speculation; the practice of speculation is already tied to systems that have been destructive of what has called itself "humanity" and "the environment." There is, then, something promissory,
disabling, and *enabling* in the inscriptive conditions of the present:
any certainty we are given occurs by way of processes of synthesis
and sense that, by their very capacity to generate truth, also promise
to be—possibly—other than what they are. I would conclude this
strand on the relation between deconstruction and transcendental
disability by making a counterdeconstructive remark. The radicalism and force of deconstruction lie in the opening of any putative
closure: any actuality must—by virtue of the forces that bring it
into being—have the possibility of being otherwise. No concept can
ever be articulated fully; no event can be exhausted without promising or being haunted by an unforeseeable futurity. The "perhaps"
is necessary as a potentiality that may open any closure; but the
"perhaps" does not necessarily open, promise, or generate futurity.
If deconstruction emphasizes the openings, misfirings, anarchic
movements, and future promissory dimensions of traces, a counter-
deconstruction would be a gentle reminder that such movements
are necessarily possible, but not necessary. And in a world that has
perhaps paid too much attention to the notion that anything is possible, one might want to think about a world in which everything
we thought might be possible is erased. Deconstruction locates conditions of impossibility in possibility: what makes speech, communication, and understanding possible (systems of repeatable inscription) preclude anything like full, unmediated, and stable meaning.
By extension, what makes life, environments, organisms, and ecology possible is an exposure, volatility, fragility, and dependence that
preclude anything like a fully sustainable ecology. To go through
time, to live, and to exist in relations of dependence *and mastery*
is to be subjected to the time of extinction. The more one expands
human ability and capacity, the more complex one's technological
and cultural ecology becomes; not only is such complexity exposed
to more and more risk, the very man who achieved such world-

transforming capacity is, today, beginning to recognize himself and all his possibilities as self-destructive and self-disabling.

One might say that what enables any inscription of the future to be read otherwise and to promise incalculable horizons is a certain *mobility or ability* of inscriptive systems. Derrida theorizes writing as a general condition, precluding any naïve opposition between the immediacy of speech and the corruption of writing systems that would be (supposedly) distanced from the full intent of the speaker. It would follow that *any* appeal to a proper, unmediated, pristine, or original ground—humanity, nature, the environment—would have to disavow its conditions of possibility: the nature that one imagines as prior to all techne, inscription, and corruption is not only known and given through inscriptive systems; it is also fully technical. This would be one main difference between Heidegger and deconstruction: Heidegger objected to considering *physis* as *techne*, as though nature unfolded according to a blueprint or thing made rather than unfolding from itself.[5] But the very "self" in "self-unfolding" occurs by way of forces that can never be fully commanded. Again we are returned to transcendental disability; any body comes to be what it is only through a relation or supplement that will expose it to risk. Nature is the outcome of an ongoing and dynamic relation among forces; what has come to be known as the stable hospitable nature before climate change was produced through forms of agriculture that increasingly destabilized what is now lived as the climate. Man and nature emerged by way of a complex history of capacities ranging from the basic syntheses of consciousness to the complex technologies that allowed for high-consumption, high-production culture. Just as one might want to observe that there is something ultimately disabling about this man of high capacities, one might also want to rethink the ability, mobility, and promissory dimension of Derridean writing. One of the dominant motifs of terms such as *"écriture,"* "trace," or *"différance"* in the early phase of Derrida's work was not only play, but anarchic and untamed force. If there is such a thing as an ethical turn in Derrida's later work, this capacity for writing in general to disinter, solicit, disturb, or disable becomes increasingly tied to an ethical vocabulary of futurity. Rather than the incalculable being presented as anarchy or chaos, it is now characterized as opening to a future. Rather than writing operating like a machine in the absence of voice and willing, writing can say anything and be figured as justice or democracy per se. Derrida insists, tirelessly, that the forces that open a text and that preclude determination are not *indeterminate* and that what is required is not an abandonment to "anything goes," but a more acute pragmatics, beyond intentionality. Taking this seriously therefore requires accepting deconstruction's demonstration of open futurity and radical promissory forces alongside counterforces where anarchy and genesis cannot be translated or figured as a "perhaps" in any human terms.

What if, as Derrida also occasionally suggests, there were something counterpromissory or disabling in inscriptive indelibility? If, as Derrida argues in "Biodegradables," the ideal of a truly great and eternal text is that it would be assimilated into the life of a culture without material or inscriptive remainder—the text becoming pure spirit, utterly indelible because no longer relying on literal inscription—one might think of another possibility. To imagine that a text has promise and a future is to insist on the possibility of future readings. What if, to take a literary example, the text of *Ulysses* were to become increasingly unreadable by an increasing attrition of a certain type of ability? What if deep attention, a mastery of Western languages and literature, and a feel for high modernism and the avant-garde were to atrophy to the point where *Ulysses* were nothing more than a mark, no more enlivening or promissory than a discarded advertising flyer? One might say that the text's inscription—its necessary material instantiation—enables both an open future of ongoing readings *and* a possibility of lying dead and inert, saying nothing. A certain notion of texts (and worlds) as promissory (as promising other readings) depends upon a certain conception of *ability*. What if one were to imagine a future or a reading of the present in which one were not blessed with the capacity to read? What if the humanity of ability were to imagine itself from the point of view of blind life, a life that did not imagine grand horizons of futurity?

To make a claim about the ethics of extinction or the imperatives that follow from predicted extinction requires *both* that one appeals—as in the case of the Anthropocene—to a reading of the earth that any possible observer would share *and* that other possible readings would follow. Such imaginings require a future in which reading (as we know it) continues. It is the nature of scientific claims, as both Husserl and any contemporary philosopher of science would concede, that the condition of making a claim about the facts of the world assumes an ongoing process and community of verification. One assumes an ongoing reason and, implicitly, the value of sustaining that reason.

One might think about this promissory temporality in politically (and environmentally) open or radical terms: whatever one claims about justice or the earth, other possible claims might be made; any political or environmental claim is made within a terrain of terms and conditions that can never be closed, and so there can be no state of climate emergency declared for the sake of saving humanity, precisely because what counts as the human remains open to contestation. There can be no imperative for survival at all costs. One might also think about this promissory quality *conservatively* insofar as science and claims to fact must presuppose an ongoing earth that remains the same and a community of observers for whom the same truths would pertain. If one claims that geological inscription promises *either* a dire future that commands that we must act now for the sake of the species *or* that this same inscription might be read otherwise and that humanity might find another (utopian and possibly inhuman) mode of being, one nevertheless—as Kant's theory of reflective judgment and Husserl's history of reason as the constitution and incarnation of sense argued—appeals to a presupposed community of ongoing and rational readers who would intuit in the world the same truth, sense, promise, and temporality that one now reads in the present.

So here I would conclude by asking two questions, and then putting forward the claim for a hyperbolic and improper extinction. I will do so by way of deconstructing the proper: what appears as secondary or parasitic needs to be accounted for *not* as an accident that befalls what appears as life, but as the very potentiality that has come to be known as life.

First question: What is the human such that it can extend itself or establish relations that can operate *without return*? (Against a conception of nature as symbiotic, with the earth as a dynamically self-creating living system, one might think of the human as a tendency to relate to the environment as just that: as nothing more than that which environs and that has no potentiality of its own.)

Second question: What is the earth such that it generates life that does not appear as ecology or that appears as nonrelational? The human would perhaps be all too human, all too subjective, and all too Cartesian after all: capable of closing itself off from anything other than itself. What sort of reversal would allow us to confront the twenty-first century without a sense of the proper, the proximate, or the *oikos* of self-regulating harmony?

As we face a future of resource depletion and of a series of choices regarding what sort of world we leave for the future, we rely upon normative concepts of the person. The question of what sort of life we will be able to lead and the question of what forms of life we wish to sustain are intensified by questions of climate change. Rather than see disability as an added concern when thinking about the future of the environment, I would argue that both the humanities and traditional forms of ecocriticism are already intertwined with normative conceptions of life. The supposed crisis of saving the humanities is not so much a disciplinary war (where the humanities have been overtaken by the hard sciences, marketing, or the social sciences) as it is a varied attempt to avert human self-extinction. *Both* sides of the wager—neoliberal rationalization and its opposite—are discourses of human self-maximization. Whereas straightforward rationalization operates in a flagrantly quantitative manner by arguing that we should orient ourselves to practices that allow us to make the most of our selves and our world (produce more, live longer, manage the environment in order to be sustainable) and in its extreme form argues for the use of any means of technology available to enhance and extend human excellence, counterdiscourses argue that the self is properly more than maximization of what it already is and ought to be oriented to some broader and more complex whole. (Such counterdiscourses would include notions of the human that are intrinsically oriented to empathy, care, justice, and nonhuman, or less than fully rational, forms of life.)[6] What is assumed by both rationalization (or hyperhumanism) and critical models that seek to redirect the course of human history that has lost its proper trajectory is that humans are capable of not realizing their potential and that possessing a potential generates an imperative: "I can, therefore I ought."

In its liberal and post-Kantian mode, this anxiety regarding the human capacity *not* to achieve realization has been expressed most recently and most vociferously by Bernard Stiegler: the capacity for humans to develop their reason, arrive at maturity, and to flourish has been threatened by the weakening of the role of the humanities. When neoliberal imperatives for market efficiency overtake the universities and when digital humanities projects do not take part in the ongoing care of the self, reason becomes nothing more than rationalization, and we fall back into stupidity. For all the radicalism of Stiegler's "pharmakology" (in which the potentiality of reason is tied essentially to a failure of realization), it is nevertheless human realization that he declares *must* come to fruition.[7] Here, in its Stieglerian mode at least, the tendency for deconstruction to emphasize futural ability and promise resonates with less critical ideologies of ability and capacity. What liberalism and neoliberalism share, especially when they seem to be starkly opposed, is a normative conception of self-maximization that is directed against a counterhuman or disabling tendency that is at once admitted to be a potential and yet is nevertheless warded off as improper. What we might want to ask is whether we might think not so much of a humanity that is threatened by its own tendency to extinguish and disable itself, but a capacity for extinction and incapacity that has—in the epoch of global climate change—created what has come to be known as man and that is perhaps the only being that has worked against incapacity and extinction so intensively as to create an accelerated complex of contradictory forces. It is precisely the drive to sustain and enable, to maximize what comes to call itself "human," that generates a war on disability and extinction, the two of which are intertwined.

Notes

1. David Wood, "Specters of Derrida: On the Way to Econstruction," in *Ecospirit: Religions and Philosophies for the Earth*, ed. Laurel Kearns and Catherine Keller (New York: Fordham University Press 2007), 286–87.

2. Timothy Clark, "Some Climate Change Ironies: Deconstruction, Environmental Politics and the Closure of Ecocriticism," *Oxford Literary Review* 32, no. 1 (2010): 134.

3. Paul de Man, *Aesthetic Ideology*, trans. Andrzej Warminski (Minneapolis: University of Minnesota Press, 1996), 127.

4. Bernard Stiegler, *What Makes Life Worth Living*, trans. Daniel Ross (Cambridge: Polity, 2013).

5. Martin Heidegger, "On the Essence and Concept of *Physis* in Aristotle's *Physics B*, I (1939)," in *Pathmarks*, ed. William McNeill (Cambridge: Cambridge University Press, 1998), 183–230.

6. Martha C. Nussbaum, *Frontiers of Justice: Disability, Nationality, Species Membership*, Tanner Lectures on Human Values (Cambridge, Mass.: Harvard University Press, 2006).

7. Stiegler, *The Re-Enchantment of the World: The Value of Spirit against Industrial Populism*, trans. Trevor Arthur, Philosophy, Aesthetics and Cultural Theory (London: Bloomsbury, 2014).

Environmental Ethics

An Eco-Deconstructive Account of the Emergence of Normativity in "Nature"

Matthias Fritsch

How "Value" Comes into the World

Environmental philosophy takes as its object the natural world with a view to the ontological and normative place of humans and other living beings in it, especially in the context of modernity's increasing alienation between nature and history or culture. As such, environmental philosophy comes to the fore with an increasing awareness of the destructive force of modern technology and capitalism and the revelation of the fragility of both nature and humanity. Thus understood, it is hardly surprising that the early Frankfurt School's blending of the Romantic philosophy of nature and the Marxist critique of capitalism yields, especially in Horkheimer and Adorno's *Dialectic of Enlightenment*,[1] one of the high points of environmental philosophy. Another center emerges in the wake of Heidegger's rethinking of human existence as mortal being-in-the-world, in particular in Merleau-Ponty's reflections on "the flesh of the world" and his lectures on "The Concept of Nature"[2] and Hans Jonas's *The Phenomenon of Life* and his later *The Imperative of Responsibility*,[3] with both Jonas and Merleau-Ponty explicitly raising the question of where and how normativity emerges in life and nature.

From this point of view, the beginning of environmental ethics in the English-speaking world of the 1970s could be seen as a late start, even if many accounts of the history of environmental philosophy reverse this chronology. Nonetheless, it may be helpful to pick up some of the terms of the latter in an effort not only at bridge building, but also at articulating the potential usefulness of Continental-European resources in thinking a problem that, as urgently global as it is, calls for, if not a common language, then at least sustained dialogue across traditions.

What is more, from the beginning in the 1970s, English-speaking philosophers of the environment have elaborated valuable accounts of the scope of moral value and have thus been concerned with the question that will be the focus here—namely, that of the emergence of normativity: how does "value" come into the world[4]? In exploring a deconstructive account of moral normativity in what follows, I will be much more concerned with the latter question (regarding the origin of normativity) rather than with the former (the exact scope of moral consideration), though these issues are closely related.

Among the avenues that have been pursued in answering the question concerning the emergence of value, three crucial ones stand out.[5] Each of these three may be formulated as a question:

(V) Can an entity value itself?
(S) Can it suffer?
(H) Can the value be predicated of environmental wholes (species, ecosystems, the earth)?

I will begin by briefly presenting what I see as the main intuition behind these approaches, followed by my "eco-deconstructive" account of normativity and how this account relates these three to each other in a common but internally differentiated approach.

Self-Valuing

Many environmental ethicists have given great weight to (V) self-valuing or self-affirmation. The intuition is that if a being values its own life, value emerges in this self-relation and is to be respected by others in at least some way—above all, of course, by humans capable of moral valuation. And if an entity values itself and thereby creates intrinsic value in the world, it may be seen to confer instrumental value on other things in its environment, such as its shelter, food, and so on. Holmes Rolston III's "value-ability" may be the most aptly titled contribution here, provided we understand it precisely not as a potential to be valued from the outside.[6] What we track in circumscribing moral scope, on this view, is whether an entity has the ability to value itself. (For Rolston, this ability famously extends not only to animals and even plants, though to a lesser degree, but to species and even ecosystems, even if in a highly qualified way.)[7]

Another account that makes self-valuing central is to be found in the work of Hans Jonas. On the basis of what he calls "an existential interpretation of biological facts,"[8] Jonas argued already in the 1960s that if we follow the phenomenological approach to yield an "empathic study of animal existence,"[9] we will see that, in response to being exposed to death, all living organisms exhibit (Heideggerian) transcendence. Their being is a being-ahead of themselves in caring for, and affirming, their lives.[10] With the help of Merleau-Ponty and Maturana's biology of autopoiesis, Evan Thompson recently elaborated the core idea (while, however, downplaying the role of mortality).[11] Here, too, value emerges in the world with the self-affirmation of living organisms.

Suffering or Vulnerability

Regarding (S) suffering, we may recall, with Peter Singer, Bentham's famous question, "Not, Can they *reason*? nor, Can they *talk*? But, Can they *suffer*?"[12] The intuition here is that vulnerability, however understood, confers moral status. For Singer, as is well-known, many nonhuman animals can feel pleasure and pain, though in differing degrees: mammals, birds, vertebrates, to a sentience-determined cut-off point "somewhere between a shrimp and an oyster."[13]

Relatively little work has been done on how to combine (V) value-ability and (S) vulnerability. There is, however, a strong tendency in our anthropocentric and mechanistic tradition to think of life itself—that is, life without consciousness or reason—as a mere machine. The capacity for "true" suffering is then taken to be dependent on rationality or consciousness, which is capable of converting a mere bodily existence into a sovereign subject that affirms itself over time. (S), the capacity for value-conferring suffering, would then depend on (V), the (here, rational) capacity to value oneself as oneself over time. Infamous in this rationalist tradition is Descartes's claim that lacking reason, animals are mere machines, without consciousness of suffering, thus without true suffering.[14] Less crudely but nonetheless still discernibly rationalist, Tom Regan's well-known work also makes (S) vulnerability depend on (V) the capacity to value one's own life. For it is this capacity that, as in Rolston, though construed much more restrictively in Regan, confers claim rights onto an entity. If humans have such rights, so do many animals who are "subjects of a life": to be such means that this life is "better or worse" for them, "logically independently of anyone else's valuing [that life] or finding [that life] useful."[15] An entity suffers only to the extent that someone interferes with that entity's valuing its own life; raw pain, as it were, has no moral traction.

For his part, Singer conceded this dependency of vulnerability on self-valuing ability to some extent when confronted with the "problem of replaceability": for instance, if we could kill cows for meat painlessly while replacing each killed cow with another one to equalize total utility, there would be no discernible reason a standard utilitarian calculus would not permit, even encourage, such killing. This problem has led Singer to distinguish irreplaceable self-aware beings who have a notion of persisting into the future and so prefer to go on living (humans, some mammals such as apes and whales) from replaceable but suffering beings.[16] Thus, even here, suffering comes to be seen to be of greater value if tied to the ability to value one's own life, an ability of valuing extended into the future of one's life; what counts is the ability to prefer to go on living.

Beyond these problematically hierarchical accounts, however, the two camps, those beginning with (V) value-ability and those beginning with (S) vulnerability (for instance, Kantians and utilitarians), remain opposed to each other. Despite the fact that both intuitions seem eminently plausible, it seems unclear how we could account for both in one approach.

Holistic Interdependence

In addition to value-ability and vulnerability, it has also been urged, and with good reason, that an adequate environmental ethics must be able to understand value as emerging with holistic entities such as species, ecosystems in habitats, biospheres, perhaps even the earth itself, understood itself as a kind of superorganism.[17] The intuition behind this third desideratum is that, in environmental ethics in particular, it makes little sense to value an individual entity, whether for reasons of its own self-valuing or to respect its vulnerability, when, ontologically or ecologically speaking, that entity can be what it is only in a network of relations. We ought to do justice to the massive fact of holistic interdependence that questions the individualism of (V) and (S).[18] If individual entities can emerge only with wholes to which they belong, then prior to (or at least alongside) valuing individual entities, according to the holists, we must value wholes and properties of wholes, such as ecological stability, diversity, integrity, beauty, and so on (Aldo Leopold's land ethics may be most famous in this regard).[19] Holistic interdependence must bear in some way on self-valuing and suffering, for ecological wholes both sustain the life of individual living beings and reabsorb those entities upon their death.

The reference to death here is meant to flag the following complication. If we value both individuals (whether for reasons of value-ability or vulnerability) and the networks that sustain them, as seems reasonable, then we cannot just value life for the sake of life,[20] but must take into account death. In this spirit, for example, Callicott criticizes egalitarian biocentric individualism (most famously, deep ecology) for its failure to recognize ecosystemic structures and functions at the heart of evolutionary processes, to which belong competition, predation, and death.[21] He writes, "Struggle and death lie at the heart of natural biotic processes, both ecological and evolutionary. An adequate biocentric axiology for environmental ethics could hardly morally condemn the very processes which it is intended to foster and protect."[22]

The conclusion to be drawn here is that in rethinking the moral relation between individuals and environmental wholes, we must also rethink the relation between life and death in the context of the emergence of value. We must allow that life is not just the opposite of death, but sustained by it in a larger, interdependent whole.

But as concerns the relation between individualism (V, S) and holism (H), it has seemed very difficult, and for good reasons, to integrate both intuitions, however plausible they may seem when taken in isolation. Thus the understandable tendency on the part of individualist environmental ethicists to *supplement* their accounts, often in a more or less ad hoc fashion, with holistic points, and vice versa.

For example, Leopold's holistic land ethic has often been seen to have a hard time accommodating individualist intuitions. Callicott attempts to help out by distinguishing between "mixed communities" in which individualism prevails and more holistic, ecocentric biotic communities in which the mixed ones are "nested."[23] The distinction between mixed and biotic communities, however, remains tenuous and overlapping. Similar problems can be raised from the other side, that of individualism. For instance, Paul Taylor's egalitarian biocentrism, which claims that all living things have intrinsic value simply in virtue of living, seeks to incorporate holist intuitions by viewing individualist biocentrism as merely a part of a more general "respect for nature" that recognizes ecological inter-dependence. A problematic or at least unclear reconciliation may be detected in Rolston, too, who argues that value-ability depends on what he calls "systemic" value. For Rolston, ecosystems possess systemic value by engendering value-able species and individuals, which then value themselves intrinsically and others instrumen-tally. Systemic value thus answers to the question of value's ori-gin: something has systemic value if it produces values that can be valued functionally, instrumentally, and intrinsically by entities in the system. Furthermore, the systematicity of the value comes to the fore in the fact that an ecosystem converts intrinsic into instru-mental value and vice versa.[24]

However, it remains unclear how intrinsic value and systemic value are related. Rolston admits as much when he writes, "Intrin-sic value, the value of an individual 'for what it is in itself,' becomes problematic in a holistic web."[25] Nonetheless, Rolston seems to me to gesture in the right direction—that is, to recognize the way in which self-valuing proceeds, as it were, from what he calls "the ongoing planetary natural history in which there is value wherever there is positive creativity," a history to which humans (as well as the higher animals) must be considered latecomers.[26] Ecological interdependence is to be brought to bear on the valuing self-relation, the outside on the inside, in the sense of the former provoking the latter, making the self-relation a *response* to systematic relationality in the environment without merely being caused by it.

To travel a bit further in this direction, I turn to what I see as an eco-deconstructive account of normativity.

Derrida's Double Affirmation

In recent years, Heideggerian and "poststructuralist" texts, from Foucault's biopolitics and Deleuze's "becoming animal" to Derrida-inspired "eco-deconstruction" (the one to be launched in this book), have been mobilized in relation to environmental philosophy. The texts have typically been taken to contribute to a better ontology, not a better account of environmental value or normativity. And this for good reasons, often based in a rejection of the subjectivist projectionism that arrogates to itself an objective standpoint from which the "rational" subject can determine what has value. The question of how normativity comes about, however, does not disappear with this rejection. Accordingly, in this essay I would like to take these nonsubjectivist reasons to heart but still show that Derrida's work on the (quasi)ontological emergence of normativity in general, and on "animal ethics" in particular, if read in the right way, suggests a promising way to relate (V-value-ability), (S-sufferability), and (H-holism) to one another—again, not primarily in view of answering here the normative question of moral scope, or a question regarding the content of environmental duties, but the normative-cum-ontological question of the emergence of "value" in the world. In response to this question, and that is my primary goal here, I will above all be concerned with the issue as to how to think (V), (S), and (H) together, even if that ensemble will not be free from tensions (for essential reasons that we will discuss).

In brief, I will argue that Derrida's concept of double affirmation shows that a living being must affirm both itself and its others as a result of each living entity being constitutively and differentially related to other entities in its life context. Put in the terms used previously, (V) self-affirmation or self-valuing (what, in *The Animal That Therefore I Am*, Derrida calls auto-bio-graphy, auto-affection, and self-presentation)[27] is a response to both (S) the fundamental vulnerability of living things and (H) the dependence on a life-sustaining context that always threatens with death and reabsorption into a larger context. Vulnerability (S) is here understood to lie in the relational spatiotemporality of a mortal being: the fact that an entity does not originally coincide with itself means that it must (V) perpetually strive to reaffirm or reidentify itself by (H) drawing on its context, welcoming its others that it must also keep at bay. This account of the emergence of value in a vulnerable self-affirmation, I will conclude, suggests new and productive avenues for thinking environmental ethics beyond the two stand-offs outlined previously—that is, beyond (V) vs. (S) and beyond the stand-off individualism vs. holism. The goal, then, is not to achieve a better, more harmonious system of environmental ethics, but to reconceptualize each element (V, S, and H) such that their interrelations stand out.

To begin with deconstruction "in a nutshell," its point of departure is to view an object as emerging out of its differential relations to others. These relations of difference are taken to be constitutive of the object. That is, the object is not seen as preexisting its context but as owing itself to the environment of its emergence and being. Difference, as de Saussure had it, is prior to identity. (This priority may already counsel a certain caution with regard to the language of entity and context, self and environment, a caution well-known in environmental philosophy—see Arne Naess's critique of the "man-in-environment" model.)[28] But the context is not itself exhaustively analyzable, as if we could list all of its elements in a complete list. A complete list would stabilize identity despite its constitutive relationality, because the relations would be bounded and always travel along the same lines. By contrast, the claim here is that the context is inexhaustible, not because the list would be too long to itemize, but rather for the more essential reason that the context is itself undergoing change as it constitutes the elements of which it is made up.[29] Each element in the context is in a similar position of changing with its context, the context changing with them, so that no element can bounce off a stable identity. Further, and for the same reason, the dependence on a constitutive context is not fully determining for an element, for it can be recontextualized; an entity can, indeed cannot but, shift from context to context.

It is these two moments of differentiation and recontextualizability (or iterability) that Derrida sought to capture economically with the neologism *"différance."* The term is to encompass difference and deferral—that is, situation in context but without final determinability, anticipation of future environments (for not anything goes), but also exposure to the open-ended future the elements in an ongoing system cannot know. And despite having first developed it primarily in the context of structuralist accounts of language and culture, it is the notion of *différance* that Derrida sees as "coextensive" with mortal life (AW 108; FW 63/106–7). He has from the beginning insisted that it holds wherever there are elements in a more or less holistic system—for instance, DNA or organisms in an environment.[30]

An organismic identity, then, does not come into the world ready-made, or nonrelationally, identical to itself. It must strive for that identity, for its time and its space. With this striving, normativity—the "must" or *"il faut"*—comes into the world. If the identity of an element, such as an organism, is spaced-out, a little mad in not already coinciding with itself, it can be itself only by being identified, again and again (hence the deferral), with itself. Its identity must rely, and from the beginning, on a repetition that intends the same. In the case of a living entity capable of thus "intending" or repeating itself, we may speak of a self-affirmation that expects to come back to itself despite its need to refer to and appropriate from others in the context into which *différance* will have cast it. Others thus hover between friend and enemy.[31] But if self-affirmation requires differentiation from its environment, then this context precedes the affirmation and is affirmed first of all. An identity must always already have affirmed the preceding context from which it is in the process of appropriating to be self-affirmative in the first place.

For these reasons, we may say that affirmation is double, a duplicity that also doubles itself.[32] Affirmation must affirm the self and the other, the one to affirm the other, but also, at the same time and in the same moment, repeat itself—that is, affirm its future repetition. In parsing out the duplicity, we can try to analytically separate what is inseparable in affirmation:

First, affirmation must affirm, along with the self, the other-than-self; otherness here ranges from the future self, to other identities, to the open-ended context. Affirmation is a response to preceding contexts.[33]

Second, the self-reference must "from the beginning" (but the beginning is already a response) affirm to repeat the affirmation. The generation of the "self" in reference to alterity is never complete, never settled but always to be repeated. The (more or less spatial) self-other affirmation is duplicated by the (more or less temporal) affirmation of past as future. It helps if we see this future as itself split into two futures to be affirmed.

Third, affirmation must affirm the future of its self-repetition (hence, a future in which its self is predicted, anticipated, foreseen) as well as an open-ended future, a future it cannot anticipate.[34] The open-ended future-to-come is not a mere accident but is (quasitranscendentally) necessary and thus must be affirmed as enabling of the self.[35] It indicates the inexhaustibility or indeterminability of context and hence the possibility of relaunching toward another context. This possibility is affirmed along with, or in and through, the affirmation of the preceding context. However, this future-to-come is also the essential possibility of alteration, contamination, and death.

In other words, identity must of necessity be repeated, for an identity is never given once and for all. Each affirmation is already promised to a repetition that brings in the other once more (hence we should speak of iteration), so that the promise of return to oneself cannot ever be finally fulfilled; it is a promise made to a future forever to come. As the "first time" that is to be repeated never took place as such either, life is a "living-on" suspended between the "absolute past" and "future to come," neither of which took place or will take place as such. The structure of living-on, of sur-vival (*sur-vie*) as the condition of any event, involves inevitable change in the suspension between past and future. Double affirmation affirms the self as constitutively and differentially related to others and thus also to its mortality. For this reason, the fundamental injunction to affirm self and other is necessarily contradictory (SM 7/27): it affirms the self as self, but also as other; it affirms the self as living, but also as dying and being reborn against the background of life's mortality.

I will now try to see how double affirmation can help us to see the interconnections among individual self-valuing, vulnerability, and holistic interdependence.

Conclusions for Relations among (V), (S), and (H)

On the basis of double affirmation, how do we think anew the relations among (V) value-ability, (S) vulnerability, and (H) holistic connectedness? Recall that many environmental ethicists recognize, at least to some extent, the validity or importance of these three values and the associated accounts of their emergence. However, typically, one is prioritized, resulting in an (often unreflective) tension between (V) self-valuing and (S) suffering and between their individualism and (H) holism. On the basis of double affirmation, I'd like to suggest new (and perhaps surprising) ways of tying them together, thereby rethinking each one. Given the intrinsically unstable and conflictual nature of double affirmation, however, the goal in this rethinking is not to eliminate tensions in the links, among the three desiderata (V), (S), and (H)—in environmental ethics above all, we must beware of harmonious value axiologies, the environment living off its tensions and, in fact, its own dying and rebirth.

Let us then begin to explore the connections and complications double affirmation suggests between (V) value-ability, (S) vulnerability, and (H) holistic connectedness.

(V) Value-Ability and (H) Holism

(1a) First, on this account the relation between (V) and (H) becomes much more intrinsic, but also conflictual. Due to the constitutive (always-already) relation between an organism and its context, between the inside and the outside, self-valuing willy-nilly also affirms another self or entity and other beings in an indefinite and open-ended fashion. The reason self-valuing is doubled in this way, as we saw, is that once the outside context is seen as co-constitutive of the self, the self to be valued can no longer be thought of as given, with predefined boundaries. Self-affirmation thus inevitably slips away or drifts into the context, in fact in an ultimately erratic, indefinite, open-ended way. This is not an accident: given that this open-ended future, the indeterminability of the context, is enabling the self (and not just undermining its boundaries), affirming the future to come in the here and now is part of the very being of an organism.

The distinction between intrinsic and instrumental value, widespread in environmental ethics, thus becomes problematic, as well. For a self cannot be supposed to be able to clearly distinguish between its own intrinsic value and the environmental instruments needed to maintain that value. Further, the indeterminability of the "whole" also means that holism has to be rethought, for not only can we not extricate ourselves from it, but we also cannot exhaustively grasp it and comprehend it. The recognition that nature is not in the end wholly objectifiable and graspable has important consequences, as many environmental philosophers have pointed out.[36]

As an example of the way in which self-valuing drifts off into the environment, consider the way in which biochemist Michael Fischbach has suggested we rethink the relation between the human body and intestinal microbes:

> We used to think the immune system had this fairly straight-forward job. All bacteria were clearly "nonself" so simply had to be recognized and dealt with. But the job of the immune system now appears to be far more nuanced and complex. It has to learn to consider our mutualists [that is, resident bacteria] as self too. In the future we won't even call it the immune system, but the microbial interaction system.[37]

The absence of constructive engagement between microbes and the immune system (particularly during certain windows of development), Fischbach argues, could be behind the increase in autoimmune conditions in the West. Keeping the immune system productively engaged with microbes (and with intestinal worms)—exposed to lots of them in our bodies, our diet, and our environment—turns out to be critical to human health.

(1b) This reference to the immune system as that which seeks to negotiate frontiers between self and nonself helps us to also rethink the relation between (organismic) self, environment, and death. Recall that we said that to value holistic interdependence is to rethink, and revalue, the role of death in life. Now Derrida uses the biomedical case of autoimmunity to do just that. He generalizes autoimmunitary "diseases" to the very condition of the self in its struggle with the environment:

> The living ego is auto-immune. . . . To protect its life, to constitute itself as unique living ego, to relate, as the same, to itself, it is necessarily led to welcome [*accueillir*] the other within (so many figures of death: différance of the technical apparatus, iterability, non-uniqueness, prosthesis, synthetic image, simulacrum, all of which begins with language, before language). It must therefore take the immune defenses apparently meant for the non-ego: the enemy, the opposite, the adversary and direct them at once *for itself against itself*. (SM 177/224)

In the last ten years or so of his writing, Derrida often used the biological structure of autoimmunity to express the complication of a self-affirmation that must affirm its other and its alteration. Autoimmunity is another way to state this impurity, the death in life, the conflictual coimplication of the one in the other and the second time in the first.[38] Autoimmunity results from the noncoincidence of the self with itself: to protect itself from the other, it would have to already know itself and its borders.[39] But this "itself" is generated only in the reference to the other, a relation that is both welcoming and protective, hospitable and hostile, unconditional and conditional. For to relate to itself, to constitute itself as itself, a living being has to welcome the other within. Breathing and eating, it has to extract from its environment and convert into energy that which it absorbs. It has to relate to other entities, of the same and of other species, by distinguishing itself from them. It has to relate to them but such that it keeps them at a sufficient distance, the right measures of distance and closeness, of intake and output, never being given in advance. Self-affirmation cannot know in advance of such affirmations where its borders begin and end. If self-protection does not in advance know what the self is and what or who the enemy is, but comes to constitute it only in repeated acts of protection, and if, further, the self consists in relation to its outside, its environment, and other entities, then the protection will necessarily direct itself against the living self as well; this is what Derrida means by speaking of a "generalized" autoimmunity.

In thus welcoming the other to affirm its life, the self also affirms death. In affirming an open-ended context along with itself, the self affirms that which sustains it, but also that which claims it (as its "own" in some way) upon its death, and in fact is always already in the process of reabsorbing it. To affirm the self is to affirm a "whole" in which the self lives and dies, and in fact lives by dying and being reborn at every moment. That is why I said we have to rethink and revalue death in its relation to a larger context of life. Val Plumwood has done so admirably. In response to being attacked, as food, by a crocodile, she proposes a "chain of reciprocity" with the earth: we appropriate from the earth, but we also return to it—for instance, in the form of food for other living beings.[40]

While selves cannot be entirely removed from nature, their iterated exappropriations and futural recontextualizations cannot but seek independence from it. Hence, we may say, the human fear of being devoured by the animals, the earth, or the sea. Derrida explored this fear most extensively in his final seminar, now published as the second volume of *The Beast & the Sovereign*. The seminar discusses attempts at human sovereignty in relation to the decision regarding the inhumation or cremation of one's future corpse and so in reference to death in life. It also devotes considerable attention to Robinson Crusoe's fear that wild animals, cannibalistic savages, the trembling earth, or the engulfing sea will swallow him alive, his life becoming the living death that it seems to have already been on the island, in the sea. Not unlike the account of the enlightened self in Horkheimer and Adorno's *Dialectic of Enlightenment*, Robinson's imperialistic and colonial claim to ownership of the island, his making himself king on it, is presented in these seminars, I would argue, as a certain (and certainly questionable) response to this fear, a fear that also reflects the reality of mortal, embodied life.

While self-affirmation as self-separation from holistic webs is ontologically necessary, singular, possessive, and human-centered sovereignty need not be understood as the only possible response. Crusoe's imperialism is not to be ontologized and depoliticized. There are different ways to attempt to stay "sovereign" in the face of nature. Aiming for mastery over the elements—"knowledge itself is power," Bacon said[41]—or seeking to avert the return of one's corpse to the earth or the sea are only possible responses. Much is at stake in how we respond to a fear-inspiring nature into which we are given over by the autoimmune complication of double affirmation. Not the least question may thus be the one Derrida asked in *Glas*: "What is it to make a gift of a corpse?" (G 143/163).[42]

(S) Vulnerability and (H) Holism

Now that we have rethought the relation between (V) and (H), let us turn to the adjacent (better: coimplicated) relation (H) holism—(S) vulnerability. Following the discussion of autoimmunity, we could now say that, in a certain sense, (S) vulnerability lies in this exposure to indeterminable others in a holistic (H) context. This exposure is not secondary to, but comes from the very inside of, (V) self-affirmation. Vulnerability now names first of all this dependence on a larger context that sustains but also reabsorbs an entity. The capacity for suffering is related to the interrelations among beings in nature. Perhaps the young Schiller put it best when he wrote that "physical pain" is "a sounding gold chord struck on the lute of nature."[43]

The context on which life depends is not just outside but nestled into the very self-difference that alone allows a living thing to be itself (or better, to become itself in an ongoing, incompletable process). Vulnerability is thus closely related to mortality, for the "whole" in which a being exists constantly "nibbles" at it, claiming it as "food" or resources for others. Death gives back to nature, as it were, such that mortal temporalization bespeaks the presence of the other in the self.

On this conception of vulnerability (S), whether an entity can suffer is no longer generated by a nonrelational, intrinsic property such as rationality, consciousness, sentience, possession of a central nervous system, and so on, though these retain a newfound significance. Rather, vulnerability lies first of all in the fact that an entity is relationally, differentially constituted. A self given in advance, completely closed in on itself, could not be affected by others, and so would not be vulnerable. In truth, as a result of having the other within itself, a being is porous and affective, or what we might call "affectable": always already open to an influence that is no longer merely external. This relational affectivity is the ground on the basis of which a nervous system, for example, can first of all make a difference in amplifying and "fleshing out" that relational vulnerability. How we *value* these different amplifications is of course yet another question (including the question as to whether we can be taken to value collectively at all). The point here is, first of all, to put into question an axiological ethics in which a being does or does not have moral standing on the basis of empirical properties that we can or cannot ascertain. This "desire for a because," as Cora Diamond puts it, evades our exposure to a being sufficiently similar (or better, implicated, involved) to demand a moral response, but sufficiently other to disallow that I "settle" my attitude in some categorical way.[44]

A similar point is made in Derrida's reading of Levinas in *The Animal*. Levinas argues that ethics begins with a call to responsibility that comes from alterity. Alterity here means, among other things, that the call is without a because whose range reason (which is always the reason of "man") masters, and that means, determines, judges, agrees, and corresponds with. Ethics thus begins in unmastered exposure, one's own but also, and primarily, that of the other. The other's vulnerability is figured by Levinas as a nonresponse, the death in life that shows the other's "im-power" to respond, there where there are bodies that will be corpses. But if ethics begins with this mortal alterity, then the moral call cannot be restricted to the human face. The call should not be seen to be "extendable" to animals, but should in fact "privilege," as Derrida puts it provocatively, the command to responsibility of the animal, not because animals are more valuable than human beings, but to hear differently the call of the human being herself. The call will then no longer come from the human, the humanity of man, but from an even more alterior place, both ulterior and other, a call that "calls within us outside of us, from the most far away, before us after us, preceding and pursuing us in an unavoidable way" (AA 113/156).

The attempt to understand differential relationality, in the difference from oneself as well as the mortality that it foregrounds, as the root of vulnerability, should be seen as part of an attempt to pluralize limits among living beings that share death. Viewing relationality as the ground of suffering is not to deny that, for example, as an entire tradition from Descartes to (yes, even) Singer stresses, consciousness of suffering may alter it, and indeed, increase its negative value at least for us (the claim it has on us to prevent it). Pluralizing limits is to recognize that different ways of drawing a limit may be salient in different contexts and with different criteria of valuing coming into play. With Derrida, this would be an attempt to think life in an "infinite plural" (BS2 198/279) in a way that multiplies the limits (AA 29/51).

Precisely because holistic interdependence makes each being vulnerable to the other, the death of each one affects all others, even across species:

> Every time it [the organism] dies [*ça meurt*], it's the end of the world. Not of a world, but of the world, of the whole of the world, of the infinite opening of the world. And this is the case for every living being: from the tree to the protozoa, from the mosquito to the human, death is infinite; it is the end of the infinite [*la fin de l'infini*]. The finite of the infinite [*le fini de l'infini*]. . . . It's an end of the world that is without equivalent, that has so little equivalent that, with regard to the death of the least living being, the absolute end of the world or, if you prefer, the singular destruction of earth and of earthly humanity, changes nothing, makes not the least bit of difference, remains in any case incommensurable. (DP2, 81/118–19)[45]

This thought of the infinity of finitude may be seen as an attempt to think the sharing of mortality and the experience of vulnerability *nongenerically*. To think the "possibility of sharing the possibility of this nonpower [to suffer], the possibility of this impossibility, the anguish of vulnerability" on the basis the very "finitude of life" (AA 28/49) means to think a sharing without attributing a common genus or species (such as "all mortal beings") that would be in itself unaffected by the death of its mere particulars. It also means to think it not vertically, but horizontally or laterally: I experience the suffering of the others only by way of the suffering of the more proximate other, here, too, constitutive relationality being the condition of relating to distant others.

(V) Value-Ability and (S) Suffering or Vulnerability

Finally, let us say a few things about the relation between (V) value-ability and (S) vulnerability. The first thing to note is that we might have to reverse a dominant hierarchy in their relation, one that sees vulnerability as dependent on the ability to value oneself (though we will seek to find the truth of that dependency, as well). For in some sense, self-affirmation should be understood as *a response to* a more fundamental vulnerability, mortality, dependence, and difference of the self from itself, the way we have thought it here. As we saw, on this view it is *because* the self is not itself from the get-go that it has to seek to affirm itself.

In a certain sense, then, a passivity is prior to an activity, an exposure prior to affirmation. We may also say that the pair "active/passive" begins to become less useful, given that affirmation is co-constitutive of a being in the first place, but is also not an expression of the activity of (free) will, circulating through a being to make it what it is. This is the context in which we should consider what I see as Derrida's references to Bentham's prioritizing of suffering over reason. In view of a passivity prior to the active-passive distinction, Derrida attempts to radicalize Bentham's "can they suffer," which is still thought a bit too actively:

> "Can they suffer?" amounts to asking "Can they *not be able*?" . . . Being able to suffer is no longer a power, it is a possibility without power, a possibility of the impossible. Mortality resides there, as the most radical means of thinking a finitude that we share with animals, the mortality that belongs to the very finitude of life, to the experience of compassion . . . [to] the anguish of this vulnerability, and the vulnerability of this anguish. (AA 28/49)[46]

Now that we have seen that (V) is a response to (S) and remains so, and if we do not misunderstand (V) too quickly as an activity, an ability, a power or faculty possessed by (isolated) subjects, we can perhaps better understand why I said I also wanted to discover a new truth of the reverse dependence, that of (S) vulnerability on (V) self-valuing. For mere relationality, one may object, is not sufficient for vulnerability; it may be a necessary but not a sufficient condition. Not even on this account, one may say, is the stone vulnerable in the relevant sense, although the relations in which it is situated may of course destroy it. The difference between a living being and a nonliving being must, however, be taken into consideration in thinking vulnerability, especially one so closely tied to mortality.[47] To put this in other terms, vulnerability demands that a being has "a good of its own," as Rolston put it, the good minimally including the attempt to affirm one's life against threats, and so seeks to avoid harm.[48] For as we saw, what is distinctive of a being suspended between life and death, one that lives death as rebirth at each moment, is that it responds to this vulnerability by seeking to protect itself from it, albeit in an "autoimmune" fashion. This self-protection against death is an attempt to seek to control and to negotiate the ordeal of contextual imbrication by redrawing the lines between self and other anew on each occasion.

Conclusion

To conclude, we see that all three elements of environmental value—self-valuing, suffering, and interdependence—can be seen to play in an ensemble that, qua ensemble, gives rise to normativity. We should not lose sight of the connections among these three elements, and any hierarchization among them, for this or that purpose, would have to carefully respect their connectedness. Self-affirmation is a form of vulnerability in that it responds to the fact that an entity only emerges within a larger context from which it can never quite extricate itself, and into which it is in the process of returning.

Notes

1. Max Horkheimer, *Gesammelte Schriften: Dialektik der Aufklärung une Schriften 1940–1950*, ed. Gunzelin Schmid Noerr (Frankfurt am Main: S. Fischer Verlag GmbH); trans. Edmund Jephcott as Max Horkheimer and Theodor W. Adorno, *Dialectic of Enlightenment: Philosophical Fragments* (Stanford: Stanford University Press, 2002).

2. Maurice Merleau-Ponty, *La nature: Notes de cours du Collège de France*, ed. Dominique Séglard (Paris: Seuil, 1995); trans. Robert Vallier as *Nature: Course Notes from the College de France* (Chicago: Northwestern University Press, 2003).

3. Hans Jonas, *The Phenomenon of Life: Toward a Philosophical Biology* (Evanston, Ill.: Northwestern University Press, 2001); Jonas, *Das Prinzip Verantwortung: Versuch einer Ethik für die technologische Zivilisation* (Frankfurt am Main: Suhrkamp Taschenbuch, 1984); trans. Hans Jonas and David Herr as *The Imperative of Responsibility: In Search of Ethics for the Technological Age* (Chicago: University of Chicago Press, 1986).

4. The quotation marks around "value"—they will soon come off for better readability—are meant to flag concerns regarding the frequent subjectivism or projectivism in talk of "values." See, for instance, Heidegger's critique of the very notion of value as betraying a humanist "subjectivization" of beings: what counts on this view is never the being itself but only its subjective evaluation. See, for example, Heidegger's "Letter on Humanism" (in *Pathmarks*, ed. William McNeill [Cambridge: Cambridge University Press, 1998], 239–76). I do not think that the eco-deconstructive account of normativity I offer here is humanist or subjectivist in this sense;

see in particular the section "Conclusions for Relations among (V), (S), and (H)" in this chapter. On this topic, see also Chapter 13, Dawne McCance's chapter in this volume.

5. I don't mean to imply that these three provide an exhaustive list of what objects or states of affairs have been proposed to be of value, or of "intrinsic" value. In addition to self-valuing and suffering, the list would include consciousness, rational autonomy of will, sentience, and being alive itself (as in some versions of biocentrism, such as Paul Taylor's). In focusing on these three (V, S, H), my point is to say that very often, though perhaps not always, (V) self-valuing and (S) suffering come to play an important role when we ask further what makes the properties or states of affairs (intrinsically) valuable. For example, we may say consciousness or rationality are valuable because they allow for self-valuing, or sentience is morally important because it permits the feeling of pleasure and pain. When it comes to the ontological connectedness, my third desideratum (H) seeks to address those states or qualities that can be predicated of environmental wholes, such as diversity, equality, "integrity, stability, beauty" (in the words of Leopold's land ethic); cf. Aldo Leopold, *A Sand County Almanac: With Essays on Conservation* (Oxford: Oxford University Press), 189.

6. Holmes Rolston III, "Value in Nature and the Nature of Value," in *Philosophy and the Natural Environment*, ed. Robin Attfield and Andrew Belsey (Cambridge: Cambridge University Press, 1994), 13–30.

7. Ibid.; for his fuller view, see Rolston, *Environmental Ethics: Duties to and Values in the Natural World* (Philadelphia: Temple University Press, 1988).

8. Jonas, *Phenomenon of Life.*

9. In chapter 3 of *The Phenomenon of Life,* Jonas shows that a mathematically conceived divine mind (the god of the philosophers) would not perceive life; 78–79. As Thompson put it later, "Only life can understand life"; namely, life as lived, as feeling and self-valuing. See Evan Thompson, *Mind in Life: Biology, Phenomenology, and the Sciences of Mind* (Cambridge, Mass.: Harvard University Press, 2000), 162ff.

10. Jonas, *Phenomenon of Life,* 83–84.

11. Thompson, *Mind in Life.*

12. Jeremy Bentham, *Introduction to the Principles of Morals and Legislation* (Oxford: Clarendon, 1907), chap. 17, n. 122.

13. Peter Singer, *Animal Liberation* (New York: Harper Collins, 2009), 174. Note that ability still plays a role here, for suffering is first and foremost conceived as an ability, the capacity for an activity, rather than primarily as a passivity and receptivity. We will come back to this point.

14. Cf. AA 81/115, on Descartes's claim that perhaps animals can suffer but not truly.

15. Tom Regan, "Animal Rights, Human Wrongs," *Environmental Ethics* 2, no. 2 (1980): 116.

16. Peter Singer, "Taking Life: Animals," in *Practical Ethics* (Cambridge: Cambridge University Press, 2011).

17. See, for instance, James Lovelock's Gaia hypothesis, in Lovelock, "Gaia as Seen through the Atmosphere," *Atmospheric Environment* 6, no. 8 (1972): 579–80, but also Rolston, *Environmental Ethics*, and others.

18. Strictly speaking, of course, we are dealing with different kinds of individualism: in the approaches that prize self-valuing, we focus on organisms, extended by Rolston to species though considered as unified entities; in the case of utilitarianism, we are dealing with an individualism of experiences—we value experiences of pleasure and disvalue pain, with the organismic individual coming in only in response to the problem of replaceability, which is triggered by the value attributed to total experiences without regard to the entity having the experiences.

19. Apart from holists such as Leopold, *Sand County Almanac*, and J. Baird Callicott, *In Defense of the Land Ethic: Essays in Environmental Philosophy* (Albany: SUNY Press, 1989) and *Thinking Like a Planet: The Land Ethic and the Earth Ethic* (Oxford: Oxford University Press, 2014), at least one attribute of environmental wholes—diversity—is also valued by policy-makers and global publics. The World Charter for Nature (1982) http://www.un.org/documents/ga/res/37/a37r007.htm, and the Global Biodiversity Treaty, https://www.cbd.int/convention/text/, the latter signed by approximately 160 nations at the Rio Conference in 1992, both explicitly state that biological diversity is an intrinsic good.

20. Cf. Paul Taylor, *Respect for Nature: A Theory of Environmental Ethics* (Princeton: Princeton University Press, 1986).

21. Similarly, ecofeminist Val Plumwood argued against egalitarian biocentrism by asking whether we should intervene to protect the sheep against wolf, thus depriving the wolf of his food; see Plumwood, "Nature, Self, and Gender: A Critique of Rationalism," in *Environmental Ethics: The Big Questions*, ed. David R. Keller (London: Wiley-Blackwell, 2010), 303.

22. Callicott, "Non-Anthropocentric Value Theory and Environmental Ethics," *American Philosophical Quarterly* 21, no. 4 (1984): 299–304.

23. Callicott, "Animal Liberation and Environmental Ethics: Back Together Again," in *In Defense of the Land Ethic*.

24. Rolston, *Environmental Ethics*, 186–91.

25. Ibid., 217.

26. Rolston, "Value in Nature and the Nature of Value," 29; cf. Rolston's more recent *A New Environmental Ethics: The Next Millennium for Life on Earth* (London: Routledge, 2012).

27. In *The Animal That Therefore I Am* (AA), Jacques Derrida refers to self-affirmation chiefly under the title of "auto-biography," by which he means the desire for self-positioning, and "auto-deicticity," as what is often seen as the essence of the living (AA 49/75, 50/76, 52/79): quot-

ing oneself, originary repetition in ipseity, the "I" affirming itself doubly, but so as to do so *before* another, promising to tell the truth about oneself (57/84), and in this self-other-affirmation, responding without responsibility (56/84) and exhibiting iterability-automaticity (AA 125/172).

28. Arne Naess, "The Shallow and the Deep, Long-Range Ecology Movement," *Inquiry* 16, nos. 1–4 (1973): 95–100.

29. Cf. Vicki Kirby, in this volume, on evolution and the environment as a force that speciates and individuates itself *in* the organism.

30. Derrida, "Eating Well," in *Points* (PI 268–69/282–84, et *passim*); see also TS 76–77. As commentators have pointed out, some of Derrida's metaphors and analogies in the last ten years of his writing, beginning in particular with "Faith and Knowledge," are also drawn from biology (e.g., autoimmunity). For an elaboration, see Francesco Vitale's recent work on what he calls "bio-deconstruction"; cf. Vitale, "The Text and the Living: Jacques Derrida between Biology and Deconstruction," *Oxford Literary Review* 36, no. 1 (2014): 95–114, and Vitale, "Life Death and Différance: Philosophies of Life between Hegel and Derrida," *CR: The New Centennial Review* 15 (2015): 93–112.

31. Here, we should be reminded of what Derrida's *Politics of Friendship* says about the indiscernibility, even the convertibility, of friend and enemy (PF 32/51, 71/90, 88/107, 163/188, 174/198, 216/244); cf. Matthias Fritsch, "Antagonism and Democratic Citizenship (Schmitt, Mouffe, Derrida)," *Research in Phenomenology* 38, no. 2 (2008): 174–97.

32. See P 112/129, 122ff./140ff., 140–41/161–62, 182/208; NII 247; see also "A Number of Yes," in P2.

33. This insertedness or "inscription" of affirmation in contexts to which it first of all responds is the reason double affirmation is not a structure we can merely describe from a neutral standpoint (see the beginning of the passage I cite later: NII 247). In describing it we must also perform it from where *we* are, in our historical context, to which we thus respond with a promise. The welcome extended to the other is necessarily a response to a prior welcome by a preceding other. As Derrida puts in *Adieu*, "The welcoming *of* the other (objective genitive) will already be a response: the *yes to* the other will already be responding to the welcoming *of* the other (subjective genitive), to the *yes* of the other" (AE 23/51). In affirming itself, a "living ego" thus affirms, in what Derrida does not hesitate to call a "blind submission" and "obedience" (SM 7/28), its inherited context, including, I would suggest, biological evolution and ecological environments.

34. Allow me to cite a long passage in which Derrida makes these points: "What I call double affirmation . . . is neither a descriptive observation nor a theoretical judgment; it is precisely an affirmation, with the performative characteristics that any affirmation entails. The 'yes' must also be a reply, a reply in the form of a promise. From the moment that the

'yes' is a reply, it must be addressed to the other, from the moment that it is a promise, it pledges to confirm what has been said. If I say 'yes' to you, I have already repeated it the first time, since the first 'yes' is also a promise of this 'yes' being repeated. To say 'yes' is to acquiesce, to pledge, and therefore to repeat. To say 'yes' is an obligation to repeat. This pledge to repeat is implied in the structure of the most simple 'yes.' There is a time and a spacing of the 'yes' as 'yes yes': it takes time to say 'yes.' A single 'yes' is, therefore, immediately double, it immediately announces a 'yes' to come and already recalls that the 'yes' implies another 'yes.' So, the 'yes' is immediately double, immediately 'yes-yes.' This immediate duplication is the source of all possible contamination—that of the movement of freedom, of decision, of declaration, of inauguration by its technical or technical double. Repetition is never pure"; NII 247.

35. For more on the idea of the "quasi-transcendental" condition of possibility in Derrida, permit me to refer to Matthias Fritsch, "Deconstructive Aporias: Both Quasi-Transcendental and Normative," *Continental Philosophy Review* 44, no. 4 (2011): 439–68.

36. See, e.g., John Passmore, *Man's Responsibility for Nature* (London: Duckworth, 1974), and Kenneth M. Sayre, *Unearthed: The Economic Roots of Our Environmental Crisis* (Notre Dame, Ind.: University of Notre Dame Press, 2010).

37. Quoted in Michael Pollan, "Some of My Best Friends Are Germs," *New York Times Magazine*, May 15, 2013, http://www.nytimes.com/ 2013/05/19/magazine/say-hello-to-the-100-trillion-bacteria-that-make-up -your-microbiome.html?ref=magazine&_r=0.

38. Perhaps the most well-known example of autoimmunity is Derrida's discussion in *Rogues* of the 1992 elections in his native Algeria. The elections were suspended to prevent suspected non-democrats from coming to power (R 28–41/51–66). In welcoming its other (e.g., non-democrats), democracy must also protect itself from them. Democracy can sustain itself only by promising to return to itself in relating to an other and in becoming other to itself.

39. Cf. Michael Naas, *Derrida from Now On* (New York: Fordham University Press, 2008), especially chapters 7, 8; Elizabeth Rottenberg, "The Legacy of Autoimmunity," *Mosaic* 39, no. 3 (2006): 1–14; Martin Hägglund, *Radical Atheism: Derrida and the Time of Life* (Stanford: Stanford University Press, 2008). A similar passage in *Rogues*, one that insists on the interminability of democracy due to its constitutive relation to *différance* as to the deferral to the future to come, goes so far as to claim that the reference to the alterity of the other cannot but be denied by the self that wishes to come back to itself: "For what is also and at the same time at stake—and marked by this same word in *différance*—is différance as reference or referral [*renvoi*] to the other, that is, as the undeniable, and I underscore *undeniable*, experience of the alterity of the other, of het-

erogeneity, of the singular, the not-same, the different, the dissymmetric, the heteronomous. I underscore *undeniable* to suggest *only deniable*, the only protective recourse being that of a send-off [*renvoi*] through denial" (R 38/63)—hence, the reference to the phantom, the spectral, and the simulacrum in the passage from *Specters* cited in the body of the text. For an analysis of this necessary illusion of oneself, see Naas, "Comme si, comme ça: Phantasms of Self, State, and a Sovereign God," *Mosaic* 40, no. 2 (2007): 1–26, and Fritsch, "Antagonism and Democratic Citizenship."

40. Plumwood, *In the Eye of the Crocodile*, ed. Lorraine Shannon (Canberra: Australian National University Press, 2012), 19.

41. *The Works of Francis Bacon*, ed. James Spedding et al., 15 vols. (Boston: Brown, 1860–64), 14:95. It would be useful in this context to consider Theodor Adorno and Max Horkheimer's analysis of this famous claim, made at the beginning of the modern epoch (1597), in their *Dialectic of Enlightenment* (see note 2).

42. I treat these claims more extensively in "Interring: Earth and Lifedeath in Derrida," in Fritsch, *Taking Turns with Earth* (book manuscript in preparation).

43. Friedrich Schiller, "Versuch über den Zusammenhang der thierischen Natur des Menschen mit seiner geistigen [An Essay on the Connection between the Animal and the Spiritual Nature of Human Beings]," 1780 dissertation, in *Sämtliche Werke*, Band 5 (Berlin: Hanser, 2004), §9, "In Verbindung [In Relation]", 297; my translation.

44. See Cora Diamond, "The Difficulty of Reality and the Difficulty of Philosophy," in *Philosophy and Animal Life*, by Stanley Cavell et al. (New York: Columbia University Press, 2008), 71; see also Cary Wolfe, "Exposures," in ibid., 13.

45. Note that calling each finite mortal living being an "infinite" is another way of saying that on account of the self-difference that makes it up, it is not exhaustively describable. This "infinity" is thus part of its finitude and mortality.

46. Cf. BS2, 243–44/338–39, regarding "the first possibility as nonpower that we share with the animal, whence compassion" stems from "power as non-power" (i.e., animals are *able* to suffer in that they are *able* to *not* be able).

47. More should be said about the relation of the living to the nonliving, a boundary that must be made porous and multiplied as well; not by accident does Derrida, when speaking of the relation of life to death, use the examples (exemplary rather than mere examples) of eating and breathing. In *Beast & Sovereign*, vol. 2, Derrida also parallels two aspects of *partager* (both sharing and partitioning) among the living: what they have in common is that they die (but each singly) and they share (so as to divide) a habitat. The habitat, including the inorganic, is that which reabsorbs the living, itself nibbling at them, exemplarily in receiving the corpse. To say

that death is not just external to life is to also say that the nonliving are not just exterior to the living. For this reason, Derrida at times associates the troubling but also necessary dividing line between the human and the nonhuman with an equally necessary troubling of the distinction between living and nonliving; see, e.g., BS1 108–11/154–59).

48. So here we value not mere sentience of pain, but the ability to respond by seeking to avoid it in view of some kind of telos; cf. Jonas, *Phenomenon of Life*; Christian Kummer, "Pflanzenwürde: Zu einem Scheinargument in der Gentechnik-Debatte," *Stimmen der Zeit* 138, no. 1 (2013).

Opening Ethics onto the Other Shore of Another Heading

Dawne McCance

In his essay collection *A Likely Story*, Canadian novelist, poet, and critic Robert Kroetsch, born and raised on a homestead near Heisler, Alberta, and educated during the 1930s in a one-room Catholic school, recounts the occasion on which he lost his boyhood faith. On a bitterly cold Sunday morning well into the depths of a prairie winter, the local church, which stood empty and unheated during all other days of the week, had scarcely warmed in time for Mass. As the congregation filed out of the frigid church, young Kroetsch pulled his right hand out of his overcoat pocket and, as was the custom, reached up, hand over head, to dip his fingers into the holy water fount—discovering, to his utter astonishment, that the holy water was frozen, thus that saintly Father Martin's words during the autumn ritual of blessing "had not rescued ordinary water from its ordinariness."[1] This seemingly prosaic story of a boy's experience of "betrayal" by a religion's promise to indemnify what it sanctifies[2]— to elevate water out of the merely natural into the realm of the holy (never should drops from one's hand be let fall on a church floor)— recalls Derrida's musing in "Faith and Knowledge" as to whether a discourse on religion can be *"dissociated from a discourse on salvation: which is to say, on the holy, the sacred, the safe and sound, the unscathed* (indemne), *the immune* (sacer, sanctus, heilig, *holy, and their alleged equivalents in so many languages*)?" (FK 42/9). In so many languages and so many traditions, this discourse on the unscathed prevails: in India, for example, the Ganges River, along with the Yamuna that flows into it, ranks among the most polluted rivers in the world, used for laundry, defecation, and disposal of industrial wastes, cremation ashes, and partially decomposed human and animal corpses. Nevertheless, Hinduism associates the Ganges with the goddess Ganga, regarding it as holy water that, notwithstanding its state of despoliation, cleanses, purifies, grants health and salvation to pilgrims who bathe in it and drink from it.[3] Such is "the grand disjunction"[4] that in a number of religious traditions separates nature from what is deemed holy, the story of salvation from the fate of the natural world.

Treating the question of religion in "Faith and Knowledge," Derrida refers to an *autoimmune* drive that would seek to safeguard and preserve intact that which is deemed holy from whatever is perceived as alien and contaminating—such as modern technoscientific reason, which the Roman Catholic Church has inveighed against at least since Vatican I (1869–70); the Council, not incidentally, included in its decrees *both* a statement on the "twofold" sources and objects that separate "divine faith" from "natural reason" *and* a declaration of papal infallibility, the sovereign "full and supreme power" bestowed by God on the Bishop of Rome. As Derrida observes, however, in opposing technoscientific reason, the church attacks, in autoimmune fashion, its own systems of sustenance and protection, *"digital culture, jet* and *TV* without which there could be no religious manifestation today, for example no voyage or discourse of the Pope" (FK 62/39–40).[5] The church, acting autoimmunely, opposes the very teletechnological reason that is internal to its own working and that it so often exemplifies. On the latter point, Derrida comments, in a footnote on the marketing and communication technologies expertise of Pope John Paul II (pontiff at the time of "Faith and Knowledge"), about "the unprecedented speed and scope of the moves of a Pope versed in televisual rhetoric" whose "last encyclical, *Evangelium vitae*, against abortion and euthanasia, for the sacredness and holiness of a life that is safe and sound—unscathed, *heilig*, holy—for its reproduction in conjugal love—sole immunity admitted, with priestly celibacy, against human immuno-deficiency virus (HIV)—is immediately transmitted, massively 'marketed' and available on CD-ROM" (FK 62n17/40n13).

As explained in *Evangelium vitae* (March 25, 1995) and a number of other encyclicals on the subject of modernity and technology that the Holy See has issued in recent years—all of which, by the way, are digitalized and distributed on the Internet (see, for example, www.vatican.va)—life, *human* life, is of inestimable value precisely because it is not only corporeal but also spiritual, hence in Derrida's words, "worth *more than* life" (FK 87/78). Human life "is sacred, holy, infinitely respectable only in the name of what is worth more than it and what is not restricted to the naturalness of the bio-zoological" (FK 87/78). Thus, in presenting the church's "anthropological vision," the February 22, 1987, encyclical, *Instruction on Respect for Human Life in Its Origin and on the Dignity of Procreation: Replies to Certain Questions of the Day*, explains that "the human body cannot be considered as a mere complex of tissues, organs and functions, nor can it be evaluated in the same way as the body of animals," and it follows from this that "marriage possesses specific goods and values in its union and in procreation which cannot be likened to those existing in lower forms of life." Confined to a heterosexual union in marriage, human reproduction, unlike animal copulation, procreates (cocreates) life that is worth *more than* life, that is infused with a spiritual soul from its very beginning. Acting according to a divine plan that is known and interpreted by a sovereign pontiff who speaks on God's behalf, the church, then, in both *Evangelium* and the *Instruction*, outlaws contraception for blocking the cocreative potential to which every conjugal act must be open and condemns as morally unlawful a range of reproductive technologies.

In his superb study of "Faith and Knowledge" in *Miracle and Machine: Jacques Derrida and the Two Sources of Religion, Science, and the Media*, Michael Naas remarks that "Derrida's thought always develops through a mise-en-scène of other texts and other voices,"[6] which I take to mean, for one thing, that Derrida's writing always proceeds as a work of inheriting tradition, "Faith and Knowledge" being a case in point, and not the least as it pertains to the "anthropo-theological" (FK 87/79) concept of sovereignty we inherit. Derrida's analysis in the essay of the "religious" postulation of a life "worth *more than* life" continues, as it were, in the first volume of his 2001–3 seminar, *The Beast & the Sovereign*, in pages that dwell on the scene of a famous dissection conducted at the decadent Palace of Versailles under the reign of King Louis XIV, his royal menagerie providing the corpse of an elephant to be autopsied under the gaze of the Sun King himself. The monarch presiding over the carving up of the great animal has, Derrida suggests, "a double body," one that is corporeal, earthly, and mortal and another that is spiritual, "celestial sublime, eternal, that of the function of the majesty, of the royal sovereignty supposed to survive eternally" (BS1 285/382). Even the physicians attending to the ailing Louis le Grand looked after him as two bodies at once, "the king's body both as the body of a respected, admired, venerated, feared, all-powerful, and omniscient God, and as the objective, objectified, coldly regarded and inspected body of an animal with irresponsible reactions" (BS1 286/383). For my purposes here, the important point is that the structure of sovereignty always presupposes this double body, which, Derrida notes, the Revolution did not put to an end by decapitating Louis XVI, but only transferred from the king to the individual, now the "free" and "equal" subject of a political fraternity (BS1 285/382, 290/388).[7]

What interests me in this essay, I should have stated by now, is *water* and the profound failure as yet on the part of us academics to develop what might be called, albeit tentatively, a water ethics. This failure, I suggest, is owed at least in part to the still-widespread dismissal of Derrida's work within the academic institution, a characterization of it as already over and gone, and/or, not unrelated to these two attitudes, an indifference to it as engaging the task of inheritance.[8] Indispensable in the context of ethics is Derrida's sustained analysis and critique of the "double-body"—of the life "worth *more than* life"—that is constitutive of the concept of sovereignty bequeathed to us by the Western tradition, an inherently autoimmune logic that prevails not only in the teachings of Roman Catholicism, but also in legal, political, and *ethical* discourses that remain Christian-theological in their thinking and rhetoric.[9] *Rogues*, then, should be studied by every would-be theorist of ethics who is attempting to grapple with the fundamental "*I can*" that holds sway today in ethics and that ties moral worth to power and capacity. Although not elaborated in *Rogues* with reference to ethics per se, Derrida's analysis of the sovereign subject of democracy is absolutely relevant to a study of the *autos* of ethics, and what Derrida intends to suggest in *Rogues* with the term "ipseity" is especially applicable: "the power that gives itself its own law, its force of law, its self-representation, the sovereign and reappropriating gathering of self in the simultaneity of an assemblage or assembly" (R 11/30). In major streams of contemporary ethics, much as in the order of the political, the assemblage or assembly, a fraternity, admits only one's likes, determining moral worth by calculating the degree to which one is similar to the subject-author of ethics, the latter auto-represented in accord with the seventeenth-century ideal of *homo rationalis*, hence a "double body," at once corporeal and sublime. In relation to this, what Derrida in *Rogues* poses as one form that the question of "democracy-to-come" might take is also a question for ethics, for the "to-come" of ethics: "What is 'living together'?" And especially, "What is a like, a compeer (*semblable*)," "someone similar or semblable as a human being a neighbor, a fellow citizen, a fellow creature, a fellow man," and so on? Or even, must one live together only with one's like, with someone semblable?" (R 11/31). From "Faith and Knowledge," six words provide ethics with a point of departure for the task of pondering what "living together" might mean: "Axiom: no to-come without heritage" (FK 83/72).

From the beginning to the end of his life, Derrida remained close to water. He was born and grew up in Algiers, a city on the edge of the Mediterranean Sea, a marginalized place (for Derrida, in more ways than one) colonized by France and separated from it by vast expanses of ocean water.[10] Not until he was nineteen did Derrida leave Algiers for France, "Jackie's first real trip: the first time he had left his parents, the first time he had taken the boat," Benoît Peeters writes. "The crossing, on the *Ville d'Alger*, was hellish, with a terrible seasickness and twenty hours of almost uninterrupted vomiting."[11] Yet every "heading off" can entail disorientation, a heightened awareness of vulnerability and of the need for change, perhaps the need to change ways of thinking and living, as is required of humans today. Derrida acknowledges this in *The Other Heading: Reflections on Today's Europe*, where, writing about response and responsibility, he notes that the expression "The Other Heading" can suggest the need to change destinations and direction,

> "to change goals, to decide on another heading, or else to change captains, or even—why not?—the age or sex of the captain. Indeed it can mean to recall that there is another heading, the heading being not only ours [*le nôtre*] but the other [*l'autre*], not only that which we identify, calculate, and decide upon, but *the heading of the other*, before which we must respond." (OHG 15/20)[12]

The *heading of the other* before which we must respond: citing a World Health Organization report, Richard Chamberlain in *Troubled Waters* notes that, as of the year 2000, 1.1 billion people, one-sixth of the world's population, lacked access to quality drinking water and 2.4 billion people, two-fifths of the world's population, lacked adequate sanitation.[13] For 2001, he goes on to point out, WHO estimated that over 2 million deaths from infectious diseases would be attributable to water and sanitation concerns (some 5,843 deaths daily); as of 2006, *The United Nations World Water Development Report 2* stated that 1.4 million children under the age of five die annually from such water-related diseases.[14] As factors contributing to this situation, Chamberlain lists overpopulation, water pollution, inadequate or no water treatment facilities, agriculture and industry demands, especially those of the computer industry, global warming affecting Arctic and Antarctic ice, and changing water temperature; and, as he allows, again citing a United Nations report, the situation is less dire in the developed world, where a child typically consumes thirty to fifty times more water resources than does a child born in an underdeveloped country.[15] That said, in the water-rich country of Canada where I live, many First Nations communities remain without access to clean drinking water and water for sanitation. In my home province of Manitoba, the capital city of Winnipeg pipes its water from Shoal Lake, located in First Nations land just across the Ontario border. When it arrives in the city, the water passes through a state-of-the-art treatment plant, while back at the lake, the First Nations band Shoal Lake 40 has been under a boil-water advisory for eighteen years. In other words, many disadvantaged populations live where there is water, but without affordable access to the water they need.[16]

In keeping with the growing privatization of water and appropriation of it, largely by the First World, as a commodity to be marketed at a price (which can be considerable where it is in short supply), allocation decisions "are no longer governed by a sense of the common good of communities but rather by private property rights."[17] Prominent in ethics today, the discourse of rights serves as an indicator of the challenges involved in changing destinations or direction. For although a rights framework has been promoted by a number of scholars as the way forward from here, the post-Cartesian genealogy ties rights to some formulation of "I think, therefore I am" that—along racial, sexual, and speciesist lines—has colonized and disenfranchised those deemed to lack the essential rationality that, in another version of the "double body" concept of sovereignty, the (historically male) subject claims for himself. For reason of this heritage, Derrida asks in *The Animal That Therefore I Am* whether rights is the pertinent concept, whether we must resort to the discourse of rights when grappling with the political, legal, and moral status of nonhuman animals, indeed whether the discourse of rights might "at bottom share the axiom and founding concepts" through which violence against animals has been exercised (AA 88–89/123–25). Although Derrida's particular focus in this text concerns violence against animals, his overriding concern, in his own words, is with "our responsibilities and our obligations vis-à-vis the living in general" (AA 27/48). Moreover, it is impossible to separate today's water crisis from the violence humans have unleashed on animals over the past two hundred years. Industrialized ("factory") farming practices, for instance, involve untold brutality to animals. They also contribute significantly to the pollution of water systems, thus to the contamination and death of aquatic creatures, to widespread loss of biodiversity, and with the vast quantities of manure these factories produce and the methane this waste releases, to global warming and increasing desertification, a phenomenon with which the state of California is now contending.

What Derrida refers to in *The Animal That Therefore I Am* and analyzes in the philosophical heritage from Descartes through Heidegger and Lacan is, again, ipseity, the "power of the 'I'" that Western culture upholds, "the power to make reference to the self in deictic or autodeictic terms, the capability at least virtually to turn a finger toward oneself in order to say 'this is I'" (AA 94/132). While Tom Regan, a leading ethical theorist, promotes rights discourse as a nonpatriarchal, nonandrocentric, and even nonspeciesist orientation, his publications seem to me to reinstate this problematic autoreferentiality, which in contemporary ethics results in taking "I can" to be the universal standard of moral worth. Although Regan makes the case for extending moral rights to some nonhuman animals, then those he would include in the moral assemblage are not merely living but "subjects of a life" who "resemble humans" in what he calls "morally relevant ways"—that is, who have "a variety of sensory, cognitive, conative, and volitional *capacities*" equivalent to that of "mentally normal mammals of a year or more."[18] That the "mentally normal" standard is inapplicable to water and to most water species is implicitly acknowledged by Regan with his definition of rights, in line with the post-Cartesian tradition, as adversarial demands that grant "negative freedom" to the holder (freedom from, rather than freedom for, the other), something like a "No Trespassing sign."[19] Regan insists on the "*individualistic* nature" of rights, deploring what he calls the "environmental fascism" that would compromise individual rights in the name of achieving some greater good for others or for the biotic whole.[20] Concerned exclusively with allocating "I can" powers to *individuals* who qualify as "subjects of a life," Regan's ethics is not about the rights of species to anything, including survival. My point here is not to suggest that "water possesses rights," but that Regan's rights logic, rooted in anthropocentric individualism, is a symptom of, rather than a solution to, the current ecological crisis. With his strong dismissal of ecological thinking and his seeming indifference to the social and ecological breakdowns to which an atomistic social physics has led, Regan's approach suggests why, in Judith Butler's terms, life has become "precarious." "The structure of address is important for understanding how moral authority is introduced and sustained," Butler writes, "if we accept not just that we address others when we speak but that in some way we come to exist, as it were, in the moment of being addressed, and something about our existence proves precarious when that address fails."[21] Life proves precarious when ethics fails to "challenge the very notion of ourselves as autonomous and in control."[22]

Although Roman Catholicism in *Evangelium vitae* condemns "the utilitarian motive of avoiding costs which bring no return," utilitarianism thrives in ethics today as an alternative to the discourse of rights and as an approach that, basically, determines the morality of an action by assessing whether its consequences will yield a surplus of benefits over costs. At least according to Peter Singer, a leading proponent of utilitarianism, the best outcome in any case is that which minimizes suffering and maximizes pleasure, or freedom from pain, for those with "like interests"—that is, those who possess the "mental capacities" on which, he argues, the "capacity for suffering" depends.[23] His utilitarian framework is non-speciesist, Singer argues, in that those animals matching the mental capacity of a "normal adult human,"[24] such as himself, can be counted, by virtue of this resemblance, as members of the moral assemblage. Possibly, due to the suffering that shortage of water implies for humans (with the requisite mental capacities), Singer's approach, developed along the lines of his treatment of vegetarianism,[25] could lead to a human-centered water ethics, one consistent nonetheless with the "I can" double-body logic that Derrida opens to question. But it makes no sense to apply Singer's utilitarian framework to water itself, any more than it does to suggest, according to an approach such as Regan's, that water possesses rights.[26] Indeed, what I take from these two leading theorists is the sorry conclusion that ethics today has become a "discourse of domination," a situation in which,

as Derrida puts it in *The Animal That Therefore I Am*, domination
is exercised "through the forms of protest that at bottom share the
axioms and founding concepts in whose name the violence is exercised" (AA 89/125).

It is important to consider the understanding of *ethics* implied by
both the rights and utilitarian frameworks previously outlined: a discourse that, having ostensibly left religion behind, rests nonetheless
on an anthropotheological concept of sovereignty, hence, that counts
spiritual, "rational" life as "worth *more than life*" that is "restricted
to the naturalness of the bio-zoological" (FK 87/78); a discourse that
foregrounds *capacity*, the "I can" of the white male subject in post-
Cartesian culture; a discourse in which the male subject represents
himself as both the sovereign author-ity as to what is right and
wrong and the standard of moral worth against which admissibility
in the moral assemblage is to be measured; a preprogrammed calculus that could in principle be carried out by a machine. On the latter
point, Cary Wolfe suggests that, "by relying upon a one-size-fits-all
formula for conduct," an ethics that concerns itself with calculating interests or rights "actually *relieves* us of ethical responsibility"
in favor of "an application that, in principle, could be carried out
by a machine."[27] On the one side, then, *Evangelium vitae* and the
Instruction, and on the other side, the calculation of interests and
rights: ethics here would seem to be deadlocked in the opposition
between, to use Michael Naas's title words, *miracle* and *machine*.

Derrida delivered the first draft of "Faith and Knowledge" in February
1994 on the island of Capri, completing the essay in April 1995 at
Laguna Beach, California,[28] two places, if not two sources, for "Faith
and Knowledge" where it is not easy to lose sight, or thought, of water. Heading off "on the boat that brought us from Naples to Capri,"
Derrida may or may not have pondered the degree of contamination
in the water he was crossing—the heavy metals, pesticides, and
PCB concentrations in Bagnoli Bay, the Gulf of Naples, and the Tyr-
rhenian Sea—but he was reflecting, at least implicitly, on ethics
when he told himself, "silently, that one would blind oneself to the
phenomenon called 'of religion' or of the 'return of the religious'
today if one continued to oppose so naïvely Reason *and* Religion,
Critique or Science *and* Religion, technoscientific Modernity *and*
Religion" (FK 65/45). Derrida's remarks might be read, it seems to
me, with reference not only to the Catholic Church's opposition to

the technoscience that supports and sustains it or to the "return
of the religious" in utilitarian and rights theories informed by an
anthropotheological concept of sovereignty, but also to the potentially fatal autoimmune drive that, in every context, an oppositional logic harbors. Again on the matter of inheriting tradition, I
might recall here Derrida's reading in "To Speculate—On 'Freud'"
(originally in his *La vie la mort* seminar), of the nonpositional, non-
oppositional structure of Freud's *Beyond the Pleasure Principle*,
where, Derrida points out, the pleasure principle (the drive to *ipseity*) that would retain unbroken self-proximity, mastery, or sovereignty is not renounced in oppositional terms, but rather detoured,
deferred, by the reality principle, so that the return to (one)self is
not absolute and fatal. In Derrida's words, Freud's reality principle
"belongs to the same economy, the same house" (PC 282/301)—the
same psychic ecosystem—as does the pleasure principle, "a master
who is sometimes hard to educate" (PC 282/301).[29] This suggests,
at least to me, that if it is on the basis of the mastery exercised by
what Freud calls the "pleasure principle" that "one can determine
any mastery at all, figuratively or literally" (PC 392/419), then the
human, rather than elevating himself as life worth *more than* life,
might need to acknowledge that an oppositional logic has proven
not conducive to "living together," not conducive to the survival of
one's likes, any more than to the survival of the planet. For we all
occupy the same ecosystem, the same house—or, if you will, we are
all in the same boat.

Such is the reality depicted by Bill Reid in his magnificent bronze
sculpture *The Spirit of Haida Gwaii*, a dugout canoe filled beyond
full with thirteen passengers shown in their myth-image forms and
symbolizing the entire family of living beings.[30] Although Raven is
steering and other figures, including Dogfish Woman and the Ancient Reluctant Conscript, are paddling, the boat's destination is
not known: perhaps it is one of the islands in the archipelago on
the northern coast of British Columbia once occupied by the Haida
people; perhaps it is a destination always "to-come." In any case,
central to Reid's work of inheriting Haida tradition through art are
the two worlds of land and sea, with the amphibious Frog creature
(he's in the canoe) being one who can hold the two worlds together.
Reid's Haida ancestors learned from the Frog the lesson of respecting
and being responsible to both land and water, for although popu-
lating the Haida Gwaii Islands, they could not reach out to each

other without crossing the sea that provided their livelihood. And no doubt, when they crossed the water from one island to another in the archipelago, they discovered not so much the sameness as the diversity that had developed among them.

Derrida's final seminar on "The Beast & the Sovereign" has much to say about water and islands—for instance, the Island of Despair on which Robinson Crusoe washed up after being shipwrecked; the island over which he soon autorepresented himself as sole master and sovereign, Crusoe's sovereignty as undivided, unshared with other living beings, as is Heidegger's in *The Fundamental Concepts of Metaphysics: World, Finitude, Solitude*, the text Derrida reads in this seminar alongside Defoe's novel. For according to Heidegger, man is alone in "that *solitariness* in which each human being first of all enters into a nearness to what is essential in all things, a nearness to world."[31] Man alone has access to the essence of "the world," or as Heidegger puts it in the three theses that guide him in *The Fundamental Concepts*, "the stone (material object) is *worldless*; the animal is *poor in world*; and man is *world-forming*."[32] Derrida argues, however, in the second published volume of his remarkably rich and complex *The Beast & the Sovereign* seminar, that "the world" is and can only be a construction, a phantasm that, if it facilitates human claims to sovereignty in the political, legal, or ethical domains, does so only by refusing to acknowledge the infinity of differences that prevails between and among all living beings. There is no world *as such*, Derrida submits. "There is no world, there are only islands" (BS2 9/31).

Not an apocalyptic statement, Derrida's "there is no world, there are only islands" allows for multiple modes of sharing, of *cohabiting*, that are prohibited by the exclusivity of Heidegger's "the world *as such*." And as Derrida may suggest in citing the Paul Celan passage, "The world is gone, I must carry you" ["*Die Welt ist fort, ich muss dich tragen*"], or "The world is gone, I can only carry you" (BS2 9/31–32), relieved of recourse to "the world"—or to the life *worth more than life*—humans may yet change moral direction and begin responding to the heading of the other.[33] Not apocalyptic, Derrida's statement is not perfectionist, either, not any more than is Bill Reid's *The Spirit of Haida Gwaii*, with all the biting and griping, all the give and take, that is shown going on between the passengers in the dugout canoe. One thing is certain: no passage, no carrying, is possible without the water on which all life depends.

Notes

1. Robert Kroetsch, *A Likely Story: The Writing Life* (Red Deer, Alberta: Red Deer College Press, 1995), 59.

2. Ibid., 58.

3. Gary L. Chamberlain, in *Troubled Waters*, points out that, among many Indian pilgrimages and bathing festivals, "perhaps the best-known rituals are performed at the junction of the Ganga and the Yamuna during the Great Kumba Fair, the six-week religious festival Maha Kumba Mela. The festival gathers over thirty million people who come to bathe in the sacred waters, visit the hundreds of holy people, and scatter the ashes of their loved ones into the waters to mingle with the purity and divine wisdom symbolized in the rivers. The experience of bathing in the waters brings a spiritual awakening and heightened reality for millions"; see Chamberlain, *Troubled Waters: Religion, Ethics, and the Global Water Crisis* (Lanham, Md., New York, and Toronto: Rowman & Littlefield, 2008), 19.

4. Ibid.

5. In biomedical discourse, an autoimmune disorder results from a malfunction of the body's immune system that causes the body to attack its own tissues, misrecognizing and targeting them as foreign. While noting in "Faith and Knowledge" that his own use of the term draws from the biological domain, Derrida states as well that he feels "authorized to speak of a sort of general logic of autoimmunization" that "seems indispensable to us today for thinking the relations between faith and knowledge, religion and science, as well as the duplicity of sources in general" (FK 72–73n27/67–68n23). See also Derrida's discussions of autoimmunity in *Rogues* (R) and in "Autoimmunity: Real and Symbolic Suicides" in *Philosophy in a Time of Terror* (PT).

6. Michael Naas, *Miracle and Machine: Jacques Derrida and the Two Sources of Religion, Science, and the Media* (New York: Fordham University Press, 2012), 10.

7. On the fraternal order in the history of the political, including democracy, and with relevance for the ethical, Derrida's *Politics of Friendship* (PF) is a crucial resource.

8. For according to Derrida in *Specters of Marx*, and as all his work demonstrates, we are heirs: "The *being* of what we are *is* first of all inheritance, whether we like it or know it or not" (SM 54/94).

9. Derrida suggests in "Faith and Knowledge" that nothing could be more problematic than an attempt "to dissociate the essential traits of the religious as such from those that establish, for example, the concepts of ethics, of the juridical, of the political or of the economic," for these latter, even as they "pretend" to isolate themselves, "remain religious or in any case theologico-political" (theologico-juridical, theologico-ethical) (FK 63/43).

10. Of water, one must recall that it is both vulnerable and formidable, both sustenance and threat: I write this (April 20, 2015) only two days after the sinking off the Libyan coast of a boat overloaded with desperate migrants seeking refuge in Italy, as many as 850 now reported missing and presumed drowned in the Mediterranean Sea.

11. Benoit Peeters, *Derrida: A Biography* (Cambridge: Polity, 2012), 35.

12. My title is taken from Jacques Derrida's *The Other Heading: Reflections on Today's Europe* (OHG 76/74).

13. Chamberlain, *Troubled Waters*, 81.

14. Ibid.

15. Ibid., 81–82.

16. Every reader will have his or her own story. For all intents and purposes, Lake Winnipeg, not many kilometers north of my home and one of the largest inland lakes in the world, has been pronounced dead, while the Winnipeg River system that flows into Lake Winnipeg from northwestern Ontario, and on which First Nations populations depend on for their livelihood, has been poisoned with mercury and other industrial pollutants.

17. Chamberlain, *Troubled Waters*, 132.

18. Tom Regan, *The Case for Animal Rights* (1983; repr. Berkeley and Los Angeles: University of California Press, 2004), xvi; my emphasis.

19. Regan, *Empty the Cages: Facing the Challenge of Animal Rights* (Oxford: Rowman & Littlefield, 2004), 38–39.

20. Regan, *Case for Animal Rights*, 361–63.

21. Judith Butler, *Precarious Life: The Powers of Mourning and Violence* (London: Verso, 2008), 130.

22. Ibid., 23.

23. Peter Singer, *In Defense of Animals: The Second Wave* (Oxford: Blackwell, 2006), 5–6. While Singer acknowledges Jeremy Bentham as the source for his utilitarian ethics and its "I can" calculus, Derrida, in *The Animal That Therefore I Am*, receives the Bentham proposal—that the issue is not to know whether animals think, reason, or speak, but whether they suffer—very differently. Bentham's dictum "changes everything" (AA 27/48), Derrida maintains, in that it replaces the standard "think, reason, or speak, etc." (AA 27/48) with the question of suffering—a "sufferance" or "passion" that issues not from some power or capacity, the "*can-have-the-logos*," but from incapacity, from what he refers to as "passivity," a profound "not-being-able" (AA 27/49). Passivity is not a power or capacity, but an *impouvoir*, a "nonpower at the heart of power" (AA 28/49), a heightened sense of one's vulnerability before—and incapacity, finally, *to know*—an other. Derrida refers to the passivity that disturbs Bentham's question—"Can they suffer?"—as "a possibility without power, a possibility of the impossible" (AA 28/49), and for him, morality, in every case, "resides there, as the most radical means of thinking the finitude that we share with animals, the mortality that belongs to the

very finitude of life, to the experience of compassion, to the possibility of sharing the possibility of this nonpower, the possibility of this impossibility, the anguish of this vulnerability, and the vulnerability of this anguish" (AA 28/49). He holds out passivity as an experience of the impossible, and at the same time, as the possibility of an ethics that exceeds, goes beyond, calculation and preprogrammed rules; as an ordeal that an ethical ethics must suffer in every response to the "wholly other they call 'animal'" (AA 13/30).

24. Singer, *In Defense of Animals*, 6.

25. See for this, Singer, "Utilitarianism and Vegetarianism." *Philosophy & Public Affairs* 9, no. 4 (Summer 1980): 325–37.

26. Chamberlain, *Troubled Waters*, 136–37.

27. Cary Wolfe, "Humanist and Posthumanist Antispeciesism," in *The Death of the Animal*, ed. Paola Cavalieri (New York: Columbia University Press, 2009), 53.

28. Naas, *Miracle and Machine*, 23.

29. Much work remains to be done on the resources that Derrida's inheritance of Freudian psychoanalysis offers to our understanding of sovereignty, in ethics and in philosophy, for instance, to cite Derrida from a discussion of Adorno in *The Animal That Therefore I Am*, on Kant's "hate for the animality of the human," and on "the whole Freudian problematic of the religions of father and son" (AA 103/142–43).

30. A wealth of material is available on Bill Reid's *The Spirit of Haida Gwaii* sculptures, two of which he created: the Black Canoe, held at the Canadian Embassy in Washington, D.C., and the Jade Canoe, on display at the Vancouver International Airport in British Columbia, Canada. For a photographic record of Reid's sculpting process, see Ulli Steltzer, *The Spirit of Haida Gwaii: Bill Reid's Masterpiece* (Vancouver: Douglas & McIntyre, 1997).

31. Martin Heidegger, *The Fundamental Concepts of Metaphysics: World, Finitude, Solitude*, trans. William McNeill and Nicholas Walker (Bloomington: Indiana University Press, 1995), 6.

32. Ibid., 177.

33. As I have attempted to suggest in this essay, it is precisely response and responsibility that are elided in frameworks of ethics that, on one side, on the side of "religion," dictate a pregiven rule to be followed in every case, and on the other side, on the side of "reason," reduce decision-making to a preprogrammed calculation. Not only does Derrida challenge the oppositional logic that underlies this impasse in ethics, he also argues that ethics—not as the discourse of a first-person addressor (author, authority) but as the attempt of an addressee to receive and respond to a call—cannot bypass the ordeal that a decision, to be free and to be ethical, necessarily entails.

CHAPTER

14

Wallace Stevens's Birds, or, Derrida and Ecological Poetics

Cary Wolfe

One of Wallace Stevens's last and most famous poems, "Of Mere Being," begins:

> The palm at the end of the mind,
> Beyond the last thought, rises
> In the bronze décor.
>
> A gold-feathered bird
> Sings in the palm, without human meaning,
> Without human feeling, a foreign song.[1]

Stevens would seem here to take his place in a long line of poets who (as is well-known to generations of school children) work the bird/bard trope, where the bird—like Shelley's skylark, Hopkins's wind-hover, Keats's nightingale, or Poe's raven (just to name a few) brings news from another, unearthly realm, infusing the poet's own song with something not exactly known but understood. An inordinate number of Stevens's most important poems turn on the presence of a bird and/or its song, but I want to suggest that Stevens here takes these matters—"beyond the last thought," "without human meaning, / without human feeling, a foreign song"—more seriously, indeed one might even say more literally, than any poet I can think of.

To begin to make headway on these questions, we have to understand that at the core of Stevens's poetics as it develops over the course of his career is a two-stage progression. First, as Simon Critchley puts it, "Stevens's poetry allows us to recast what is arguably the fundamental concern of philosophy, namely the relation between thought and things or mind and world, the concern that becomes, in the early modern period, the basic problem of epistemology." At the heart of Critchley's reading of Stevens is the claim that his poetry "recasts this concern in a way that lets us cast it away," that Stevens's poetry "shows us a way of overcoming epistemology."[2] The question then becomes how to do this nondialectically, as it were (how not to be Whitman, if you like—which Stevens had not yet figured out in a poem such as "Sunday Morning"), which leads to the second stage of the progression of Stevens's poetics, his contention that "the theory / Of poetry is the theory of life, / As it is, in the intricate evasions of as."[3] This "overcoming" is hence "performed in the specific poem insofar as that poem concerns itself with some real particular, with some object, thing or fact," which in turn means, as he asserts, that "things as they are" (to use Stevens's phrase from "The Man with the Blue Guitar") "only are in the act that says they are."[4] Thus, for Stevens, poetry "reveals the idea of order which we imaginatively impose on reality. . . . The fact of the world is a *factum*: a deed, an act, an artifice."[5]

We are now in a better position to understand the rhetorically curious procedure of many of Stevens's most famous poems, such as "On the Road Home":

> It was when I said,
> "There is no such thing as the truth,"
> That the grapes seemed fatter.
> The fox ran out of his hole.
>
> You . . . You said,
> "There are many truths,
> But they are not parts of a truth."
> Then the tree, at night, began to change,
>
> Smoking through green and smoking blue.
> We were two figures in a wood.
> We said we stood alone.[6]

The poem goes on in this vein for another three stanzas—which begin, in turn, "It was when I said," "It was when you said," and "It was at that time"—but what I want to emphasize here is simply Stevens's undeniable insistence on the performative *factum* of "saying." I mean, when you think about it, how odd it is to say, "We *said* we stood alone."[7]

I am dwelling on this rhetorical slackness, which is also a kind of heightened artifice that Stevens repeatedly insists upon, to under-score the fact that the poem doesn't just foreground the contingency of the fact that "there is no such thing as the truth," that "there are many truths,/But they are not parts of a truth." And it doesn't just tell us that it is precisely this fact that reveals the *thereness* or *quidditas* of the world to us—that makes the grapes fatter, makes the fox run out of his hole. It also and most importantly insists on its *own* contingent performativity—"It was when I said," "It was at that time"—to prevent everything we just said from flipping over into its own form of idealism—that is, "the only universal truth is that there is no universal truth." Stevens's antirepresentationalist poetics here makes available to us a logic that is "heterogeneous" to idealism, as Derrida puts it, one that "entails the necessity of think-ing *at once* both the rule and the event, concept and singularity," marking "the impossibility of idealization as such" (LI 119). And thus, Stevens stages his complex inheritance of Emerson by fore-grounding the contingency of observation and the performativity of utterance in the *factum* of the poem itself—features of the Emer-sonian legacy that begin with the opening pages of the essay *Nature* but really only become constitutive and definitive in texts from the first and second series of *Essays* such as "Experience":[8]

> It was when I said,
> "Words are not forms of a single word,
> In the sum of the parts, there are only the parts.
> The world must be measured by eye"[9]

This eye—which takes up in echo not only the Emersonian "I/eye," but also as iterative and performative "I" of "I said"—is at the center of Stevens's poetic universe, and it's the same eye that pops up in a key bird moment (one of the most famous of all in Stevens) halfway through *Notes Toward a Supreme Fiction*, where Stevens writes:

> Eye without lid, mind without any dream—
>
> These are of minstrels without any minstrelsy,
> Of an earth in which the first leaf is the tale
> Of leaves, in which the sparrow is a bird
>
> Of stone, that never changes.[10]

By the logic of the poetics I'm trying to articulate here, the eye needs a lid for the very same reasons that Derrida is so interested in blinking, how the *Augenblick* figures a hiatus or "spacing" that holds the sound of one's own voice at a distance of a before and an after—"It was when I said"—that divides the "I" (as in Emerson's "transparent eye-ball"—"I am nothing, I see all") from itself. And it is in that space of "between"—"between that disgust and this," as Stevens puts it in "The Man on the Dump"[11]—that the "mind" can have its "dream," that imagination can enact its "purifying change."[12]

Stevens in much of his poetry is almost obsessed with this dynamic, and it often takes the form of the *Augenblick* of the before and after, as in the iconic poem "Thirteen Ways of Looking at a Blackbird":

> I do not know which to prefer,
> The beauty of inflections
> Or the beauty of innuendoes,
> The blackbird whistling
> Or just after.[13]

This "just after" captivates Stevens in all sorts of important heard bird poems— "Autumn Refrain" is one example, "Not Ideas about the Thing but the Thing Itself" is another—but I want to note in closing out "On the Road Home" that, for these very reasons, Stevens's insistence, "The world must be *measured* by eye," is precisely right, since "measuring" is only a particularly schematic instance of what is always already true of the "spacing" of iteration and the movement of protention and retention that it generates, creating the strange loop of temporality, of the "before" and "just after," that Stevens very much insists on with all those "It was whens" in "On the Road Home."

This inheritance and then radical reworking of the Emersoninan "I"/eye has been noted by many critics—indeed it is the most familiar story of the Emerson/Stevens legacy transmitted to us by critics like Harold Bloom[14]—but explaining how Stevens does this through a logic that is "heterogeneous" to idealism (as Derrida characterizes it) is what has been missing in the critical literature. I've been giving you the deconstructive version of what that heterogeneous logic looks like, but let me now briefly redescribe it in the more "naturalistic" and pragmatic terms of systems theory to more fully contextualize how Stevens's intense interest in the "epistemological" problems attending the relation between mind and reality eventuate, as Critchley put it earlier, in "overcoming epistemology."

Epistemology is often understood, these days, as a branch of what Graham Harman and other object-oriented ontologists call "correlationism," but this charge misunderstands a couple of quite fundamental postulates of systems theory (and, I think, of deconstruction). First of all, for systems theory, the question is not epistemological but *pragmatic*. As both Niklas Luhmann and Humberto Maturana make clear, the veracity of the systems-theoretical analysis is not about epistemological adequation to some pregiven state of ontological affairs (whether conceived in realist *or* idealist terms), but is rather based on its *functional* specificity. In other words, what we think of as "fact" is here rewritten precisely as "*factum*." Contrary to the understanding of autopoietic systems as solipsistic, the operational closure of systems and the self-reference based upon it arise as a practical and adaptive necessity precisely because systems are *not* closed—that is, precisely because they find themselves in an environment of overwhelmingly and exponentially greater complexity than is possible for any single system. What this means is that systems are characterized by a kind of *finitude* that can be formalized as a complexity differential; they maintain themselves and achieve their autopoiesis, in a sense, against all odds. To put it another way, systems have to operate selectively and "blindly" (as Luhmann puts it) not because they are closed, but precisely because they *aren't*, and the asymmetrical distribution of complexity across the system/environment difference is in fact what *forces* the strategy of self-referential closure.[15] Indeed, the "second-order" turn, as I have argued elsewhere,[16] is to realize that the more systems build up their own internal complexity through recursive self-reference and closure, the *more* linked they are to changes in their environments to which they become more and more sensitive—which is why a bat or a dolphin—or a bird!—can register a higher degree of environmental complexity than an amoeba that responds only to either gradients of light or dark, higher or lower sugar concentrations, and so on. Or as Luhmann puts it in one of his more Zen-like moments, "Only complexity can reduce complexity."[17]

To put it this way is to realize that "epistemological" isn't the *opposite* of "environmental" but in fact *means* "environmental" in this very specific sense: you can't take embodiment seriously, of *whatever* form of life, without also taking epistemological questions seriously, because if epistemology is precisely the study of how a being knows things, then those modes of knowledge and experience of the world depend directly on the embodied "enaction" (to use Maturana and Varela's phrase),[18] which is a product of the recursive loop between an organism's wetware and how it gets rewired by external interactions, environmental factors, semiotic systems, cultural inheritances, the use of tools, and much else besides. There are, to be sure, flows of energy, toxins, climate change, etc., across those autopoietic boundaries on the level of structure, and indeed, such flows are the drivers for the adaptive strategy of autopoietic closure.[19]

It ought to be obvious by now that this is not at all about correlationism as it is usually described—the idea that "being exists only as a correlate between mind and world"—and this is the case not least of all because "mind" is no more a constitutive element for systems theory (for whom observers can be nonhuman and even, in Luhmann's work, inorganic communication systems) than it is for deconstruction.[20] Quite the contrary; it is about systems *not* being free to just think whatever they will at their whimsy about objects and things. To put it another way, there is nothing to stop you on *epistemological* grounds from thinking that humors or ethers exist; indeed, the history of philosophy and science make that abundantly clear. But there is plenty to stop you on *pragmatic* grounds, even if, as Bruno Latour reminds us, bad ideas can work well enough under the right constraints for long periods of time.

All of this is crucial, I think, for understanding and contextualizing the precise mechanics by which Stevens moves away from an "idealist" rendering of mind and imagination toward one that ceaselessly foregrounds how the paradoxes that are attendant upon "overcoming epistemology" are actually, pragmatically *productive* of a meaningful relationship between mind and world. And we are thus in a better position to make sense of the "new knowledge of reality" put forth in another important heard bird poem, the last piece in *The Collected Poems*, "Not Ideas About the Thing But the Thing Itself," which handles this problem in a manner close to what we might call the "weakness" or "slackness" of "On the Road Home":

> At the earliest ending of winter,
> In March, a scrawny cry from outside
> Seemed like a sound in his mind.
>
> He knew that he heard it,
> A bird's cry, at daylight or before,
> In the early March wind.

The sun was rising at six,
No longer a battered panache above snow . . .
It would have been outside.

It was not from the vast ventriloquism
Of sleep's faded papier-maché . . .
The sun was coming from outside.

That scrawny cry—it was
A chorister whose c preceded the choir.
It was part of the colossal sun,

Surrounded by its choral rings,
Still far away. It was like
A new knowledge of reality.[21]

Here—to put it very telegraphically—what cannot be missed is how Stevens's insistence on this "new knowledge of reality" from exposure to the "outside" is, as it were, "blinking" at sunrise, how the scrawniness of the bird's cry is matched (weakness for weakness, you might say) by the scrawniness of the poem and its heavily mediated immediacy—its im-mediacy—by the phalanx of prepositions Stevens walks us through: "at," "in," "of," "from," and, most crucially, "by" and "like." And we need, moreover, to be told (again, oddly enough) that "he *knew* that he heard it." This makes us wonder, with the speaker of the poem, where did it come from, this sound, this "new knowledge of reality" that is only "*like* a new knowledge of reality," this "c" that is also a "pre-c," that precedes and, as it were, defers the Apollonian "colossal sun" that is "still far away" (and now, we know, always will be).

Here, I think, we find a bridge between animal studies and ecology in Stevens's poetry, but Stevens shows us that ecology in literature is not always where you think it is. For Stevens, it is not a matter of representing and, as it were, cheering for nature, but rather showing how nature must be replaced by that better term "environment," in the theoretical sense that we have sketched previously, for this very simple reason: that the term "environment," rather than "nature," reminds us that what counts as nature is always a product of the contingent and selective practices deployed in the embodied enaction of a particular autopoietic system. In biological terms, this realization reaches back, of course, to Jakob von Uexküll's theories of human and animal *Umwelten* and forward to ecological work in

the philosophy of mind and consciousness by thinkers such as Alva
Noë's, who argues that "the locus of consciousness is the dynamic
life of the whole, environmentally plugged-in person or animal."[22]
As his work shows, recent research in the biology of consciousness
makes it clear that these questions do not neatly break along lines
of human versus animal, inside versus outside, brain versus world,
or even, for that matter, organic versus inorganic. Mind is not brain,
in other words, precisely *because* it is ecological, a recursive loop
between the inside of the system, including its biological wetware,
and the outside of things like semiotic systems, technologies, and
cultural practices (among many others).[23] As Noë puts it, "It is not
the case that all animals have a common external environment,"
because "to each different form of animal life there is a distinct, corresponding, ecological domain or habitat," which means, in short,
that "all animal live in structured worlds."[24]

From here it is but a short step indeed to the question Derrida
raises early in the second volume of *The Beast & the Sovereign*
seminars—"the question is indeed that of the world" he writes—
which is framed by his entire engagement of Heidegger on the issue,
"What do beasts and men have in common?" In reply, Derrida offers
this quite remarkable passage made up of a movement through three
possible theses, finished off with a meditation on the last:

"1. Incontestably, animals and humans inhabit the same world,
the same objective world even if they do not have the same experience of the objectivity of the object.

"2. Incontestably, animals and humans do not inhabit the same
world, for the human world will never be purely and simply identical to the world of animals.

"3. In spite of this identity and this difference, neither animals of
different species, nor humans of different cultures, nor any animal
or human individual inhabit the same world as another, however
close and similar these living individuals may be (be they humans
or animals), and the difference between one world and another will
remain always unbridgeable, because the community of the world
is always constructed, simulated by a set of stabilizing apparatuses,
more or less stable, then, and never natural, language in the broad
sense, codes of traces being designed, among all living beings, to
construct a unity of the world that is always deconstructible, nowhere and never given in nature. Between my world . . . and any
other world there is first the space and time of an infinite difference, an interruption that is incommensurable with all attempts to

make a passage, a bridge, an isthmus, all attempts at communication, translation, trope, and transfer that the desire for a world . . . will try to pose, impose, propose, stabilize. There is no world, there are only islands" (BS2 8–9/30–31).

Now it is the first thesis that is usually taken to be ecological, but my point here is that by the peculiar logic of Derrida, Stevens, and second-order systems theory, it is actually the *third* thesis that is the most radically ecological.[25]

To show why, we must track Derrida as he moves rapidly in the next moment of the seminar from an equally bracing phrase taken from Daniel Defoe's *Robinson Crusoe*—the phrase, "I am alone"—to Paul Celan's memorable line, "The world is gone, I must carry you."[26] The world is "gone" for precisely the reasons marshaled in Derrida's third thesis—reasons that we have seen elucidated in more naturalistic terms by Maturana and Varela, Noë, and others. More than this, Derrida contests Heidegger's assertion that humans are "worlding" while the stone is "without world" and the animal "has a world in the mode of not having." First of all, as Derrida argues in a number of places, there is no generic entity called "the animal" about which such a blanket claim could ever be made as anything other than a "dogma";[27] and second, humans themselves, in fact, *do not* have a world in Heidegger's sense. Why? Because, for "all animals who live in structured worlds" (to remember Noë's phrase) the very thing that makes the world available to us—the *grille* or *gramme* or *machinalité* of semiotic code or program in deconstruction,[28] the blind spot of the contingent self-reference of observation in systems theory that dictates that "reality is what one does not perceive when one perceives it"[29]—is also and at the same time what makes the world *unavailable* to us.[30] Or as Stevens puts it memorably in "The Idea of Order at Key West," "It was her voice that made / The sky acutest at its vanishing."[31] And this fact—that "the world is gone," and not just for nonhuman life but also for humans, thus linking human and nonhuman life in their shared finitude (indeed, in the finitude of their finitude)—is precisely where ethics and ecological responsibility begin. As Derrida puts it in a later session that year, picking up the thread:

> We could move for a long time, in thought and reading, between *Fort und Da, Da und Fort*, between. . . these two *theres*, between Heidegger and Celan, between on the one hand the *Da* of *Dasein* . . . and on the other hand Celan's *fort* in "Die Welt

ist fort" . . . the world has gone, in the absence or distance of
the world, I must, I owe it to you, I owe it to myself to carry
you, without world, without the foundation or grounding of
anything in the world, without any foundational or fundamental mediation, one on one, like wearing mourning or bearing a
child, basically where ethics begins. (BS2 105/160)

Here, I think—"the world is far, the world has gone," "it would
have been outside," "it seemed like a sound in his mind"—we can locate the particular kind of ecological poetics that we find in Stevens,
particularly in his later work, a poetics that makes his work quite
different from the more conspicuously ecological poetics of a Gary
Snyder or a Robinson Jeffers. In these poems, it's as if Stevens insists
on a scene of address, a mise en scène, that in so many words says to
the reader, "*You*: it's on you, you are the one who has to look, listen,
receive, and think," along the lines that Critchley has in mind when
he writes that Stevens's poetry is about "the calm that comes from
learning to look at things, being there with things in a way that does
not seek to dominate them or appropriate them to the understanding."[32] More than that, however, it's as if Stevens says, "*You*: not the
TV, not the cell phone, not the computer, not the iPad: *you*. I am
alone. With you"—what Derrida, on the very first page of the second
set of seminars, calls "a sentence that is still more terrifying, more
terribly ambiguous, than 'I am alone'" (BS2 1/21). Hence the significance, in our hypermediated landscape, of what I have characterized
as not just the quietness and calmness of Stevens's poetics, but its
weakness, especially in the sparse and Spartan late poems, leaving
us (in Derrida's words from the very closing pages of that same seminar) "without foundational or fundamental mediation," without a
"world" that is "simulated by a set of stabilizing apparatuses" (BS2
105/160, 8/31). In that light, this weakness is part of its strength,
indeed it is key to what we might call the "carrying capacity" or
"carrying ecology" of Stevens's poetry. Stevens's poems put us on an
island, as it were, where we discover, "I am alone. The world is gone
and I'll have to carry you." Or, as Stevens puts it, again:

> You . . . You said,
> "There are many truths,
> But they are not parts of a truth."
> Then the tree, at night, began to change,

Smoking through green and smoking blue.
We were two figures in a wood.
We said we stood alone.[33]

With this movement between *"Fort und Da,"* in Derrida's text, the question of what is near and what is far, we move to a particular conception of ecology that we might, following Michel Serres, call "topological" rather than "topographical." The terminology of folds, knots, and paths that we often find in Serres's work is drawn from mathematical topology and its opposition to geometry—an opposition that is not just logical for Serres but also ethical. As Steven Brown summarizes it, while geometry "rests upon clear notions of identity and distinction, topology, and the mathematics which underpins it, is concerned with transformation and connection."[34] Or as Serres himself puts it in a series of conversations with Bruno Latour,

> If you take a handkerchief and spread it out in order to iron it, you can see in it certain fixed distances and proximities. If you sketch a circle in one area, you can mark out nearby points and measure far-off distances. Then take the same handkerchief and crumple it, by putting it in your pocket. Two distant points suddenly are close, even superimposed. If further, you tear it in certain places, two points that were close can become very distant.[35]

As Paul Harris points out, "Topology poses problems of spatial relations through questions such as: 'What is closed? What is open? What is a connective path? What is a tear? What are the continuous and the discontinuous? What is a threshold, a limit?'"[36] Now, without getting into an extended investigation of the lines of relation and difference between this sense of the topological and the very large amount of Derrida's writing on what he calls "spatialization" and "the becoming-space of time,"[37] I want simply to note here how these questions animate the very large amount of space given in the second set of seminars to the questions such as the following: "Does solitude *distance* one from others? What am I saying when I say 'I am alone?'" Derrida responds that, "to begin to reply to these questions or even to elaborate them as questions, we would need to begin by agreeing as to what *coming closer* or *distancing* means," by acknowledging that "the proxim*ity* of the close is not, for its part, necessarily close" (BS2 62/103).

Now in the most literal sense, of course, a topological vs. topographical rethinking of ecology can help us understand how, on a deeper level of relation, spaces that seem on the surface dissimilar and distant may in fact be very tightly bound to each other in other, more complex ways. For example, the province of Alberta and the state of Texas are quite distant from each other topographically, and could not seem, in many ways, more dissimilar from each other in terms of geology and landscape, flora and fauna, and so on. Topologically, however, Texas and Alberta are very close indeed in being bound tightly together by the infrastructure of the global petrochemical industry and its associated capital and transportational flows, whose most identifiable icon is the Keystone XL pipeline that connects the oil sands of Fort McMurray with the massive refinery infrastructure of Port Arthur and Houston, Texas.

On the more philosophical terrain mined by Serres and Derrida, however, the general and epistemological point is that we find a fundamentally different underlying *logic* at work in thinking topographical relations, one that folds and twists, if you like, the observation made famous by Gregory Bateson (borrowed from Korzybski), that "the map is not the territory."[38] This topological reorientation helps us understand more fully much of what is going on with Stevens's birds—where they're coming from, you might say—and even helps shed light on how we are to understand the peculiar nature of the very trees in which they roost, as with, "The palm at the end of the mind,/Beyond the last thought" in "Of Mere Being," which "stands on the edge of space."[39] We might ask, in tune with Derrida's investigations of "world" and the "-imity" of "proximity," what does it mean to mark "the end of the mind"? How is something that is the most distant, we may ask with Derrida, also something that is closest? (BS2 62–63/103–4, 72–73/116–18). And we might ask, moreover, since when does space have an "edge"?

One of Serres's answers to this question is figured in the topological process of the baker's kneading of dough, "a certain folding of half a plane of dough over the other half, repeated indefinitely according to a simple rule," which "produces a design precisely comparable to the flight of the fly or wasp," an implicating and explicating movement—literally a "folding" and "unfolding" (*pli*) of space and time[40]—that also characterizes the formal operations of

Serres's chief topological figure, Hermes: the "free mediator" (as Latour calls him) "who wanders through this folded time and who thus establishes connections" between what look at first glance like distant and unrelated phenomena (as in, for example, La Fontaine's fables and information theory in Serres's book *The Parasite*). "We must conceive or imagine how Hermes flies and gets about," Serres writes, "when he carries messages from the gods—or how angels travel. And for this one must describe the spaces situated between things that are already marked out—spaces of *interference*. . . . This god or these angels pass through folded time, making millions of connections."[41]

What this means, as Serres puts it, is that *"the semiotic is above all a topology,"*[42] and "in the linguistic field," as Harris puts it, "topological relations are most visibly expressed in prepositions. Topology in general and prepositions in particular share concerns with modes of linkage, and are therefore intrinsic to figuring the space-between."[43] Here, I think we find a useful template for understanding the peculiar poetic space that Stevens builds in poems such as "Of Mere Being," and we might well be reminded of my discussion earlier of Stevens's "weak" poetics in poems like "On the Road Home" and, especially, "Not Ideas About the Thing But the Thing Itself," with its succession of prepositions—"at," "in," "of," "from," and, most crucially, "by" and "like"—but also its figuration (one of the most famous in all of Stevens) of what we could call, following Serres, this Hermes figure of the bird whose song comes "from outside," delivering something "like a new knowledge of reality."[44] In fact, as Serres observes of the Hermes operator, "As soon as this intermediary comes to rest on a spot, he sometimes finds himself far off but also sometimes very close to foreignness."[45] "The messenger," he concludes," always brings strange news; if not, he's nothing but a parrot."[46]

Now, as it happens, a parrot *does* in fact figure centrally in the second set of seminars on *The Beast & Sovereign*—specifically, Poll the Parrot in Defoe's *Robinson Crusoe*. Here's the passage from Defoe's novel that fascinates Derrida:

> For I was very weary, and fell asleep: But judge you, if you can, that read my story, what a Surprize I must be in, when I was wak'd out of my Sleep by a Voice calling me by my Name several times, *Robin, Robin, Robin Crusoe*, poor *Robin Crusoe*,

where are you *Robin Crusoe?* Where are you? Where have you been? (quoted in BS2 86/135)

Crusoe "thought I dream'd that some Body spoke to me," Defoe continues,

> But as the Voice continu'd to repeat *Robin Crusoe, Robin Crusoe*, at last I began to wake more perfectly," and "no sooner were my Eyes open, but I saw my *Poll* sitting on the Top of the Hedge; and immediately I knew that it was he that spoke to me; for just in such bemoaning Language had I used to talk to him, and teach him; and he had learn'd it so perfectly, that he would sit upon my finger, and lay his Bill close to my Face, and cry *Poor* Robin Crusoe, *Where are you? Where have you been? How come you here?* (quoted in BS2 86/135)

What we find here, Derrida writes, is a rather "uncanny" moment that remains, nonetheless, "a circular auto-appellation, because it comes from a sort of living mechanism that he has produced, that he assembled himself, like a *quasi*-technical or prosthetic apparatus, by training the parrot to speak mechanically so as to send his words and name back to him, repeating them blindly" (BS2 86/136).

As Derrida notes, this circular or wheel-like movement of the selfsame and the self-named is unavoidable; "one could say of every autobiography, every autobiographical fiction . . . that it presents itself through this linguistic and prosthetic apparatus—a book—or a piece of writing or a trace in general, for example the book entitled *Robinson Crusoe*, which speaks of him without him . . . without the author himself needing to do anything else, not even be alive" (BS2 87/136). What we find here, then, in this autoappellation that proceeds—and can only proceed—by means of the heteroappellation of the prosthetic and technical apparatus that is the trace, language, what Michael Naas calls (making much of the pun) "a sort of originary Poly-semy or Poly-graphy that expropriates right from the start what we might believe to be a live, spontaneous, self-naming voice"[47]—a polygraphy that may be said to be marked by the deformation of "Robinson" to "Robin" in the bird's enunciation.[48] But Derrida's point, of course, is that this generalized exappropriation could ever only be controlled or mastered—could ever be turned into a perfectly "circular auto-appellation"—in the realm of fantasy: that is to say, on a "desert island," where Crusoe explicitly figures his place in relation to the lands and creatures around him, as Krell notes, in terms of sovereignty. "Then to see how like a King I din'd too all alone," Defoe writes, "attended by my Servants, *Poll*, as if he had been my Favourite, was the only Person permitted to talk to me."[49]

While I cannot pursue within the limited space of this essay the entire topos of sovereignty that Derrida explores in such rich detail in both sets of seminars, I do want to conclude by noting that this discussion of "circular auto-appellation" marks both an important point of convergence and an important point of difference between Serres and Derrida's topological rendering of ecological thought—specifically, with regard to the role of the performative and ethical stakes for thinking ecology that derive from that role. Serres, like Derrida, "strives for a more radical rethinking of subject and object" in which "we have to think of sensation (and cognition) as neither directed outwards from the recesses of our bodies, nor as flowing into us from the outside world, but rather as an ever ramifying and branching network that blooms into life in the middle of worldly engagements."[50] The point of these itineraries, as Paul Harris puts it, is "to weave together the fabric of knowledge into a 'pattern that connects' (as Bateson called it) humans to the world."[51] As the invocation of Bateson here suggests, one advantage of the system's theoretical rearticulation of this relationship is to provide a more robust environmental account of these itineraries than we are likely to receive from Derrida—to underscore that these itineraries are always a matter of *specific, embodied forms of life* that have their own autopoietic system/environment relationship in ways that are far from generic. "Knowing things," as Serres puts it, "requires one first of all to place oneself between them. Not only in front in order to see them, but in the midst of their mixture, on the *paths* (emphasis mine) that unite them."[52]

Derrida's rendering of these "paths" and "itineraries," however, emphasizes not just our finitude (the particular, nongeneric form of our embodied being and the vulnerability and mortality to which it subjects us) but also how the performativity of autopoietic self-reference constitutes *the finitude of our finitude*, its nonappropriability—a fact that Derrida often in the seminars treats under the rubric of (auto)immunity. As Derrida writes,

everything that can happen to the *autos* [of autoaffection, of autobiography, of autonomy, and so on] is indissociable from what happens *in the world* through the prosthetization of an ipseity which at once divides that ipseity, dislocates it, and inscribes it outside itself *in the world*, the world being precisely what cannot be reduced here, any more than one can reduce *tekhne* or reduce it to a pure *physis*. The question, then, is indeed that of the world . . . and what I want to say is that there is no ipseity without this prostheticity in the world, with all the chances and all the threats that it constitutes for ipseity, which can in this way be constructed but also, and by the same token, indissociably, be destroyed. (BS2 88/138)

Where Derrida is reaching with such formulations, I think, is in the direction of the ecological described by both Serres and systems theory, an understanding that moves us out of what might sometimes seem a rather thin epistemological understanding of such questions. In Serres's work, however—and Latour presses him on this point repeatedly in the *Conversations*[53]—it is never really clear where and how the reality of these topological paths and itineraries are to be located; sometimes Serres speaks as if *we* do it, sometimes as if we merely find ourselves in the midst of a topological space that has its own antecedent reality, whose topological quality does not depend upon our activity (as his recourse to the mathematical grounding of topology often suggests).

Now from the pragmatist point of view I have already invoked, one might well say—and sometimes Serres himself *does* say, in so many words—"Who cares, since both realist and idealist forms of this charge are, from the standpoint of an anti-representationalist philosophy, both unprovable and empty?"[54] Or as Michael Naas puts the Derridean version of this itinerary, "There is thus no solipsism that needs to be overcome by putting worlds together or by demonstrating that we in fact belong from the beginning to a shared or common world," because it is "only through the self's detour through the world as originary prosthesis—that the self, that any ipseity or identity, is first constituted."[55] As we have seen, however, Derrida is quite decisive on this point in a way that Serres is not. It is only *because* "the world is gone" that the ethical responsibility of "I'll have to carry you" is possible—a responsibility that, in turn, we can only take on performatively, not constatively. Or to put it another way, what is "near" and what is "far," what is close to us and concerns us and what does not, is something that is *made*, not given—by mathematics or anything else.

Here again, a systems-theoretical redescription of Derrida's point will help it seem less counterintuitive to those who are not deconstructively inclined. As Maturana and Varela put it, "We do not see what we do not see,"[56] and this is so not only for epistemological reasons but also for ecological and evolutionary ones. As Hans-Georg Moeller points out, such a perspective denies, with Derrida, "transcendental agency and intentionality" with regard to the determination of "world." For Derrida's "performative" (in the words of J. Hillis Miller), "iterability disqualifies the requirement that a felicitous performative must depend on the self-consciousness of the ego and its 'intentions.'"[57] For systems theory, because "an ecosystem has no center . . . the very condition of seeing something is not to see everything. The ability to observe, paradoxically, also implies limitations, and thus inabilities, of observation. The partial blindness that comes with evolution also implies a certain ethical and pragmatic blindness. Since it is impossible to see everything, it is also impossible to see what is good for all."[58]

And this returns us, then, to the quintessentially Derridean point that I take up in some detail elsewhere around Derrida's thinking of conditional and unconditional hospitality: that ethics is, in this sense, a matter of doing the impossible. (And in this connection, it is perhaps not surprising that the central scene of ethical responsibility in the second set of seminars—"The world is gone. I'll have to carry you"—is staged in terms of an uncanny hauntology: "I'll have to carry you," as in "wearing mourning or bearing a child" [BS2 105/160], marking our responsibility to those spectral ones who are not here, either already departed or not yet arrived, those whose alterity can never be done justice, and structurally so.) We have to choose, we have to act, and yet we do so without grounds, without foundations, without, precisely, "world"—without what we care about being *given* to us by nature.[59]

It is exactly here, however, in the closing pages of the second set of seminars, that Derrida locates a special role for art and poetry, one *predicated* on the fact that "the world is gone." As he puts it in the tenth and last session, "If *Die Welt ist fort*, if we think we must carry the other . . . this can only be one of two things." Either we "carry the other out of the world" and "share at least this knowledge without phantasm that there is no longer a world" or "make it that there be precisely a world . . . do things so as to make *as if* there were just a world, and to make the world come into the world," to "make the gift or present of this *as if* come up poetically" (BS2 268/369–70).

Unlike Kant's version of the "as if" (the *als ob*) in the form of a "regulative idea" of the world that considers phenomena "as if they were the arrangements made by a supreme reason of which our reason is a faint copy" (BS2 269–70/370–72), this "poetic" "as if" does not deny or repress its fictive nature, but acknowledges and embraces the fact that "the community of world," as Derrida puts it in the opening pages of the first session, "is always constructed" (BS2 8/31). If there is to be a "world," a "shared world," in other words, *we* must make it, without any taking for granted of who or what this "we" might be.[60]

What I am calling the "ecological poetics" that Stevens and Derrida share comprises a three-part progression: first, for both, "the question is indeed that of the world," as Derrida puts it, but for both, "the world is gone" (to borrow Celan's line once again) in the sense that the very thing that makes the world available to use—the performative for Derrida, the *factum* of the poem for Stevens, the contingency of self-reference in systems theory's terms—is also the very thing that makes the world "as such" (in Heidegger's sense) *unavailable* to us. As we have seen, another term for this fact is "environmental complexity." Second, this absence of world *as such* is precisely where responsibility begins, here (or shall we say, more precisely, a "there" that must be made a "here"), in the fact that the world is not given but is rather *made* by a set of "stabilizing apparatuses" that creates what we call a "shared world." As many critics have pointed out, this fact was at the center of Stevens's poetic universe; from beginning to end, his entire struggle was against what he called, variously, "the malady of the quotidian,"[61] "romantic tenements/Of rose and ice,"[62] and "the A B C of being"[63]—all those inherited worlds and "rotted names" that had to be made new in a way that could sustain belief,[64] with the famous early long poem "Sunday Morning" announcing a first grand struggle against the grandest inheritance of all, the Judeo-Christian legacy. Or as Joan Richardson, one of the best readers of Stevens we have, puts it, ideas like "truth" and "reality" remained, for Stevens, what philosopher Donald Davidson calls "*an unexplained primitive*, like the word for 'truth,' *a-letheia*," in the sense captured by the title of Stevens's late poem, "Reality Is an Activity of the Most August Imagination."[65] It is out of this struggle, of course, that Stevens's famous doctrine of "supreme fiction" as the chief duty of poetry arose. "The final belief," he writes, "is to believe in a fiction, which you know to be a fiction, there being nothing else. The exquisite truth is to know that it is a fiction and that you believe in it willingly."[66] And third and finally, for Derrida and Stevens alike, art has a special and specific role to play in this process of "making the world come into the world"—for both, a poetics, a *poiesis*, a *making*, that is "ecological" not in the sense of being about rocks and streams, but rather in the sense of the world we will have and live in, the possibility of sharing a world, *depending on us*. And what could be more "ecological" than the assertion that the world depends on us?

For Stevens as for Derrida, however, the point is not so much epistemological as it is pragmatic and ethical. Or, as Stevens put it in *The Necessary Angel*, meditating on the "social obligation" of the poet,

> What is his function? Certainly it is not to lead people out of the confusion in which they find themselves. Nor is it, I think, to comfort them. . . . I think that his function is to make his imagination theirs and that he fulfills himself only as he sees his imagination become the light in the minds of others. His role, in short, is to help people to live their lives.[67]

As Stevens reminds us time and again, however, the poet does so by returning us to the dense situatedness of where we are, here and now, the only place where the world may be found. Or, as he puts it in a later long poem,

> And out of what one sees and hears and out
> Of what one feels, who could have thought to make
> So many selves, so many sensuous worlds,
> As if the air, the mid-day air, was swarming
> With the metaphysical changes that occur,
> Merely in living as and where we live.[68]

Notes

1. Wallace Stevens, *The Palm at the End of the Mind*, ed. Holly Stevens (New York: Vintage, 1972), 398.

2. Simon Critchley, *Things Merely Are: Philosophy in the Poetry of Wallace Stevens* (London: Routledge, 2005), 4.

3. Stevens, *Palm at the End of the Mind*, 349.

4. Critchley, *Things Merely Are*, 19.

5. Ibid., 58.

6. Stevens, *The Collected Poems of Wallace Stevens*, ed. Holly Stevens (New York: Vintage, 1990), 203.

7. I don't have space here to rehearse the details of Derrida's rendering of the performative, so I will simply refer readers to Michael Naas's able summary of J. Hillis Miller's articulation of this concept in Miller's book *For Derrida*; see Naas, *The End of the World and Other Teachable Moments: Jacques Derrida's Final Seminar* (New York: Fordham University Press, 2015), 105–7.

8. For a more detailed account of this movement in Emerson's work and how it plays out in Stevens, see chapters 9 and 10, "Emerson's Romanticism, Cavell's Skepticism, Luhmann's Modernity" and "The Idea of Observation at Key West," in Cary Wolfe, *What Is Posthumanism?* (Minneapolis: University of Minnesota Press), 239–82.

9. Stevens, *Collected Poems*, 204.

10. Ibid., 394.

11. Ibid., 202.

12. Ibid., 202. For Derrida on blinking, see, for example, "La loi du genre," in Jacques Derrida, *Parages* (Paris: Galilée, 1986); trans. Tom Conley et al. as "The Law of Genre," in *Parages* (Stanford: Stanford University Press, 2010; also in Derrida, *Acts of Literature*, ed. Derek Attridge (London: Routledge, 1991), 230–31. Emerson's "transparent eye-ball" occurs in the essay "Nature," in *Emerson's Prose and Poetry*, ed. Joel Porte and Saundra Morris (New York: Norton, 2001), 29.

13. Stevens, *Collected Poems*, 93.

14. See, for example, Stevens, *Wallace Stevens: The Poems of Our Climate* (Ithaca: Cornell University Press, 1980).

15. See Niklas Luhmann, *Social Systems*, trans. John Bednarz Jr., with Dirk Baecker, introduction Eva M. Knodt (Stanford: Stanford University Press, 1995), 12–58.

16. Namely, in Wolfe, *What Is Posthumanism?*, xx–xxv.

17. Luhmann, *Social Systems*, 26.

18. Humberto Maturana and Francisco Varela, *The Tree of Knowledge: The Biological Roots of Human Understanding*, trans. Robert Paolucci, foreword J. Z. Young, rev. ed. (Boston: Shambhala, 1998), 29.

19. For more on this point, see Maturana and Varela, *Tree of Knowledge*, 93–117.

20. This fairly standard characterization of correlationism belongs to Ian Bogost, in *Alien Phenomenology, Or What It's Like to Be a Thing* (Minneapolis: University of Minnesota Press, 2013), 5, summarizing Quentin Meillassoux's definition of the term in his book *After Finitude*.

21. Stevens, *Collected Poems*, 534.

22. Jakob von Uexküll, *A Foray into the Worlds of Animals and Humans*, with *A Theory of Meaning*, trans. Joseph D. O'Neil, introduction Dorion Sagan (Minneapolis: University of Minnesota Press, 2010), and Alva Noë, *Out of Our Heads: Why You Are Not Your Brain, and Other*

Lessons from the Biology of Consciousness (New York: Hill and Wang, 2009), xiii.

23. For more on how these questions cross-pollinate with Derrida's work, see Wolfe, *Animal Rites: American Culture, the Discourse of Species, and Posthumanist Theory* (Chicago: University of Chicago Press, 1998), 78–94, and, more recently, *Before the Law: Humans and Other Animals in a Biopolitical Frame* (Chicago: University of Chicago Press, 2013), 60–86.

24. Noë, *Out of Our Heads*, 43.

25. My colleague Timothy Morton has made his own version of this argument that ecological thinking *begins* with what he calls "the end of the world"; see, for example, Morton, *The Ecological Thought* (Cambridge, Mass.: Harvard University Press, 2012), and *Hyperobjects: Philosophy and Ecology after the End of the World* (Minneapolis: University of Minnesota Press, 2013).

26. The translation I use here is from *Poems of Paul Celan*, trans. Michael Hamburger (New York: Persea, 2002), 275.

27. See, for example, *L'animal que donc je suis*, ed. Marie-Louise Mallet (Paris: Galilée, 2006); trans. David Wills as *The Animal That Therefore I Am* (New York: Fordham University Press, 2008), 29–30, 34.

28. For more on these concepts, see Derrida, *De la grammatologie* (Paris: Minuit, 1967); trans. Gayatri Chakravorty Spivak as *Of Grammatology* (Baltimore: Johns Hopkins University Press, 1998), and "Signature, événement, contexte," in *Limited Inc.*, ed. Elisabeth Weber (Paris: Galilée, 1990); trans. Samuel Weber et al. as "Signature, Event, Context," in *Limited Inc.*, ed. Gerald Graff (Evanston, Ill.: Northwestern University Press, 1988).

29. For an expanded discussion of what falls between the dashes here—and in relation to Stevens's poetry—see Wolfe, *What Is Posthumanism?*, chapter 10.

30. Indeed, as I've argued elsewhere, "having a world in the mode of not having" is as good a definition of *Dasein* as we are likely to get; see Wolfe, *Before the Law*, 63–86.

31. Stevens, *Collected Poems*, 129.

32. Critchley, *Things Merely Are*, 5.

33. Stevens, *Collected Poems*, 203.

34. Steven D. Brown, "A Topology of the Sensible," *New Formations* 72 (Jan. 2011): 165.

35. Michel Serres, with Bruno Latour, *Conversations on Science, Culture, and Time*, trans. Roxanne Lapidus (Ann Arbor: University of Michigan Press, 1995), 60.

36. Paul Harris, "The Itinerant Theorist: Nature and Knowledge/Ecology and Topology in Michel Serres," *SubStance* 26, no. 2 (83): 44; further references are given in the text.

37. But for a very brief summary, see Wolfe, *What Is Posthumanism?*, 91–93, 132–33, and 292–95.

38. Gregory Bateson, *Steps to an Ecology of Mind* (New York: Ballantine, 1972), 449; further references are given in the text.

39. Stevens, *Palm at the End of the Mind*, 398.

40. Serres and Latour, *Conversations*, 65.

41. Ibid., 64.

42. Quoted in Harris, "Itinerant Theorist," 45.

43. Ibid.

44. Stevens, *Collected Poems*, 534.

45. Serres and Latour, *Conversations*, 65.

46. Ibid., 66.

47. Naas, *End of the World and Other Teachable Moments*, 95.

48. See David Farrell Krell's amusing discussion of this moment in *Derrida and Our Animal Others: Derrida's Final Seminar, "The Beast & the Sovereign"* (Bloomington: Indiana University Press, 2013), 49.

49. Krell, *Derrida and Our Animal Others*, 50.

50. Brown, "Topology of the Sensible," 64.

51. Harris, "Itinerant Theorist," 37.

52. Quoted in Brown, "Topology of the Sensible," 164.

53. Serres and Latour, *Conversations*; see, for example, 89–102.

54. For a more detailed handling of this issue in relation to pragmatism, see Wolfe, *What Is Posthumanism?* 244–45.

55. Naas, *End of the World and Other Teachable Moments*, 98.

56. Maturana and Varela, *Tree of Knowledge*, 242.

57. Quoted in Nass, *End of the World and Other Teachable Moments*, 106.

58. Hans-Georg Moeller, *The Radical Luhmann* (New York: Columbia University Press, 2012), 70–72.

59. For more on this question in Derrida in relation to ethics and non-human animals, see Wolfe, *Before the Law*, chapters 7 and 8.

60. See Kelly Oliver's foregrounding, which is somewhat different in emphasis from my own, of this aspect of the seminars in "The Poetic Axis of Ethics," *Derrida Today* 7, no. 2 (2014): 121–36, esp. 125, 129.

61. Stevens, *Collected Poems*, 96.

62. Ibid., 238.

63. Ibid., 288.

64. Ibid., 183.

65. Joan Richardson, *Pragmatism and American Experience: An Introduction* (New York: Cambridge University Press, 2014), 7.

66. Stevens, *Opus Posthumous*, ed. Milton J. Bates, rev. ed. (New York: Vintage, 1990), 189.

67. Stevens, *The Necessary Angel* (New York: Vintage, 1969), 29.

68. Stevens, *Collected Poems*, 326.

15

Earth: Love It or Leave It?

Kelly Oliver

> Dreaming of islands . . . is dreaming of pulling away, of being already
> separate. . . . An island doesn't stop being deserted simply because it
> is inhabited.
> —*Gilles Deleuze, "Desert Islands," in* Desert
> Islands: And Other Texts, 1953–1974, 9–14

After the first mission to the moon, Astronaut James Lovell told
Time magazine, "What I keep imagining is that I am some lonely
traveler from another planet. What would I think about the earth at
this altitude? Whether I think it would be inhabited or not."[1] The
view of earth from space conjures an alien perspective, seeing our
planet for the first time. What would an extraterrestrial see when he
saw earth? Would he think it was inhabited?

Centuries before the Apollo moon missions, Kant imagined an
alien vantage point on earth, through which we see ourselves as
earthlings, united on the limited surface of the same planet. It seems
as if we can only see ourselves united or whole by imagining we
are someone, or something, else. For example, we see ourselves as
a species only when compared to other species. Perhaps this is why
John Berger says that without other species on earth, humans would
be very lonely—that is, if it were even possible to imagine the existence of humans without other species.[2]

For Kant, it is only by imagining what we look like from the
perspective of another rational species that we might finally see
ourselves as rational beings among other rational beings, citizens of
the universe. We see ourselves as inhabitants of one planet only by
imagining seeing ourselves through the eyes of an extraterrestrial,

from a vantage point off-world, so to speak. Thus, we fragment ourselves into an "us and them" to see ourselves as one. This need to see
ourselves from the outside, from an impossible perspective, manifests what Jacques Derrida calls an "autoimmune logic" whereby
that very system that is supposed to protect us, to make us whole,
turns against us, even to the point of killing us. Only by taking an
impossible, unsustainable, and even deadly perspective do we see
ourselves as a species united on our home planet. Only from this
impossible "God's-eye" vantage point do we see ourselves whole.
Furthermore, seeing the image of the whole earth brought with it
fantasies of the annihilation of the earth. With the Apollo photographs from space, we imagined, and still imagine, that we have
finally seen the "whole" earth, "as it truly is," an "island" floating
in space. And, as an island it provokes ambivalence, wanting to flee
and wanting to fortify it.

In his last seminar, Derrida asks, "Why do some people love islands while others do not love islands, some people dreaming of
them, seeking them out, inhabiting them, taking refuge on them,
and others avoiding them, even fleeing them instead of taking refuge on them?" (BS2 64/106). What is it that one flees or seeks when
one escapes from, or takes refuge on, an island? And why do we
find these seemingly contradictory desires in the same person, even
in "the same desire?" (BS2 69/112). We are both attracted to and
repulsed by islands, caught between "insularophilia and insular-
phobia." But, what is an island that it provokes this "double contradictory movement of attraction and allergy?" (BS2 69/112). Following Martin Heidegger's discussion of solitude, Derrida finds in
islands the double senses of being alone as being lonely and isolated,
on the one hand, and being unique and singular, on the other.

The figure of an island as isolated or insular and circular or round,
floating alone as solitary and unique, becomes a figure for what Derrida calls "autoimmunity." In the case of Robinson Crusoe, this autoimmunity shows up in his desire to leave his family, particularly
to get away from his father's authority, and the stifling conventions
of his home country, only to attempt to recreate that authority and
those conventions on the desert island. Robinson Crusoe escapes
from one island, England, only to land on another, his so-called desert island. Just as he did with his first island home, he wants to flee
the second. And, after he succeeds in escaping the desert island, as
he did with his original island home, he longs to return to it. Like a
wheel turning in on itself, Robinson Crusoe becomes the very thing

he is trying to escape. Derrida associates autoimmunity with this wheel turning on its axis. Robinson Crusoe's potter's wheel, which he recreated on the desert island, becomes a symbol of the wheel of his ambivalent desire of attraction/repulsion in his movement from one island to the next. This wheel is not just any circle, because— like the earth—it turns around its own axis. And like Robinson Crusoe's wheel of ambivalent desire, planet Earth, spinning on its axis, triggers an autoimmune response.

Although it is Robinson Crusoe's island and not the earth Derrida has in mind when he asks why some people love islands and are drawn to them, while others fear islands and flee from them, his analysis of the ambivalence conjured by islands is apt when considering contradictory reactions to seeing the earth from space. Indeed, on Derrida's analysis the island itself becomes a figure for this ambivalence, for loving or leaving or, more accurately, for both loving *and* leaving. The figure of the island represents seemingly contradictory desires—namely, to love and to leave, to cling to and escape from, to stand your ground and to flee. In this regard, the figure of the island might be what Julia Kristeva would call "abject" in the sense that it is fundamentally ambiguous and thus our relationship to it is fundamentally ambivalent. An island is always in between, between land and sea, between safe harbor and threatening isolation. Like everything abject, an island both fascinates and terrifies. It fuels our ambivalent, even contradictory, desires to love *and* leave, to stay *and* to flee. Imagining the earth as an island, isolated and alone, floating against the vast sea of space, conjures the ambivalent reactions that Derrida attributes to all islands. Images of the earth as an island, beautiful and blue, floating alone in the darkness of space, are both threatening and reassuring.

On Christmas day 1968, immediately after "Earthrise," the first photograph of earth from space, was transmitted from Apollo 8, poet Archibald MacLeish wrote an article in the *New York Times* entitled, "Riders on Earth Together, Brothers in Eternal Cold," in which he claimed the significance of the moon mission as changing our very conception of earth: "The medieval notion of the earth put man at the center of everything. The nuclear notion of the earth put him nowhere—beyond the range of reason even—lost in absurdity and war. . . . To see the earth *as it truly is*, small and blue and beautiful in that eternal silence where it floats, is to see ourselves as riders on the earth together, brothers on that bright loveliness in the eternal cold—brothers who know now they are truly brothers."[3]

MacLeish speculated seeing the earth "as it truly is" would "remake our image of mankind" such that "man may at last become himself."[4] Seeing the earth "whole" for the first time would unite all of mankind, together on "that little, lonely, floating planet." Realizing we are all in this together on the precarious lovely earth alone in the "enormous empty night" of space was seen as a catalyst for our finally coming into our own as a species united as brothers. But, when MacLeish called the astronauts "heroic voyagers who were also men," we could not help but hear men as the masculine heroic space cowboys who have the power and vision to unite all men as brothers against the eternal cold of space.[5]

The spectacular images from the 1968 and 1972 Apollo missions to the moon, "Earthrise" and "Blue Marble," are the most disseminated photographs in history.[6] Indeed, "Blue Marble," the most requested photograph from NASA, is the last photograph of the planet taken from outside of earth's atmosphere.[7] Whereas "Earthrise" shows the earth rising over the moon, with elliptical fragments of each (the moon is in the foreground, a stark contrast from the blue and white earth in the background), the later image, "Blue Marble," is the first photograph of the whole earth, round with intense blues and swirling white clouds so textured and rich that it conjures the three-dimensional sphere. Even more than previous photographs of earth, the high definition of "Blue Marble" and the quality of the photograph make it spellbinding. Set against the pitch-black darkness of space that surrounds it, the earth takes up almost the entire frame. Unlike in "Earthrise," in "Blue Marble" the earth does not look tiny or partial, but whole and grand. Both of these photos from Apollo missions (8 and 17) were immediately met by surprise, along with excited exclamations about the unity of mankind on this "blue marble," this "pale blue dot," this "island earth."[8]

In the frozen depths of the Cold War and over a decade after the Soviets launched Sputnik, the first satellite to orbit earth, these images were framed by rhetoric about the "unity of mankind" floating together on a "lonely" planet. At the same time as vowing to win the space race with the Soviet Union, the United States wrapped the Apollo missions in transnational discourse of representing all of mankind. Indeed, these now iconic images ignited an array of seemingly contradictory reactions. Seeing earth from space generated new discussions of the fragile planet, lonely and unique, in need of protection. These tendencies gave birth to the environmental movement. And the Apollo missions spawned movements to unite the planet through technology. Heralded as man's greatest triumph, the moon missions led to a flurry of speculation on not just the technological mastery of the world or of the planet, but also of the universe. While seeing earth from space caused some to wax poetic about earth as our only home, it led others to imagine life off-world on other planets. While aimed at the moon, these missions brought the earth into focus as never before.

MacLeish's assessment of the "Earthrise" photograph was consistent with NASA's press releases after both Apollo missions, which included panhuman themes of uniting mankind and representing all of mankind in space outside of any national borders. For example, then NASA chief Thomas Paine told *Look* magazine that photographs of earth from space "emphasize the unity of the Earth and the artificialities of political boundaries."[9] NASA presented the Apollo 8 mission as one of peace and goodwill to all mankind.[10] And in 1969 *Time* magazine named the Apollo 8 astronauts, Borman, Lovell, and Anders "Men of the Year." The accompanying article described a New World born from their mission, one in which the human race could come together with one unified peaceful purpose as a result of the "escape from the planet that was no longer the world."[11] The world had expanded to include the universe, while the earth had shrunk into a tiny fragile ball.

Time magazine described the earth as a troubled place full of war and strife and space as the great hope to "escape the troubled planet." Again, the astronauts were seen as heroic figures conquering space: "It seemed a cruel paradox of the times that man could conquer alien space but could not master his native planet."[12] The goal is clearly to conquer, and the Apollo missions signal a great victory in escaping a troubled planet and moving beyond what appear from space as the petty disagreements between peoples. In the words of astronaut Frank Borman, "When you're finally up at the moon looking back at the Earth, all those differences and nationalistic traits are pretty well going to blend and you're going to get a concept that this is really one world and why the hell can't we learn to live together like decent people."[13] The irony is that Borman claims he only accepted the mission because as a military officer he wanted to "win" the Cold War.[14] The *Time* magazine article concluded that man will not turn into "a passive contemplative being" because he knows how to challenge nature, and in reaching for the moon he now conquered not only the seas, the air, and natural obstacles, but also space and the moon, which brings with it the "hope and promise of his latest conquest."[15] Like Borman, the American media seemed to think of the Apollo mission as a triumph for freedom and hope, paradoxically both for all of mankind and as an American victory in the Cold War.[16]

The Apollo missions are emblematic of our ambivalent relationship to the earth and to other earthlings. The reactions to seeing earth from space make manifest tensions between nationalism and cosmopolitanism and between humanism (in the sense that we are the center of the universe) and posthumanism (in the sense that we are insignificant in the universe). Indeed, the space program itself was driven by the conflicting rhetoric of nationalism and cosmopolitanism. The Apollo missions were both attempts at winning the Cold War and putting America first, and yet they were sold as missions for all of mankind intended to unite human beings across the globe. One small step for a man, one giant leap for mankind. The tension between nationalism and cosmopolitanism is clear. And it recalls the tension Derrida sees at the heart of hospitality.

Derrida contends that cosmopolitanism, or what he calls "cosmopolitics," necessarily puts into tension unconditional and conditional hospitality (CF 4–5). The rhetoric surrounding the Apollo missions manifests this ambivalence at the heart of hospitality. On the one hand, these missions were part of a military operation to secure the earth from hostile enemies; on the other hand, they were seen as benefiting all of humankind and uniting mankind as brothers who share a common world and common goals.

In terms of Derrida's analysis of hospitality, implicit in the welcoming gesture of the cosmopolitan rhetoric of uniting all of humankind, is the affirmation of the technological superiority of the United States and its concern to dominate not only the earth but also space. As the victor in the Cold War, America was in the position to extend hospitality to the rest of the world. As Derrida points out, hospitality is not only about generosity, but also always about control and mastery of what one takes to be one's own home (OHY 15/21). The ambivalence surrounding the Apollo missions and reactions to them resonates with imaging the earth as an island, as NASA did when it published the glossy photo-filled book *This Island Earth*, shortly after Apollo 17's mission to the moon.

Derrida argues that there is an internal contradiction inherent within the very notion of hospitality itself (OHY 149/131). He points to this contradiction with this question: "In giving a right, if I can put it like that, to unconditional hospitality, how can one give place to a determined, limitable, and delimitable—in a word, to a calculable—right or law?" (OHY 147–49/131). In other words, the principle grounding all conditional hospitality—namely, unconditional hospitality—is at odds with its practice, for what makes hospitality unconditional makes hostility not only possible, but also inevitable insofar as ultimately there is no calculus with which we determine how to distinguish one from the other. The threat to unconditional hospitality does not come from outside, but rather from inside. Hospitality operates according to the autoimmune logic distinctive of all appeals to the self or sovereignty. In other words, if, or insofar as, hospitality is granted by one to an other, it already comprises unconditionality. Indeed, the very terms "self" and "other" are problematic if our goal is unconditional hospitality; but these terms are required by our notion of hospitality insofar as we imagine that someone has the power to extend hospitality to another. And yet, this very power acts as a condition that prevents hospitality from being unconditional.

Thus, we must be watchful for the threat to unconditional hospitality from within our practical attempts at hospitality. We must be open to hearing from others how extending hospitality may also be extending hostility, or how extending hospitality to one entails excluding another. If it is impossible to ground unconditional hospitality on a universal principle that does not also always undermine itself, that does not operate according to an autoimmune logic, then we have only the groundless ground, which I describe in *Earth and World* as the earth itself. The earth is our home. And yet, as Derrida points out, "home" is precisely what is at stake in hospitality. The "problem of hospitality," he says, "is always about answering for a dwelling place, for one's identity, one's space, one's limits, for the *ethos* as abode, habitation, house, hearth, family, home" (OHY 149–51/133).

The ambivalence inherent in hospitality speaks to ambivalence in the concept of *home*, which is complicated, to say the least, when we consider earth as our home. If "ethics is hospitality," then ethics is about home, not only because in its Greek roots, home and ethics share a common root, *ethos*, but also and moreover because struggles over home and hospitality are at the heart of our relationship to others, particularly when considering the earth as home and our relationship to nonhuman beings for whom this planet is also their one and only "home," whatever that may mean within the limitations of their worlds. Certainly, when we consider earth as home, we encounter the uncanny at every turn, whether it is the uncanny strangeness of other animal creatures who share a singular bond to our shared planet or the uncanny strangeness of the earth itself as seen from space. And these experiences fill us with ambivalent desires.

In the case of contradictory reactions to the *Earthrise* and *Blue Marble* photographs, this ambivalent logic is operative as we acknowledge the singularity of our earthly home and at the same time attempt to escape from it to find another home. To the astronauts and subsequently the media, the earth is alone in the universe, "a planet so eccentric, so exceptional" that the mission to the moon brought the earth into focus.[17] Speaking of the Apollo 8 astronauts in a NASA publication entitled *This Island Earth*, administrator Oran Nicks echoes the cosmopolitan sentiments of the unification of mankind, stressing that from the vantage point of space, we see the "true reality" of our situation: "Their eyewitness accounts impressed millions of men with the true reality of our situations: the oneness of mankind on this *island Earth*, as it *floats* eternally in the silent sea of space."[18] Resonant with the astronauts, along with writers like MacLeish, NASA administrator George Low hopes, "By heeding the lessons learned in the last decade, and attacking our man-made problems with the same spirit, determination, and skill with which we have ventured into space, we can make 'this *island earth*' a better planet on which to live."[19]

The comparison between the earth and an island works to highlight the supposed reality of our situation: that we're all in the same boat, so to speak. Like an island, the earth is imagined alone, floating in the infinite sea of space. And, like so many fantasy islands, some see it as paradise, while others can't wait to escape from their exile here. Seeing earth from space made some appreciate earth anew, while others imagined moving further away from earth and traveling to other planets. For some, seeing the loveliness of earth "is to wish also to return to it,"[20] while for others, seeing the insignificance of earth compared to the vastness of space is to wish to leave it. In other words, these seemingly contradictory impulses, triggered within the cultural imaginary by the Apollo photographs, make manifest a deep ambivalence in our relationship to our earthly home. We feel both marooned and miraculous.

The photographs of earth from the moon continue to provoke the uncanniness of our relation to our home planet, particularly when we realize that we are "down there" somewhere, miniscule specks on that tiny "pale blue dot" floating in space. While some see the earth from space and want to protect it, others imagine escaping from earth to find our way in the galaxy, perhaps even in the universe. With environmental disaster looming large on the horizon, in recent years there is a sense among some that the earth has betrayed us or is taking its revenge on us; and rather than a safe haven, it has become a death trap and a threat to human survival.[21] For example, Buzz Aldrin recently proposed colonizing Mars and becoming a "two-planet species." "Our earth," says Aldrin, "isn't the only world for us anymore. It's time to seek out new frontiers."[22] The urge to colonize Mars or find another habitable home is getting stronger, evidenced by the *Mars One* project, which plans to start colonizing Mars in 2023, less than a decade from now, and to continue bringing people on a one-way trip to Mars every two years from then on, for a permanent, self-sustaining Mars settlement.

As we have seen, images of earth lead us to see the earth both as our amazingly singular home and a tiny, insignificant pea barely visible from space. The view of earth from space makes us both want to protect our vulnerable and fragile planet and to escape from this insignificant speck in the universe. It makes us feel simultaneously special and inconsequential. This is the double sense of the loneliness of earth; it is all alone in the universe and yet unique, a loneliness that resonates with Heidegger's notion of solitude as being alone, as in without others, and being alone, as in without equal. This double desire is the ambivalent desire or autoimmune logic that Derrida associates with islands, both wanting to take refuge and wanting to flee. Moreover, this autoimmune logic is intrinsic to the photographs themselves. For, to shoot those images, astronauts were propelled into inhospitable space in an unsustainable and precarious artificial environment where their very survival was uncertain. In other words, those images could only be taken from a vantage point where the survival of man is impossible. This extraterrestrial vista is from an impossible viewpoint, where no one could live. In this way, the photographs signal the danger inherent in the viewpoints of the people taking them.

On the one hand, these two photographs, taken by human beings rather than unmanned satellites, have more rhetorical force because they are tokens of a human eyewitness standpoint. On the other, they also signal the perilous position of these space travelers who risk their lives while taking them. The only way to get what even NASA officials called this "God's-eye view" was from an impossible point so far away from earth. The view of the whole earth could not be seen from earth, but only at a distance born out of rocket science and compared to the viewpoint of God. As creatures on the earth, we cannot see the earth; it is never a whole or total object presented to our perception. Apart from photographs, the view of the earth as a whole has been reserved for the rare astronaut who left the earth's atmosphere.

Speaking of their view of "the whole globe," as "the first humans to see the world in its majestic totality," astronaut Frank Borman exclaimed, "This must be what God sees," and many of the astronauts talked of traveling to "the heavens."[23] To see the earth whole, as it "really is," human beings must travel to the heavens to get a God's-eye view of the planet. Some earthlings even asked the astronauts whether they had seen God in space.[24] The view of the whole earth is the view of God. It is no wonder, then, that the Apollo missions sparked as much discussion of conquering and mastery as they did vulnerability and fragility.

However, what the astronauts and the media assumed they saw in the photographs, particularly "Blue Marble"—namely, the whole earth—was an illusion, for both images show only part of earth, indeed, a fraction of the earth. "Earthrise" shows an elongated piece of the top of a sphere, while "Blue Marble" shows one side of the earth; and both are rendered in the two-dimensional space of the photographic medium. In other words, we did not, and do not, see what we thought we saw.

The impact of seeing the earth whole, seeing it as it *really* is, was based on the fantasy of the whole earth, which not only was never visible in these photographs, but also, at least with current technology, never will be. The whole earth cannot be captured from any human vantage point, even one floating in a space capsule orbiting the moon, or any other point in space. For, as phenomenologists teach us, the human perspective is always only partial; there is always something that is occluded and missing from our viewpoint.[25] Perhaps this is why we imagine desert islands as uninhabited. From our perspective, nothing lives in the desert, nothing like us, anyway. Inverting the epigraph from Deleuze with which we began, we could say that just because an island is deserted doesn't mean it is uninhabited.

In "Faith and Knowledge," Derrida talks of the desert as a place wherein barrenness is a lack of predetermined meaning, language, or symbolic systems. Yet this barren desert before meaning is the intersubjective space that makes meaning possible. The barrenness of this desert is what enables the fecundity of life as we know it, which is to say, the possibility of meaningful life. In this desert, there is no dogma, no traditions, and no prejudice. There is just the primary relationality that makes communication possible. In that essay, the desert is both a hopeful place (or place before place) that allows us to presuppose the possibility of communication through a primary relatedness and a violent place where the force of law, its performative conditions, gives way to law proper. For Derrida, the desert is not only both a hopeful and a violent place, but also a fecund place that gives birth to human meaning and human law. The desert seems to be a metaphor of hope or hopelessness, another metaphor for the ambivalence at the heart of our autoimmune existence on planet Earth.

In his last seminar, Derrida suggests that the desert is never barren or pure; rather, it is always already contaminated by dogma, traditions, and prejudice. There, the desert is associated with Robinson Crusoe's desert island, and the question becomes whether the island is actually deserted. Does Crusoe share his island, or is he alone? Robinson Crusoe longs for human companionship more than anything else on his deserted island. And yet, when he discovers the footprint in the sand, he is terrified and longs for solitude. He both wants and fears company on his island. The "savages" whom he witnesses on the beach threaten the supposedly deserted nature of his island home. He is both fascinated and repulsed by these strangers and intends to kill or enslave them. He wants both to flee the island and return to the island, which represents both the terror of solitude and the blissful paradise of solitude. Robinson Crusoe's ambivalent relationship to his island is akin to the ambivalence witnessed after the Apollo missions—namely, the terror of our solitude on earth and the absolute uniqueness of our earthly paradise in the vast emptiness of space.

Speaking of islands and singularity, Derrida says, "Between my world and any other world there is first the space and the time of an infinite difference, an interruption that is incommensurable with all attempts to make a passage, a bridge, an isthmus, all attempts at communication, translation, trope, and transfer that the desire for world or the want of a world, the being wanting a world will try to pose, impose, propose, stabilize. There is no world, there are only islands" (BS2 9/31). Not only every man is an island, but also every beast, and perhaps every living being, each radically separated from every other. Emphasizing the radical singularity of each living being—not just human beings—Derrida claims that with each one's death, the world is destroyed. This radical claim evokes a sense of urgency in relation to ethics and politics. Derrida embraces the impossible intersection of ethics and politics, the demand to respond to the singularity of each living being and the demand for justice for all. This incalculable obligation to the other, even the ones whom we may not recognize, challenges all political pluralities and notions of peoples in whose names we act and brings to the fore the tension between ethical obligations that necessarily take us beyond calculation and universal principles, on the one hand, and political obligations that necessitate calculation and universal principles, on the other. It challenges the notion that there is some vantage point from which we can accumulate all worlds, all perspectives, to see *the world* as it is. In this sense, the world is always singular and never plural.

On the one hand, Derrida insists that we do not share the world and that each singular being is a world unto itself, not just *a* world, but *the* world. On the other hand, and at the same time, we are radically dependent on the other and others for our sense of ourselves as autonomous and self-sufficient, illusions that come to us through worldly apparatuses. We both do and do not share the world. And, even if we do not share a world, we do share the earth.

So, how can we reconcile these two strands in Derrida's thought? Namely, that the self is constituted through the other, on one hand, and that each singular being is unique to the point that its death is the end of the world, on the other hand. Derrida makes a certain double movement to circumnavigate the seemingly oppositional, even contradictory, claims that no man is an island and every man is an island, that we are radically interdependent and that we are radically singular, that relationality permeates every living being and that alterity inhabits every living being. Rather than either/or, Derrida gives us both/and. Every living being is fundamentally relational and interdependent, and yet at the same time, each is singular and unique. Furthermore, the singularity of each comes through the other and others. The answer to the question "Do we share a common world?," then, is both *yes* and *no*. We both share a common world and we don't. Each living being has its own world, and yet common codes (or languages in the case of humans) assume that we share a world.

In this regard, whether we are talking about life with other human beings or with nonhuman animals (and perhaps even plants and rocks), we both share and do not share a world or *the* world. As earthlings, we depend on the network of relationships that constitute earth. In addition, every living being is a world made up of diverse living organisms living together in one body, which operates as a semi-contained, but fundamentally interconnected, ecosystem or world.

Even Robinson Crusoe on his desert island is not alone. The desert, or deserted, island turns out to be an illusion, a fantasy of autonomy and self-sufficiency. The ethical obligation that stems from the singularity of each living being is not Robinson's solipsistic delusion of mastery or control over himself or the plants, animals, and ultimately other humans that both share and do not share that island. Following Derrida, we could say that we live in an age of autoimmunity where life turns against itself and leaves us with a desert island or the fantasy of one. Like Robinson's wheel and his island, the earth spinning on its axis incites ambivalence. These contradictory reactions exist within one and the same nation, within one and the same person. And yet, this ambivalence now figured by the island— this island earth—undermines the sovereignty of that nation and that person.

In Derrida's terms, islands contain within them the logic of autoimmunity that makes all autos, all returns to self, eventually turn against themselves; and then what was intended to protect the self begins to destroy it. This mythical return to self contains within it a threat to self, the threat of self-annihilation. For example, technologies developed to provide endless energy deliver nuclear destruction. Or fossil fuel–driven technologies that allowed us to leave earth's atmosphere and visit the moon contribute to the destruction of earth's atmosphere and perhaps eventually the destruction of life on earth. It is noteworthy, however, that most likely, the earth itself will survive. When environmentalists talk of saving the planet, they really mean saving ourselves. The earth has already survived an ice age.[26] As I argue elsewhere, the discourse of "saving the earth" assumes that we can master both the earth and ourselves.[27]

If our conception of earth determines our conception of ourselves and vice versa, then we must rethink the earth as island metaphor. The island metaphor suggests that the earth is isolated, insular, and self-contained. The initial reactions to the photographs of earth were illusions in part because they perpetuated this view of earth. If we conceive of the earth as a network of dynamic living relationships, however, then we also change the way that we see ourselves. No longer maverick space cowboys desperately holding on to the fantasy of the autonomous self-contained individual, species, nation, we can embrace our earthbound existence, sharing our planetary home with diverse worlds beyond our comprehension.

If earth is an island, then we must reconceive of islands as dynamic spaces constituted by their relationships to air, sea, and the elements that make them what they truly are. In other words, we must embrace the fact that we are limited creatures who are not just living on earth, but rather part of the biosphere that constitutes its very being. And, as such, the earth and other earthlings are always necessarily in excess of our world(s) and function as limits to our world(s).

Earth ethics demands that we acknowledge the ways in which we do not share a world because the earth and other earthlings refuse us. With equal force, earth ethics demands that we acknowledge the ways in which we do share worlds because the earth and other earthlings respond to us. Ultimately, earth ethics demands that we not only acknowledge, but also embrace the vital fact that although we may not share a world, we do share a singular bond to the earth. By emphasizing the radical relationality of each living being, together with our shared but singular bond to the earth, the island no longer appears deserted or isolated but rather inhabited by immense biodiversity and surrounded by oceans teeming with life, which we risk discounting at our peril.[28] Indeed, the island is the meeting of land and sea, an in-between space, which provokes uncanny ambivalence. When we disavow the uncanny strangeness and our own ambivalence toward it, then we risk reducing the island to the fantasy of barren isolation or exotic paradise, neither of which is sustainable.

The danger of the island is that it becomes a figure for isolation rather than a figure for the in-between, the meeting of land and sea. Even Derrida's invocation of the island as a figure for singularity risks overshadowing the necessary interconnections and interrelationality through which and in which habitation happens. The island metaphor, when extended to the earth itself, discounts the very features that make the earth unique—namely, the dynamic relationships between land and sea, globe and atmosphere, planet and solar system. We begin to articulate an ethics and politics grounded on the earth, not as isolated island or totalizing globe, but rather as the unknown and unpredictable source of diverse life on our uncanny earthly home. Earth ethics is grounded on the earth as the singular home to all earthlings, which inhabit a vast and plentiful plurality of worlds.

Notes

1. "Men of the Year," *Time* Magazine, January 3, 1969, 16.

2. John Berger, *About Looking* (New York: Random House, 1991).

3. Archibald MacLeish, "Riders on Earth Together, Brothers in Eternal Cold," *New York Times*, December 25, 1968.

4. MacLeish, "Riders on Earth Together."

5. See Denis Cosgrove, "Contested Global Visions: One-World, Whole-Earth, and the Apollo Space Photographs," *Annals of the Association of American Geographers* 84, no. 2 (1994): 270–94.

6. See Benjamin Lazier, "Earthrise; or, the Globalization of the World Picture," *American Historical Review* 116, no. 3 (2011): 606.

7. See Lazier, "Earthrise," 620, and Cosgrove, "Contested Global Visions," 272.

8. See Carl Sagan, *Pale Blue Dot: A Vision of the Human Future in Space*, chapter 1, "You Are Here," 1–8 (New York: Ballantine, 1994), and Oran W. Nicks, ed., *This Island Earth* (Washington, D.C.: Government Printing Office, National Aeronautics and Space Administration,1970).

9. Quoted in Robert Poole, *Earthrise: How Man First Saw the Earth* (New Haven: Yale University Press, 2008), 134.

10. Cf. Cosgrove, "Contested Global Visions," 282.

11. "Men of the Year."

12. Ibid.

13. Quoted in Poole, *Earthrise*, 133–34.

14. See ibid., 17.

15. "Men of the Year," 17.

16. Cf. Poole, *Earthrise*, 134.

17. Lazier, "Earthrise," 623.

18. Nicks, *This Island Earth*, vi; my emphasis. Upon seeing the photographs of earth from space, news anchor Walter Cronkite described the Earth as *"floating* in space"; Poole, *Earthrise*, 146; my emphasis.

19. Quoted in Nicks, *This Island Earth*, iv; my emphasis.

20. Lazier, "Earthrise," 620.

21. See ibid., 619.

22. Buzz Aldrin, "The Call of Mars," *New York Times*, June 13, 2013, Opinion/ Global Opinion, http://www.nytimes.com/2013/06/14/opinion/global/buzz-aldrin-the-call-of-mars.html.

23. Quoted in Poole, *Earthrise*, 20.

24. See ibid., 129.

25. It is noteworthy that while some of the geographers and historians who have discussed the Apollo photos quote Edmund Husserl on pre-Copernican Earth or point out that the Blue Marble only shows Africa and Asia, none linger on the fact that these photographs are actually not of the *whole* earth. For a sustained discussion, see Kelly Oliver, *Earth and World: Philosophy after the Apollo Missions* (New York: Columbia University Press, 2015).

26. David Beillo, "CO2 Levels for February Eclipsed Prehistoric Highs," *Scientific American*, March 5, 2015, http://www.scientificamerican.com/article/co2-levels-for-february-eclipsed-prehistoric-highs/.

27. Oliver, *Earth and World*.

28. Biologist Edward O. Wilson says, "There is no question in my mind that the most harmful part of ongoing environmental despoliation is the loss of biodiversity"; Wilson, "Biophilia and the Conservation Ethic," in *The Biophilia Hypothesis*, ed. Stephen R. Kellert and Edward O. Wilson (Washington, D.C.: Island Press, 1993), 35.

Contributors

Karen Barad is Professor of Feminist Studies, Philosophy, and History of Consciousness at the University of California at Santa Cruz. Barad's Ph.D. is in theoretical particle physics and quantum field theory. Barad held a tenured appointment in a physics department before moving into more interdisciplinary spaces. Barad is the author of *Meeting the Universe Halfway: Quantum Physics and the Entanglement of Matter and Meaning* (Duke University Press, 2007) and numerous articles in the fields of physics, philosophy, science studies, poststructuralist theory, and feminist theory. Her work on Derrida includes "Quantum Entanglements and Hauntological Relations of Inheritance: Dis/Continuities, SpaceTime Enfoldings, and Justice-to-Come," *Derrida Today* 3 (2010): 240–68. She is currently completing a book project entitled *Infinity, Nothingness, and Justice-to-Come.*

Timothy Clark is Professor of English at the University of Durham. His latest books are *The Poetics of Singularity: The Counter-Culturalist Turn in Heidegger, Derrida, Blanchot and the Later Gadamer* (Edinburgh University Press, 2005); *The Cambridge Introduction to Literature and the Environment* (Cambridge University Press, 2011); and *Ecocriticism on the Edge: The Anthropocene as a Threshold Concept* (Bloomsbury, 2015).

Claire Colebrook is Edwin Erle Sparks Professor of English at Penn State University. She has written books on literary theory, literary history, gender theory, queer theory, the philosophy of Gilles Deleuze, and posthumanism. She has recently completed a book on fragility and is now working on a book concerning the outside of thinking.

Matthias Fritsch is Professor of Philosophy at Concordia University, Montréal. His research interests are social, political, and moral philosophy (in particular, democratic theory and Marxism), as well as nineteenth- and twentieth-century European philosophy (especially German critical theory and deconstruction). He has published a monograph, *The Promise of Memory* (SUNY Press, 2006), and a range of articles in scholarly journals, has coedited two anthologies, and has translated authors such as Heidegger, Gadamer, and Habermas into English. He has been a Humboldt Fellow in Frankfurt and a Visiting Research Professor in Kyoto. At present, he is working on a book manuscript (for which he has been awarded federal Canadian funding) on intergenerational ethics. A second project develops a concept of deconstructive normativity in relation to metaethics, biopolitics, and environmental philosophy.

Vicki Kirby is Professor of Sociology in the School of Social Sciences, the University of New South Wales, Sydney. Books include *Quantum Anthropologies: Life at Large* (Duke University Press, 2011); *Judith Butler: Live Theory* (Continuum, 2006); and *Telling Flesh: The Substance of the Corporeal* (Routledge, 1997). She teaches and lectures regularly in Europe and the United States and was Erasmus Mundus Professor at Utrecht University in 2013. In 2015, she was Visiting Professor at the Winter School, Institute of Advanced Study in the Humanities and Social Sciences, Bern University. She has articles forthcoming in the journals *Derrida Today*, *Parallax*, and *Philosophy of the Social Sciences*. The motivating question behind her research is the puzzle of the nature/culture, body/mind, matter/ideation division, because so many political and ethical decisions are configured in terms of this opposition and its cognates. Her particular interest is embodiment and matter, and she brings feminism and deconstruction into conversation in order to shift the terms of these inquiries.

John Llewelyn has been Reader in Philosophy at the University of Edinburgh and Visiting Professor of Philosophy at the University of Memphis, Loyola University of Chicago, and Vanderbilt University. As well as translations and edited works, his publications include *Beyond Metaphysics?* (Prometheus, 1989); *Derrida on the Threshold of Sense* (Palgrave Macmillan, 1986); *Appositions of Jacques Derrida and Emmanuel Levinas* (Indiana University Press, 2002); *The Middle Voice of Ecological Conscience* (Palgrave Macmillan, 1991); *Emmanuel Levinas: The Genealogy of Ethics* (Routledge, 1995); *Seeing through God* (Indiana University Press, 2004); *Margins of Religion* (Indiana University Press, 2008); *The Rigor of a Certain Inhumanity* (Indiana University Press, 2012); and, most recently, from Edinburgh University Press, *Gerard Manley Hopkins and the Spell of John Duns Scotus* (2015). A semi-autobiographical volume by him, entitled *Departing from Logic*, has been published by Y Lolfa (2012).

Philippe Lynes earned his Ph.D. from Concordia University in Montreal, specializing in deconstruction, environmental philosophy, biopolitics, and ecolinguistics. In 2017–18, he will serve as the Fulbright Visiting Research Chair in Environmental Humanities at the University of California, Irvine. Lynes is also a translator of French philosophy, with a translation of and introduction to Jacques Derrida's *Advances* forthcoming in late 2017 with the University of Minnesota Press. His first monograph, *General Ecology: Futures of Life Death on Earth*, is currently under review, and he is now working on his second book, *Dearth: Eco-Deconstruction after Speculative Realism*, on Blanchot, Derrida, and Heidegger.

Michael Marder (michaelmarder.org) is Ikerbasque Research Professor of Philosophy at the University of the Basque Country (UPV-EHU), Vitoria-Gasteiz, Spain. A specialist in phenomenology, environmental philosophy, and political thought, he is the author or editor of twelve books. His most recent monographs include *The Philosopher's Plant: An Intellectual Herbarium* (Columbia University Press, 2014); *Pyropolitics: When the World Is Ablaze* (Rowman and Littlefield International, 2015); *Dust* (Bloomsbury Academic, 2016); *Through Vegetal Being: Two Philosophical Perspectives*, coauthored with Luce Irigaray (Columbia University Press, 2016).

Dawne McCance is a Distinguished Professor at the University of Manitoba and editor of the interdisciplinary critical journal *Mosaic*. She teaches and writes on the work of Jacques Derrida in such books as *Posts: Re Addressing the Ethical* (SUNY Press, 1996); *Medusa's Ear: University Foundings from Kant to Chora L* (SUNY Press, 2004); *Derrida on Religion* (Equinox, 2009); *Sleights of Hand: Derrida Writing* (Kalamalka Press, 2008); and *Critical Animal Studies: An Introduction* (SUNY Press, 2013); and in numerous published essays and book chapters.

Michael Naas is Professor of Philosophy at DePaul University. He works in the areas of Ancient Greek Philosophy and Contemporary French Philosophy. His most recent books include *The End of the World and Other Teachable Moments: Jacques Derrida's Final Seminar* (Fordham University Press, 2014), and *Miracle and Machine: Jacques Derrida and the Two Sources of Religion, Science, and the Media* (Fordham University Press, 2012). He is also the cotranslator of several works by Jacques Derrida, including *The Other Heading* (Indiana University Press, 1992); *Memoirs of the Blind* (University of Chicago Press, 1993); *Adieu—to Emmanuel Levinas* (Stanford University Press, 1999); *Rogues* (Stanford University Press, 2005); and *Learning to Live Finally* (Melville House, 2007). He also coedits the *Oxford Literary Review*.

Kelly Oliver is W. Alton Jones Professor of Philosophy at Vanderbilt University. She is the author of over one hundred articles, thirteen books, and ten edited volumes. Her authored books include, most recently, *Hunting Girls:*

Sexual Violence from The Hunger Games to Campus Rape (Columbia University Press), forthcoming; *Earth and World: Philosophy after the Apollo Missions* (Columbia University Press, 2015); *Technologies of Life and Death: From Cloning to Capital Punishment* (Fordham University Press, 2013); *Knock Me Up, Knock Me Down: Images of Pregnancy in Hollywood Film* (Columbia University Press, 2012); *Animal Lessons: How They Teach Us to Be Human* (Columbia University Press, 2009); *Women as Weapons of War: Iraq, Sex and the Media* (Columbia University Press, 2007); *The Colonization of Psychic Space: A Psychoanalytic Theory of Oppression* (University of Minnesota Press, 2004); and perhaps her best-known works, *Witnessing: Beyond Recognition* (University of Minnesota Press, 2004), and *Noir Anxiety: Race, Sex, and Maternity in Film Noir* (University of Minnesota Press, 2001). She has published in the *New York Times* and has been interviewed on ABC television news, various radio programs, and Canadian Broadcasting network. Her work has been translated into seven languages.

Michael Peterson is a graduate student at DePaul University, Chicago. He received an M.A. in Philosophy from Concordia University in 2014 and a B.A. Honors from the University of Alberta in 2011.

Ted Toadvine is Professor of Philosophy and Environmental Studies at the University of Oregon, where he served as Head of Philosophy from 2011 to 2014. He specializes in contemporary Continental philosophy, especially phenomenology and recent French thought, and the philosophy of nature and environment. His current research interests include the philosophical significance of deep time, the eschatological imaginary of environmentalism, the relation between geomateriality and memory, and biodiacritics. He is author of *Merleau-Ponty's Philosophy of Nature* (Northwestern University Press, 2009) and editor or translator of six books, including *The Merleau-Ponty Reader* (Northwestern University Press, 2007); *Nature's Edge* (SUNY Press, 2007); and *Eco-Phenomenology* (SUNY Press, 2003). Toadvine directs the Series in Continental Thought at Ohio University Press, is Editor-in-Chief of the journal *Environmental Philosophy*, and is a coeditor of *Chiasmi International: Trilingual Studies concerning Merleau-Ponty's Thought*. He is currently completing two manuscripts, *Eschatology and the Elements* and *Diacritical Life: Animality and Memory*, as well as coediting (with David Alexander Craig) a volume on *Animality and Sovereignty: Reading Derrida's Final Seminars*.

Cary Wolfe's books and edited collections include, most recently, *What Is Posthumanism?* (University of Minnesota Press, 2010), and *Before the Law: Humans and Other Animals in a Biopolitical Frame* (University of Chicago Press, 2013). He is Founding Editor of the series *Posthumanities* at the University of Minnesota Press, which publishes four books per year by noted authors such as Donna Haraway, Roberto Esposito, Isabelle Stengers, Michel Serres, Vilém Flusser, and many others. He currently holds the Bruce and Elizabeth Dunlevie Chair in English at Rice University, where he is Founding Director of 3CT: The Center for Critical and Cultural Theory.

David Wood is W. Alton Jones Professor of Philosophy at Vanderbilt University, where he teaches Continental and Environmental Philosophy. His published books include *The Deconstruction of Time* (Northwestern University Press, 2001); *Thinking after Heidegger* (Polity and Blackwell, 2002); *The Step Back: Ethics and Politics after Deconstruction* (SUNY Press, 2005); and *Time after Time* (Indiana University Press, 2007). *Reinhabiting the Earth* and *Deep Time* are forthcoming with Fordham. He is also an earth artist and directs Yellow Bird Art Farm.

Index

activism: 4–5, 51, 241, 244n27

Adorno, Theodor: 2, 18n2, 19nn2,3,5, 279, 291, 296n1, 301n41, 316n29

aesthetics: 150–51, 155

affect, affectivity, autoaffection, heteroaffection: 57–8, 61, 104–5, 107–8, 113, 118, 119n10, 132, 135, 285, 292, 294, 331

affirmation: 16, 101–2, 104, 107, 110–11, 114, 116, 118, 144–45, 153, 234, 280–81, 284–92, 294–96, 298–99n27, 299n33, 299–300n34. *See also* double affirmation

agency: 13, 38–40, 42, 46, 81, 85–87, 91–92, 97n28, 110, 124, 132–33, 169, 213, 262, 333

agriculture: 46, 194, 268–69, 271, 308

air: 45, 70, 72, 74, 343, 352

Alterity: 12, 16, 86, 105–7, 110, 113, 115–17, 126, 163, 177, 234–35, 287, 293, 300n39. *See also* Other, the

animals, animality: 1–6, 8–11, 17, 20n15, 23n33, 26n64, 30, 32–33, 35–36, 41–42, 45, 67–68, 70–71, 73, 85, 88–90, 96n25, 104, 106, 109–10, 124, 127, 132–34, 143, 147, 158, 193, 211, 262, 265, 281–82, 284–85, 291, 293, 295, 301n46, 304–6, 308–11, 313, 315–16n23, 323–25, 346, 350–51

animal studies: 5, 323

Anthropocene: 5, 15, 40, 81–82, 85, 88, 90–92, 94, 121, 206, 241nn1,2, 268–70, 272

anthropocentrism: 6, 10, 16, 87, 107–8, 111, 124, 134, 136, 139, 262, 281, 310

anthropologism: 8, 24n49, 36, 41, 48n23, 107, 158, 305

anthropogenic: 35, 47, 80n44, 81, 121, 132

anthropology: 77n10, 90, 135

anthropotheologism: 16, 305, 311–12

Antigone: 142, 147

antihumanism: 122–24

apocalypse: 50–51, 55–57, 60, 62–63, 67, 76, 77n8, 121, 202, 207–8, 313

Apollo missions: 17, 339–44, 346, 348–49, 353n5

aporia: 11, 16, 30–31, 37, 39, 46, 196, 262

arche-writing/archi-writing: 78n21, 175–77

archive: 52, 60, 74, 91, 188, 190, 195–96, 201, 203–4, 250, 260n11, 261, 263, 267–68

Aristotle: 23n33, 33, 147, 154, 166, 181, 230

art: 4–5, 110, 142, 149–51, 203, 312, 333, 335

Aufhebung: 114–15, 167

autoimmunity: 15, 17, 43, 45, 80n34, 85, 93, 194, 268, 289–92, 296, 299n30,

autoimmunity (*continued*)
  300nn38,39, 304, 306, 312, 314n5,
  340–41, 345, 347, 349, 351
autopoiesis: 17, 281, 321–23, 331
axiology: 283, 288, 293

Barad, Karen: 14–15, 22n28, 139n7,
  139n11, 206–48, 264–65
Bataille, Georges: 4, 12, 102, 114–15,
  152, 163
Bateson, Gregory: 328, 331
Beauvoir, Simone de: 123, 127, 139n2
Benjamin, Walter: 2, 198, 210
Bennington, Geoffrey: 24n47, 138,
  140n25
Bentham, Jeremy: 281, 295, 315n23
Bible: 4, 33, 160, 164n3
biocentrism: 283–84, 297n5, 298n21
biodegradability: 11, 14–6, 18, 143,
  187–205, 249–50, 253–55, 259n3,
  260n11, 267–68, 272
biology: 8, 17, 23n35, 25n50, 36, 41,
  52, 81, 88–89, 95n14, 96n23, 97n34,
  103, 105, 108, 117, 123, 130–32,
  140nn15,16, 147, 211, 255, 281,
  297n9, 298n19, 299nn30,33, 314n5,
  323–24, 354n28
biopolitics: 19n3, 35, 284
biosphere: 282, 352
Blanchot, Maurice: 4, 25n53, 86, 96n22,
  102–3, 106, 161–62, 164n4, 198
body, 6, 10, 43, 78n17, 83, 89, 125,
  143–44, 147, 164n2, 167, 193, 214,
  221, 229, 235, 237–40, 243n17, 262,
  268, 271, 289, 305–6, 308, 310, 314n5,
  351. *See also* embodiment
Brown, Charles S.: 3, 20n8, 21nn16,17,
  87, 94nn4,5, 95n20
Butler, Judith: 139n6, 310, 315n21

Calliott, J. Baird: 81–82, 88, 91–92,
  94n5, 283, 298n19
cannibalism: 25n55, 48n27, 291. *See
  also* kinnibalism
capitalism/capital: 18, 45, 58, 60–62,
  84, 87, 92–93, 156–57, 198, 206, 210,
  213, 244n26, 246–47n50, 247n51, 279,
  328
carnophallogocentrism: 10, 40–41

Carson, Rachel, *Silent Spring*: 5, 22n23, 51
Cartesian thought. *See* Descartes, René
Catholicism: 304, 306, 310–11
causality: 21n17, 95n16, 125, 130, 213,
  215, 236
Celan, Paul: 65, 68, 109, 198, 313, 325,
  334
Christianity: 60, 117, 167, 177, 182, 306,
  334. *See also* Bible
cinders: 194, 201
Clark, Timothy: 12, 22n26, 81–97,
  261–62, 275n2
climate change: 1, 12–5, 35, 40, 47, 54,
  77n7, 77–78n11, 78n12, 85, 121, 132,
  134, 242n3, 256, 268–69, 271, 274–75,
  322. *See also* global warming
code, coding: 7–10, 24n44, 51, 68,
  324–25, 350
Cold War: 51, 202, 207, 227, 342, 344
Colebrook, Claire: 7, 15, 23n41, 25n63,
  261–75
colonization, colonialism: 5, 14–15, 17,
  40, 206, 209–11, 213, 215, 221–23,
  228–31, 235, 237, 241, 247n51, 269,
  291, 307–8, 347
Condillac: 172–74, 176
constructionism: 82, 126, 134
contamination: 34, 92, 105, 111, 122,
  251, 287, 300n34, 304, 309, 311, 349
context: 4–8, 10–12, 14, 16, 30, 32, 57,
  76, 94n12, 126, 139n4, 167, 194, 197,
  254–59, 265, 269, 285–89, 291–93,
  295, 296, 299n33, 312, 337n28
Continental philosophy: 1–3, 6, 11,
  18n1, 19n7, 280
cosmos: 59, 63, 65–67, 69, 117
Critchley, Simon: 317–18, 321, 326

Darwin, Charles: 128–32, 140nn13,17
death: 6, 9–10, 14, 25n50, 32, 36, 38, 55,
  57, 63–72, 74–75, 79n31, 101, 103–5,
  108–9, 111–15, 117–18, 131, 138,
  145–49, 153, 161, 172, 178–79, 194,
  202, 205nn4,5, 214–15, 222, 228, 232,
  236–38, 240, 247n53, 248n58, 263–65,
  281, 283, 285, 287, 290–96, 302n47,
  350. *See also* mortality
deconstruction: 2–12, 14–17, 20n18,
  22n28, 24nn44,49, 25n53, 29–34, 37,

39, 41–47, 55, 57–58, 62, 68, 81, 86,
88, 90, 93–94, 95n12, 96n23, 101–2,
104–7, 110, 113, 115–16, 118, 125,
134, 138–39, 141–43, 149–50, 152,
154, 160, 164–73, 175–76, 181, 182,
192, 194, 198, 205n4, 213, 250, 257,
261–62, 265–66, 268, 270–73, 275,
280, 284–85, 299n30, 320–22,
324–25
Deep Ecology: 1, 3, 6, 283
Deleuze, Gilles: 2, 20n13, 86, 96n22,
119n9, 284, 339, 348
De Man, Paul: 6, 8, 24n49, 189–92,
195–96, 199, 202–3, 204–5n3, 205n5,
266, 275n3
democracy: 17, 43, 84, 107, 133, 263,
265, 271, 300nn38,39, 306, 314n7
democracy-to-come: 37, 42, 46, 107,
261, 265, 306
Derrida, Jacques: *passim*; works by:
The Animal That Therefore I Am,
6, 23n33, 36, 42, 71, 104, 110–11,
285, 293–95, 297n14, 298–99n27,
308–9, 311, 315–16n23, 316n29, 325;
Aporias, 108, 234; *Archaeology of the
Frivolous*, 172–73; *Adieu: Emmanuel
Levinas*, 106–7, 116–17, 299n33;
Advances, 101, 117–18, 119n1; "Alterities," 116; *Archive Fever*, 260n11,
263; *Arguing with Derrida*, 7, 286;
"Avowing the Impossible," 102–3,
119n13, 264; "Avec Levinas," 119n13
The Beast & the Sovereign, vol. 1, 6,
17, 23n33, 26n64, 48n25, 80n35, 106,
260n11, 302n47, 305–6; *The Beast
& the Sovereign*, vol. 2, 6, 17, 55,
57, 64, 67–72, 80n35, 109–10, 170,
260n11, 291, 293, 301nn46,47, 313,
324–26, 328–34, 340, 349; "Biodegradables: Seven Diary Fragments," 14,
187–205, 249–50, 253–4, 259n3, 267,
272; *Counterpath: Traveling with
Jacques Derrida*, 147, 159–60, 164n2;
The Death Penalty, vol. 1, 25n64,
102, 260n11; *The Death Penalty*,
vol. 2, 26n64, 64, 79n31, 260n11,
294; *Deconstruction in a Nutshell: A
Conversation with Jacques Derrida*,
116; *Demeure: Fiction and Testi-

mony*, 148, 161–62; *Derrida*, 24n47;
Dissemination, 25n56, 111; *The Ear
of the Other: Otobiography, Transference, Translation*, 22n32; "Economimesis," 149–52, 155; *Edmund
Husserl's "Origin of Geometry": An
Introduction*, 63, 79n30; "Faith and
Knowledge: Two Sources of 'Religion'
at the Limits of Reason Alone," 16,
102, 111–12, 299n30, 303–5, 307,
311, 314n5, 314n9, 348–49; "Force
of Law," 10, 16, 25n64; *For What
Tomorrow*, 7, 102–3, 118, 286; *The
Gift of Death and Literature in
Secret*, 104, 119n5, 145–46, 149,
177; *Given Time*, vol. 1, *Counterfeit Money*, 25n54, 120n20, 144–45,
154–55; *Glas*, 114–15, 148, 176, 292;
*Heidegger: The Question of Being &
History*, 107–8, 119n12; "Hospitality," 157–59; "Hospitality, Justice,
Responsibility," 119n11; "Interpreting Signatures," 22–23n32; "Jacques
Derrida," 119n2; "La forme et la
façon," 102; *L'argent*, 24n49; "La vie
la mort," 6, 22n32, 23n34, 25n50, 88,
140n15, 312; *Learning to Live Finally:
The Last Interview*, 101; *Limited
Inc.*, 24n47, 139n4, 154, 156, 254–56,
319, 337n28; *Margins of Philosophy*,
25nn51,56, 84, 108, 129, 146–47, 153,
168, 176, 179–80, 253–55; *Memoires
for Paul de Man*, 114, 191–92; *Negotiations: Interventions and Interviews
1971–2000*, 81, 104, 112, 299nn32,33,
299–300n34; *Of Grammatology*, 5–9,
21n18, 23n33, 25n51, 38, 58, 80n35,
84, 89, 105, 108, 113, 119n10, 121,
131, 135–37, 140n23, 337n28; *Of Cosmopolitanism and Forgiveness*, 344;
Of Hospitality, 344–45; *Of Spirit:
Heidegger and the Question*, 71, 110,
143; *On Touching—Jean-Luc Nancy*,
58, 64, 117; *The Other Heading:
Reflections on Today's Europe*, 307,
315n12; *Paper Machine*, 106; *Parages*,
94n12, 299n32, 336n12; *Philosophy in a Time of Terror*, 35, 80n34,
205n8, 314n5; *Points . . . Interviews*,

Derrida, Jacques (*continued*)
1974–1994, 10, 25n54, 25n55, 41,
86–87, 95n17, 102, 105–6, 108, 110,
119n4, 132–33, 258, 286; *The Politics
of Friendship*, 119n5, 299n31, 314n7;
Positions, 25n56, 115; *The Postcard:
From Socrates to Freud and Beyond*,
23n32, 112–13, 176, 312 *The Problem
of Genesis in Husserl's Philosophy*,
105, 117; *Psyche: Inventions of the
Other*, vol. 1, 32, 51–55, 67, 80n34,
202; *Psyche: Inventions of the Other*,
vol. 2, 86, 299n32; "Que faire—de la
question 'Que faire?,'" 103; *Rogues*,
70–71, 107, 257, 259, 300n38, 300–
1n39, 306–7, 314n5; *Specters of Marx*,
29–30, 40, 43, 106–7, 109, 157, 178,
288, 290, 299n33, 314n8; *Sovereignties in Question: The Politics of Paul
Celan*, 64–67, 70; *Spurs*, 267; *A Taste
for the Secret*, 8, 24nn47,49, 116,
299n30; *The Truth in Painting*, 110–
11; "Ulysses Gramophone," 95n12;
Voices and Phenomenon, 25n51, 105;
Without Alibi, 113–14; *The Work of
Mourning*, 64, 66, 70, 79n33, 80n39;
Writing and Difference, 25n51,
112–15, 119n5, 146, 153, 267
Descartes, René: 6, 17, 18n2, 23n33,
122–24, 134, 273, 281, 293, 297n14,
308–9, 311
dialectics: 3, 102, 104–5, 114–16, 168,
172, 262, 318
différance: 6, 8, 9, 72, 78n21, 86, 96n23,
103, 105, 110–12, 115, 119n11, 125,
128–29, 137, 139, 139n4, 146, 152–54,
160, 169, 176, 180, 216–17, 261, 264,
271, 286–87, 290, 300n39
diffraction: 212, 215–26, 229, 234–35
disability: 15, 127, 268–72, 274–75
double affirmation: 284–85, 288, 292,
299n33, 299–300n34
dwelling: 3, 7–9, 39, 44, 46, 69, 109, 112,
141–47, 149–63, 166, 178, 345

earth: 1–4, 12, 17–18, 21n16, 24n44,
35, 39–40, 42, 46, 49n30, 50–51, 53,
55–56, 59, 65–67, 70–71, 73–74, 82,
85–86, 91–92, 97n34, 101–4, 107,
109–14, 117–18, 131, 157, 200, 203,
207–8, 213, 215, 230, 238, 251–52,
258, 262–63, 265–66, 268–69, 272–73,
280, 282, 291, 294, 305, 319, 339–54
ecocriticism: 3, 5–6, 89, 274
eco-deconstruction: 10–14, 16, 18,
20n14, 33–34, 56, 141, 162, 188, 203,
249, 259n1, 279–80, 284, 296n4
ecofeminism: 1, 3, 5, 298n21
eco-hermeneutics: 3–6
ecology: 2, 4, 7–8, 10–17, 21n17, 25n53,
30, 42, 51–52, 58–59, 61, 67, 82–83,
85, 89–90, 97n34, 101–9, 111–18, 121–
24, 126, 129–30, 132, 134–38, 141–64,
176–79, 181, 188–91, 203, 244n27,
250–51, 253, 261–62, 268, 270, 273,
282–84, 299n33, 310, 323–28, 331–35,
337n25
economy: 7–10, 12–13, 18n2, 43–45,
54, 58–62, 84, 92, 94, 95n14, 102–6,
108–9, 112–15, 118, 134, 141–64,
166, 171, 179–80, 210, 252, 286, 312,
314n9
eco-phenomenology: 2–4, 6, 20n8,
20–21n16, 21n17, 55–58, 81, 83, 87
eco-poetics: 5, 326, 334
ecosystem: 1, 12, 16, 93, 142, 194,
280–84, 312, 351
ecotechnics: 55–56, 58–63, 78nn17,21
elements/the elemental: 23–24n44, 55,
70–75, 193, 207, 262, 291, 352
embodiment: 4, 9, 17, 97n28, 213,
222–23, 235, 240, 291, 321–23, 331
Emerson, Ralph Waldo: 319–20, 336n8
energy: 44, 59, 82, 90, 93, 96n23, 214,
218, 220, 226–27, 230–34, 236,
246nn41,44, 251, 261–62, 290, 322,
351
environmental ethics: 1, 11, 16–17, 42,
93, 115, 118, 164, 282, 285, 288–89
environmental philosophy: 1–2, 10–11,
61, 141, 279, 284, 286
epistemology: 82, 90, 220, 318, 320–22,
328, 332–33, 335
eschatology: 50–52, 54, 56–57, 67, 70,
74–76, 77n4, 182
ethics: 1, 6, 11, 14, 16–18, 42, 64–67,
77n4, 78n12, 82, 84, 87, 90, 93–94,
102–4, 106, 108–9, 111, 113–16,

118, 119n5, 132, 138, 142, 164, 174,
176–79, 181, 191, 241, 251–52, 257,
271–72, 280, 282–83, 285, 288–89,
293, 297n5, 306–11, 314nn7,9, 315–
16n23, 316nn29,33, 325–27, 331–33,
335, 338n59, 345, 350–52
event: 8, 13, 31–32, 38, 86, 107–8, 111,
113, 141–42, 144, 146–49, 151–56,
159, 161–64, 197, 199, 202, 208, 214,
220, 228, 237, 240, 255, 270, 288,
319
evolution: 6, 9, 12, 40, 82, 88, 92,
95n14, 107, 114, 118, 127–32, 283,
299nn29,34, 333
exappropriation: 9, 78n17, 112, 291, 330
exteriority: 8, 73, 75, 137, 144, 150–51,
155–56, 174–77
extinction: 1, 14–15, 36, 42, 49n30,
52–53, 121, 261, 263, 265–67, 270,
272–75

feminism: 1, 3, 5, 10, 123, 127, 181,
298n21
forgiveness: 31, 265
Foucault, Michel: 7, 284
Frankfurt School: 2, 279
friendship: 25n55, 119n5, 179n, 265
Fritsch, Matthias: 1–26, 141, 242n1,
279–96, 299n31, 300n35
Fukushima: 58–59, 226, 229, 244n27
futurity, future: 7, 11–12, 15–16, 35,
38–40, 50–57, 59–60, 63, 67, 74–75,
80n35, 86, 107, 114–17, 191, 203, 208,
210–12, 214, 222, 224, 226–27, 241,
250–66, 269–75, 282, 286–89, 291,
300n39

Gadamer, Hans-Georg: 4
Gaia Hypothesis: 33, 48n21, 298n17
general ecology: 12, 102–4, 106, 108–9,
114–15, 118
generations: 15, 36, 53, 59, 93, 222, 237,
250, 252–53, 256–57
genes, genetics: 9, 51, 88–89, 103, 122,
129, 148, 251
Genesis (biblical): 4, 33
genesis (phenomenological): 105–6
genetics: 9, 51, 88–89, 103, 122, 129,
148, 251

geography: 82–83, 212, 262, 353n25
geology: 15, 40, 49n30, 52, 54–55, 73–74,
96n23, 121, 207, 251, 257, 262, 269,
273, 328
gift: 10, 13, 29, 31, 76, 110, 144–46,
154–55, 162–63, 177, 292, 333
global warming: 35, 37–38, 43–44,
308–9. *See also* climate change
globalization: 55, 59, 245n39, 253n6
God, gods: 33–34, 60, 62, 104, 106, 117,
147, 151, 158, 215, 228, 297n9, 304–6,
329, 340, 347–48
Gould, Stephen J.: 130–31, 140n14

haunting, hauntology: 14, 37, 66, 74–76,
156–59, 167, 169, 176, 178, 188, 190,
202, 224, 226–27, 261–63, 270, 333.
See also specter, spectrality
Hayashi, Kyoko: 211–12, 216, 221–23,
228–29, 235–36, 239, 241, 244n27,
245nn28,37, 246n50, 247–48n58
Hegel, G. W. F.: 45, 114–15, 142, 147,
152, 167, 169, 176
Heidegger, Martin: 2–4, 6, 8,
19nn2,3,4,5, 20n9, 21n18, 23n33,
25n53, 29–30, 37, 41–42, 46, 57,
64, 67, 70–73, 84, 104, 107–11, 113,
119n12, 127, 141, 143, 146, 166,
168–72, 174, 178, 181–82, 198, 263,
271, 279, 281, 284, 296n4, 309, 313,
324–25, 334, 340, 347
hermeneutics: 3–4, 6, 33, 143, 163, 195
Hiroshima: 14, 32, 208–9, 221, 226–28,
242, 243n9, 245n28
holism: 6–7, 16, 282–86, 288–92, 294,
298n19
Homo sapiens: 40, 82, 269
Horkheimer, Max: 2, 279, 291, 301n41
hospitality: 15, 46, 157–59, 178, 234,
251, 268, 271, 290, 333, 344–45, 347
Human exceptionalism: 133–34, 136–37,
139, 241–42, 248n65
Husserl, Edmund: 2–4, 9, 20–21n16,
21n18, 25n53, 55 ,57, 64, 78nn15,16,
79n30, 104–6, 113, 118, 119n9, 163,
171–72, 176, 272–37, 353n25

idealism: 32, 45, 96n23, 107, 167, 175,
319–22, 332

immanence: 58, 61, 76, 103–7, 112–13, 116, 150, 152

impossibility, the impossible: 4, 42, 55, 66, 89, 102, 108–9, 113, 115–16, 142, 152, 154–55, 160–61, 164, 167, 176, 224, 246n40, 263, 266, 270, 294–95, 315–16n23, 319, 333, 340, 345, 347

indigenous peoples: 15, 35, 135, 208, 210–12, 229, 242–43n4, 248n58, 308, 315n16

Ingold, Tim: 91

inheritance: 29, 32, 38–39, 118, 251–52, 254, 256–59, 305–6, 312, 314n8

inorganic: 70, 112, 301n47, 322, 324

intentionality: 4, 21n17, 83, 97n28, 106, 122, 265–56, 272, 333

intergenerational justice: 15, 250, 252–53

interiority: 58, 91, 97n28, 123–24, 126, 130, 134, 145, 153, 155–56, 174–75

intersubjectivity: 104–5, 116–17, 349

ipseity: 111, 152, 299n27, 306, 309, 332

islands: 17, 68, 72, 75, 291, 313, 325–26, 330, 339–42, 344, 346–52

iterability: 15, 148, 154, 173, 194, 253, 256, 266, 286, 290, 299n27, 333

Jacob, François: 88–89

Jonas, Hans: 2, 19n6, 279, 281, 296nn3,9, 302n48

justice: 3–4, 11, 13–14, 16, 26n64, 30, 44, 61, 73, 102, 106, 108–9, 147, 163, 176, 181, 206, 213, 215, 222, 230, 261, 263, 265, 268, 271, 273–74, 282, 333, 350

Kamuf, Peggy: 22n32, 188, 204n1

Kant, Immanuel: 18n2, 23n33, 33, 56, 86, 149–52, 165–72, 174, 176, 178, 180–82, 273–74, 282, 316n29, 334, 339

Keats, John: 30, 47n4, 317

khōra (or chôra): 13–14, 167, 171, 178–81, 183n1

kinnibalism: 41, 48n27. See also cannibalism

Kirby, Vicki: 9, 12–13, 22n28, 25n52, 121–40, 299n29

Klages, Ludwig: 2, 19n3

Lacan, Jacques: 23n33, 26n64, 35, 79n22, 309

language: 4, 6–9, 11, 24n49, 31–33, 35–37, 41, 68, 110, 124–25, 127, 129, 134, 136–37, 141, 149, 151, 155, 169, 173, 175, 181, 192, 197, 249, 252, 254, 265, 286, 290, 303, 324, 329–30, 348, 350

Latour, Bruno: 33, 42, 47n7, 133, 140n20, 322, 327, 329, 332

Leopold, Aldo: 5, 283, 297n5, 298n19

Levi-Strauss, Claude: 135–36, 140n23

Levinas, Emmanuel: 2–3, 8, 13, 23n33, 39, 74, 103, 105–6, 108–9, 113, 116, 119n5, 152, 159, 174, 176–79, 181, 293

life: 6–10, 12–14, 16–17, 19n3, 24n44, 25n50, 26n64, 38–40, 42, 46, 55, 57, 61, 65, 67, 70–75, 77n4, 97nn28,34, 101–2, 104–5, 108–15, 117–18, 121–22, 127, 129–32, 137–38, 140n15, 146–48, 160, 179, 188, 190, 195–97, 200–1, 203, 212, 232, 237–38, 247n53, 248n58, 251, 264, 266–70, 272–74, 279–83, 285–86, 288, 290–95, 297n9, 299n30, 301–2n47, 304–7, 309–13, 316n23, 318, 321, 324–25, 331, 343, 349–52

lifedeath, or life death: 55, 86, 71, 75, 101, 104, 256, 301n42

literary theory: 5, 31

literature: 4–5, 161, 179, 191–92

living beings or living things: 1, 9, 16–17, 55, 64, 66, 68–71, 75, 79, 102–4, 107–10, 114, 279, 283, 285, 290, 292–95, 301n45, 313, 324, 349–52

living present: 9, 15, 104–7, 109, 113, 117–18, 206, 222

Llewelyn, John: 13, 50, 165–83

logocentrism: 6, 8, 10–11, 23n33, 24n49, 40–41, 124, 127, 134, 148

logos: 23n33, 141, 143–44, 148–49, 151, 153–54, 156, 160–62, 164, 179, 315n23

Lovelock, James: 33, 298n17. See also Gaia hypothesis

Luhmann, Niklas: 321–22, 336n8

Lynes, Philippe: 1–26, 101–20, 141

Lyotard, Jean-François: 34, 79n22

machine, mechanical: 9, 13, 36, 41, 48n23, 75, 90, 104, 111, 207, 210, 271, 281, 311, 325, 330

Marder, Michael: 7–8, 13, 23–24n44, 141–64, 259n1

mark: 8, 24n49, 75, 96n23, 197, 200, 241, 257, 264, 267. *See also* trace

Marx, Karl: 13, 29–31, 38, 45, 59, 88, 157, 167, 178, 261, 263, 279

masterpiece: 14, 165, 194, 198, 200–1, 205n6, 249, 251, 253–54, 256

materialism: 5, 7, 22n28, 38, 40, 46, 58, 69, 88, 167, 175, 244n26, 252, 262

matter, materiality: 7, 15, 55, 58, 71, 73, 75–76, 88, 107, 112, 125, 167, 209, 213–16, 217, 220, 223, 226, 229–33, 235–41, 243n16, 246nn44,46, 247n51, 250, 262, 266, 268

Maturana, Humberto: 17, 281, 321–22, 325, 333, 336n19

McCance, Dawne: 16–17, 20n15, 26n65, 297n4, 303–16

McKibben, Bill: 34, 51, 77n3

meat: 41, 43, 282

Meillassoux, Quentin: 23n40, 336n20

memory (re-membering): 9, 15, 23n33, 55, 74–75, 91, 137, 144, 196–97, 201–2, 205n5, 211, 213, 215, 222–23, 226, 228, 239, 258, 268. *See also* re-membering

Merleau-Ponty, Maurice: 2–3, 19n6, 76, 279, 281

messianism, the messianic: 7, 37

metaphysics: 8, 10, 12, 21nn17,18, 25n53, 37, 83, 95nn12,14, 105, 107–8, 110, 124, 143–44, 146, 157, 160, 166, 168, 170, 172, 174, 177, 211, 224, 231, 244n22

middle voice: 39, 168, 170

mortality, mortal life: 6–7, 9, 12, 16–17, 70–71, 108–11, 114, 214, 279, 281, 285–86, 288, 291–95, 301n45, 305, 315n23, 331

Morton, Timothy: 25n53, 264–65, 337n25

mourning: 55, 64, 66–67, 74–75, 147–48, 178–79, 211, 215, 222, 239, 241, 248n65, 261, 326, 333

mundus: 59, 63, 66, 69

Naas, Michael: 14, 22n32, 69, 79nn31,32,33, 187–205, 260n9, 300–1n39, 305, 311, 314n6, 330, 332, 336n7

Naess, Arne: 286

Nagasaki: 14, 32, 208, 211, 216, 221, 222, 226–29, 236, 240, 242n4, 243n9

Nancy, Jean-Luc: 6, 11, 50, 54–63, 67, 69–73, 75–76, 78nn17,21,22, 80n39, 132; works by: *After Fukushima: The Equivalence of Catastrophes*, 58–61, 63; *Being with the Without*, 80n39; *Being Singular Plural*, 58, 61–62, 69, 75; *Corpus*, 58–59, 62; *The Creation of the World or Globalization*, 58, 61, 69; *A Finite Thinking*, 58–59, 61; "Indestructible," 56, 59, 62, 76; *Noli me tangere*, 70; *The Sense of the World*, 50, 58–59, 61–63, 69, 71–73, 75

narrative: 4, 34–35, 37, 50–52, 54, 77n4, 92, 127, 130, 134–35, 161

National Socialism: 19n3; Nazism: 19n3, 36

nature: 1, 2, 4–8, 10, 14–17, 18n1, 18–19n2, 19n3, 21n17, 23n33, 33–35, 44–45, 50, 56, 58–59, 61–62, 68, 81–83, 86–87, 89, 109, 112–13, 122, 124–28, 132–33, 135–38, 150–51, 160, 167, 193–95, 216, 249, 253, 255, 261–62, 264, 268–69, 271, 273, 279, 284, 289, 291–92, 298n19, 304, 323–24, 333, 343

New International: 62

New Materialism: 5, 7, 22n28, 46, 244n26

Newton, Newtonian physics: 14–15, 212–23, 215, 226, 230–31

Nietzsche, Friedrich: 2, 4, 8, 21n17, 29, 36–38, 97n28, 111–13, 177, 182, 267

Noë, Alva: 324–25

nonhuman: 2, 6, 7, 9, 15, 34–35, 41–42, 46, 68, 77n4, 89, 92, 94, 97n34, 102, 104, 107, 111, 127, 133, 147–48, 150, 156, 158–59, 176, 228–29, 236, 238, 241, 269, 274, 281, 302n47, 308–9, 322, 325, 345, 350

nonliving: 7, 10, 57, 106–7, 133, 295, 301–2n47

normativity: 11, 16, 32, 41, 83, 102,
    257–8, 265, 274–75, 279–80, 284–86,
    296, 296n4, 300n35
Novalis: 169, 182
nuclear war: 14, 45, 51–52, 59, 67,
    200–2, 208
nuclear waste: 14–15, 60, 74, 193,
    200–2, 249–54, 256–59, 259nn2,6

object-oriented ontology: 9, 25n53, 321
oikos: 13, 102, 108, 112–13, 115, 123,
    126, 141, 143–46, 148–50, 153–54,
    156–57, 159–64, 166–67, 171, 173–74,
    178, 180, 182, 274
Oliver, Kelly: 17–18, 20n15, 26n66,
    338n60, 339–54
ontology: 2, 9, 11–12, 18, 25n53, 30,
    61, 71, 81–82, 84, 102, 104, 114–15,
    123, 126, 130, 133, 145, 159–61, 166,
    169, 178, 214, 220, 223–24, 231,
    241, 244n22, 257–58, 262, 279, 282,
    284–85, 291, 297n5, 321
organism: 7, 9–10, 12, 43, 58, 89, 91, 104,
    112, 121, 193, 195, 270, 281–82, 286,
    288–90, 294, 298n18, 299n29, 322, 351
Other, the: 8, 12, 29, 45–46, 57, 64–70,
    75, 80n35, 94, 103, 105–8, 111, 114,
    116, 136, 138, 143, 150, 152–53, 158,
    171, 174–78, 180–82, 233, 257, 287,
    290–94, 299n33, 300nn34,39, 307,
    309, 333, 350. *See also* alterity
overpopulation: 308

past: 31, 37–38, 40, 51–54, 60, 74–75,
    86, 105, 107, 114, 118, 206, 208, 212,
    222–24, 226–27, 241, 261, 264–65,
    269, 287–88. *See also* future, futurity
performativity: 37, 41, 244n22, 299n34,
    318–19, 331–32, 333, 336n7, 349
Peterson, Michael: 15, 77n8, 188, 200,
    249–60
pharmakon: 12, 86, 103, 268–69, 275
phenomenology: 2–4, 6, 9, 21nn17,18,
    31, 37, 50, 55–58, 64, 78n17, 81, 83,
    87, 97n28, 105–7, 118, 119n11, 123,
    132, 144, 163, 169, 171, 175, 180, 266,
    281, 348
physis: 50, 112, 150–51, 271, 332
plants. *See* vegetal/vegetable life

Plato: 13, 167, 169, 174–76, 178–79, 181,
    198
Plumwood, Val: 291, 298n21
poetics, poetry: 5, 17, 69–70, 118, 138,
    151, 191, 303, 317–20, 323, 326, 329,
    333–35, 337n29, 338n60, 341, 343
poiesis: 69, 335
politics: 2, 5, 14, 17–18, 33, 35, 40,
    44, 58–59, 62, 67, 82, 84, 87, 90, 94,
    95n14, 111, 113, 121–24, 127, 132–34,
    138–39, 157, 160, 178, 181, 191, 206,
    208–9, 212–15, 221–22, 227, 235, 241,
    247n51, 252, 256, 258, 263, 268–69,
    273, 291, 306, 308, 313, 314nn7,9,
    343, 350, 352
pollution: 85, 93, 194, 268, 303, 308–9,
    315n16
population: 51, 85, 93, 96n25, 97n34,
    230, 307–8
posthumanism/posthuman: 89, 101–2,
    110, 117–18, 122–24, 139n3, 262, 344
poststructuralism: 123–24, 132, 139n3,
    163, 284
power: 7, 40–41, 47, 61, 69, 110–11, 113,
    119n10, 264, 291, 293–95, 300n46,
    304, 306, 309, 315–16n23, 342, 345
presence: 6, 8, 12, 25n53, 37, 39, 52, 69,
    83–84, 86, 89, 95n12, 102, 105, 123,
    125, 143, 167–68, 172–73, 178, 224,
    231, 265–66, 292
promise: 7, 12, 16, 62–63, 101–2, 104,
    107–8, 110–11, 114, 117–18, 192, 194,
    204, 249, 257, 261, 263, 265, 269–73,
    275, 287, 299nn27,33, 299–300n34,
    300n38, 303
prosthetic/prosthesis: 58, 82, 91–92,
    103, 290, 330, 332
psychoanalysis: 31, 102, 316n29

Quantum Field Theory or QFT: 215–16,
    220, 226, 230–36, 244n25, 245n35,
    247n51
quantum physics or quantum mechan-
    ics: 12, 125–26, 128, 130, 133, 139n8,
    210, 212–18, 220–21, 223, 244n25,
    246nn41,44
quasitranscendental: 38, 45, 55, 148,
    155, 287
queer: 7, 234, 246n25, 264–65

race, racism, racialization: 34, 115, 127, 206, 211, 213, 222, 227, 241, 247n51, 308
radiation: 74, 190, 200–1, 203, 204, 214, 221, 236–37, 243n12
radioactivity: 15, 53–54, 74, 207, 211–12, 214, 227, 250–52, 258, 259n4
rationality, reason: 10, 12, 16, 90, 97n28, 148–49, 151, 167, 175, 222, 231, 252, 273–75, 281, 283–84, 292–93, 295, 297n5, 298n21, 304, 306, 308, 311, 315n23, 316n33, 334, 339, 341
realism: 7, 25n53, 58, 248n59, 321, 332
Regan, Tom: 281, 309–10
religion: 16–17, 33, 51, 62, 303–5, 311–12, 314nn3,5,9, 316nn29,33
remainder: 14, 51, 114–15, 192, 194, 199, 201, 205n5, 250, 257, 267, 272; restance, reste, 115, 164n2, 197, 250
re-membering: 213, 222, 230, 235, 239–41
repetition: 9, 32, 111–12, 122, 128, 147–48, 154, 198, 256, 263–66, 270, 286–87, 299n27, 300n34. *See also* iterability
resistance: 56, 76, 87, 164, 194–200, 203, 250, 253–54
responsibility: 13, 15–17, 42, 46, 52, 54–55, 64–67, 77n4, 94, 110–11, 115–16, 123, 132, 134, 138, 176–77, 179, 181, 206, 208, 222, 226, 241, 250–53, 257–59, 259n5, 293, 299n27, 306–7, 309, 311–12, 316n33, 325, 332, 333–34
Ricoeur, Paul: 4
rights: 35, 42, 46, 85, 134, 156, 281–82, 308–12
Robinson Crusoe: 17, 291, 313, 325, 329–30, 340–41, 349, 351
Rolston III, Holmes: 280–81, 284, 295, 313nn6,7, 298nn17,18
Romantics, Romanticism: 2, 5, 18n2, 56, 279
Rorty, Richard: 7–8, 46, 79n22

Saussure, Ferdinand de: 176, 305
Schelling, F. W. J.: 18–19n2
science: 8, 10, 21n17, 34, 52, 58–59, 62, 77n4, 85, 87, 95n14, 107, 121,

171, 191, 194, 202, 207–8, 215, 228, 231, 242n3, 243n4, 247n51, 255, 269, 272–74, 304, 311–12, 314n5, 322
secret: 14, 141, 146–50, 200
sense: 11–12, 50, 54–59, 61–65, 69–70, 73, 75–76
September 11, 9/11: 35, 38, 46, 57, 205n8
Serres, Michel: 17, 327–29, 331–32
Shelley, Mary: 50, 52
Shelley, Percy: 317
sign: 8, 34, 91, 147, 254–56
Singer, Peter. 282–81, 293, 310, 315n23
singularity: 11–13, 55, 59–60, 62, 64–66, 70–73, 75, 104, 109, 111, 114–15, 119n5, 146–49, 151, 155–56, 197–99, 241, 247n51, 250, 266–67, 291, 301n39, 319, 340, 346–47, 349–52
social ecology: 3
sovereignty: 10, 16–18, 32, 34, 37, 43, 46, 61, 91, 104, 109, 113–14, 157–58, 230, 252, 281, 291, 304–6, 311–13, 331, 345, 351
spacetimematterings: 15, 213–15, 220, 223, 226, 229–30, 234, 236, 239, 243n16
spacing: 8, 13, 56, 61, 72–73, 75, 153, 169, 176, 300n34, 320, 327
specter, spectrality: 12–14, 87, 101, 156, 159, 178–79, 222, 231, 262, 301n39, 333. *See also* haunting, hauntology
speculative realism: 7
speech: 6, 143, 146–47, 153, 162, 175, 179, 231, 268, 270–71
Stevens, Wallace: 17, 317–20, 322–23, 325–26, 328–29, 334–35, 336n8, 337n29
Stiegler, Bernard: 90–92, 96n26, 268, 274–75
stones, rocks: 34, 50, 55, 67, 70–75, 107, 193, 241, 295, 313, 325, 335, 351
subjectivity: 10, 12–13, 16–17, 18n2, 26n64, 40–41, 46, 74, 76, 86, 91–92, 95n16, 97nn28,32, 104–11, 113, 116–17, 119n9, 122–23, 125–26, 132–34, 138, 144–46, 148, 150–51, 156, 179, 213, 222, 241, 265–66, 273, 281, 284–85, 295, 296n4, 306, 308–9, 311, 331, 349

suffering: 16, 45, 109–10, 157, 280–83, 285, 288, 292–96, 297nn5,13,14, 301n46, 310, 315–16n23. *See also* vulnerability

supplement: 31, 45, 61, 86, 89, 91, 103, 115, 124, 126, 129, 156–57, 271, 283

survival: 12, 43–44, 52, 83, 85, 121, 123, 128, 174, 207, 209, 211, 226, 235–37, 243n9, 268, 273, 347, 351

survivance: 14–15, 65–67, 70, 74, 89, 101–4, 114, 168, 188–92, 194, 196–201, 204, 249, 252–54, 256–57, 259, 260n11, 265, 305, 310, 312

sustainability: 4–5, 11, 14–15, 30, 37, 39, 44, 53, 56, 59, 97n34, 190, 268–70, 274, 340, 347, 352

systems theory: 9, 320–22, 325, 333–34

technology, techne, technicity: 2, 15, 35, 55, 57–63, 78n17, 82, 84, 90–92, 103–4, 111–12, 135, 143, 193–95, 202, 208, 215, 252, 262, 264, 268–71, 274, 279, 290, 300n34, 304–5, 311–12, 324, 330, 343–44, 348, 351

technoscience: 59, 202, 208, 215, 269, 304, 311–12

terrorism: 43, 46, 202

text: 4–11, 14, 24nn44,49, 25n53, 32–33, 51–52, 89, 96n23, 121, 124, 134, 139n4, 173, 176, 192, 195–99, 249, 251, 253–55, 263–64, 266–68, 271–72

theology: 16–17, 23n33, 33, 114, 135, 305–6, 311–12, 314n9

Thompson, Evan: 281, 297n9

Thoreau, Henry David: 5

time, temporality: 8–12, 14–15, 18, 21n17, 32, 37–40, 48n16, 53–55, 60–62, 64–66, 68, 74–75, 77nn9,10, 82–86, 90, 92–93, 95n16, 104–8, 113, 118, 125–27, 130, 132, 146, 153, 157, 162, 171, 176, 180, 182, 200–1, 206–31, 233, 236–41, 243nn14,16, 246nn41,46, 247n51, 248n65, 251–52, 258, 261, 263–66, 269–70, 273, 281, 285–87, 292, 320, 324, 327–29, 349

Toadvine, Ted: 3, 9, 11–12, 50–80, 87

trace: 7–9, 12, 14, 21n18, 23n33, 24n49, 68–69, 74, 86, 96n23, 101, 103, 105, 108, 110, 114, 117, 137, 153, 158, 176–77, 189, 194, 201, 226, 228–29, 239, 250, 263–71, 324, 330

transcendence: 64, 102–3, 105–9, 112–14, 116, 123, 152, 281

transcendental: 21, 33, 38, 40, 45–46, 55, 64, 97n28, 102, 104–5, 107–9, 119n9, 145, 148, 155, 269–70, 287, 300n35, 333. *See also* quasitranscendental

translation: 32–33, 59, 64–65, 68, 103, 125, 155, 162, 199–200, 204, 325, 349

Uexküll, Jakob von: 323

uncanny (*unheimlich*): 123, 135, 142, 149, 154–55, 159, 171, 182, 330, 333, 346, 352

undecidability: 31–32, 36, 48, 153–54, 156, 180–81, 234

utilitarianism: 16, 282, 298n18, 310–12, 315n23, 316n25

value, valuation: 1–2, 16, 18n2, 61, 67, 87, 155–57, 279–85, 288–96, 296n4, 297nn5,9, 298n18, 302n48

Varela, Francisco: 17, 322, 325, 333, 336n19

vegetal/vegetable life: 8–9, 24n44, 33, 64, 71, 73, 104, 106, 109, 121, 127, 133, 147, 158–59, 162, 193, 211, 241, 281, 351

vegetarianism: 41, 49n28, 310, 316n25

Vitale, Francesco: 23n35, 88–90, 96n23, 299n30

void: 14–15, 69, 131, 215, 228, 230–32, 235–37, 239–41, 245n35

vulnerability: 16–17, 58, 69, 75, 107, 113, 197, 237, 281–83, 285, 288, 292–96, 307, 315n10, 315–16n23, 331, 347–48. *See also* suffering

war: 14, 43–45, 51–52, 29, 67, 188, 191, 200–3, 207–9, 213, 215, 226–27, 229, 239, 256, 275, 341–44. *See also* Cold War; nuclear war

waste: 14–15, 44–45, 53–54, 60, 74, 93, 188, 190, 193–94, 200–2, 229, 238, 249–54, 256–59, 268, 303, 309

water: 17, 70–73, 227, 303–4, 306–13, 314n3, 315n10

Wolfe, Cary: 9, 17, 20n15, 102, 109, 301n44, 311, 316n27, 317–38

Wood, David: 1–26, 29–49, 81, 83, 86, 95n16, 107, 141, 261

world: 4, 7–12, 15, 17–18, 18n2, 24n44, 31, 34–35, 38, 43, 45, 50–59, 61–76, 79n31, 81–83, 92–93, 95n25, 101–3, 105, 109–14, 117, 124–26, 134–35, 139n4, 142, 144–45, 147, 151, 157–58, 166, 170–72, 176, 178–79, 182, 203, 208, 211–17, 226, 229, 232–33, 238–39, 241, 244n22, 255, 264–68, 270, 272–74, 279–81, 285–86, 294, 304, 307–8, 312–13, 318–20, 322, 324–26, 328, 331–35, 337n25, 343–45, 347–52. *See also* cosmos, *mundus*

writing: 6, 8–10, 14, 14, 21n18, 30, 78n21, 84, 96n23, 129–30, 135–38, 136, 167–68, 175–77, 190, 192, 196, 254, 260n9, 260n11, 264–66, 268–69, 271, 330. *See also* arche-writing; text

gROUNDWORKS|
ECOLOGICAL ISSUES IN PHILOSOPHY AND THEOLOGY

Forrest Clingerman and Brian Treanor, series editors

Interpreting Nature: The Emerging Field of Environmental Hermeneutics
Forrest Clingerman, Brian Treanor, Martin Drenthen,
and David Utsler, eds.

*The Noetics of Nature: Environmental Philosophy and the Holy Beauty
of the Visible*
Bruce V. Foltz

Environmental Aesthetics: Crossing Divides and Breaking Ground
Martin Drenthen and Jozef Keulartz, eds.

The Logos of the Living World: Merleau-Ponty, Animals, and Language
Louise Westling

Being-in-Creation: Human Responsibility in an Endangered World
Brian Treanor, Bruce Ellis Benson, and Norman Wirzba, eds.

Wilderness in America: Philosophical Writings
Henry Bugbee, edited by David W. Rodick

Eco-Deconstruction: Derrida and Environmental Philosophy
Matthias Fritsch, Philippe Lynes, and David Wood, eds.

Milton Keynes UK
Ingram Content Group UK Ltd.
UKHW011812021123
431764UK00001B/214

9 780823 279517